Globalized Poverty and Environment

Nathaniel O. Agola · Joseph L. Awange

Globalized Poverty and Environment

21st Century Challenges and Innovative Solutions

Springer

Nathaniel O. Agola
Doshisha University
Kyoto
Japan

Joseph L. Awange
Department of Spatial Sciences
Curtin University
Bentley, WA
Australia

and

Geodetic Institute
Karlsruhe Institute of Technology
Karksruhe
Germany

ISBN 978-3-642-39732-5 ISBN 978-3-642-39733-2 (eBook)
DOI 10.1007/978-3-642-39733-2
Springer Heidelberg New York Dordrecht London

Library of Congress Control Number: 2013945277

Printed on acid-free paper

Springer is part of Springer Science+Business Media (www.springer.com)

Foreword

In the early twenty-first century, we are standing in a critical juncture. Concerted action of the humankind may be able to put an end to structural poverty for the first time in human history, enlarging opportunities for future generations to sustain their lives on the planet. If we make a wrong choice, however, poverty and inequality will persist, joblessness will violently tear up social webs, and natural assets will continue to be plundered in irretrievable ways. We are pressed to decide our course.

In 2000, the international community has shifted the helm. All UN member states agreed upon the Millennium Development Goals (MDGs) and pledged to achieve such goals as halving the proportion of people living on less than $1.25 a day by 2015. Next goals that are now being negotiated will more explicitly include environment-related agendas discussed at Rio in 2012, pursue more equitable and inclusive ways of economic development, and address the issues of democratic accountability.

In the meantime, the phenomenal growth of Asian and African economies raises serious concerns about worsening inequalities in those emerging regions as well as the long-term sustainability of todays patterns of resource consumption. In parallel, the expansion of Foreign Direct Investment (FDI) dwarfs the conventional grant aid as means of effective poverty reduction especially on the continent of Africa, while decentralized management is becoming a normative practice of aid agencies as learning organizations.

This book provides a good summary of economic theories of poverty as well as a vivid depiction of the state of environmental degradation in the world. People often work separately on different issues that are, in fact, closely intertwined. The principle of holism is that the whole is greater than the sum of its parts, and

I believe that this joint venture of two experts on poverty and environment has produced something more than a sum of two separate monographs on the issues. Various points raised in this volume are worth heeding when we think of formulation and implementation of a truly effective post-MDGs development agenda.

Kyoto, Japan 2013

Yoichi Mine
Graduate School of Global Studies
Doshisha University
Kyoto, Japan

Preface

Poverty reduction and environmental degradation challenges rank top among the most difficult problems for the global community. These problems are multidimensional, and therefore in terms of conception, and efforts aimed at providing solutions, they straddle the fields of economics, social studies, health, education, and natural science. This book reviews the key conceptions and economic theories of poverty, how it is intertwined with environmental degradation challenge—poverty-environment nexus, and offers innovative socioeconomic and scientific geospatial solutions for the twenty-first century.

This book dispels the overly simplistic view that the poor cause environmental destruction, and as such there is linear causality relationship between poverty and environmental destruction. We validate that the two issues are two sides of the same coin, and they are a function of a two-way complex cause and effect mechanism. In certain instants, the poor cause environmental degradation, and in some instances they are mere victims of other groups socioeconomic actions. Surprisingly, there are also instances where the poor perceive environmental quality as an important determinant of their earning capacity, housing quality, health, security and energy suppliers, and therefore their actions are positively aligned with environmental conservation.

We make it possible for our readers to understand poverty by giving a thorough concise review of the major theoretical economic frameworks, measures of poverty, and point out the need to understand rural–urban dichotomy of poverty. Of all the theories and measures, none stands out as faultless, therefore we point the need to treat these measures as convenient tools lacking in perfect accuracy and utmost scientific reliability. We also posit that if conceptions and measures are not beyond fault, and so are the supposedly knowledgeably crafted poverty reduction and environmental preservation solutions.

The book finally proposes a raft of innovative economic and geospatial-based solutions. The economic solutions are centered on the idea of need to go beyond the humdrum macroeconomic and policy measures targeting poverty and environmental issue to envisaging a new paradigm in which private sector and other stakeholders can create innovative and inclusive markets where value is co-created and shared.

Our concern with the need for bringing the concepts of innovation to solving poverty leads us to propose an inclusive-shared-value model based on what we call

Value Ecosystem, which requires a paradigm shift in how market value is perceived, and how judiciously costs and the accompanying returns could be willfully allocated. In the low-income countries, the most pervasive question is always where money is going to come from whenever any attempt is made to practically solve any socioeconomic problem. Affordability is therefore one of the most pressing problems when considering implementation of almost all shades of solutions conceived of by public, private sector actors, and even more so NGOs. Value Ecosystem Model proposed here is a novel approach, which starts with the idea that there is always latent value locked up in the various entities and subsystems that constitute entire economic and social systems.

Income inequality remains a pervasive problem in the current global economic system. Inequality is a problem for consumption, resource mobilization and is even a disincentive to labor motivation. The central issue here is therefore how resources can be effectively mobilized, and labor productivity improved based on high worker motivation? We think social innovation is a necessary approach to how new economic organizations can be created to deal with resource mobilization, equity, and motivation of workers to create wealth for not only individuals, but even whole communities. We validate this approach by analyzing the Mondragon Model. Mondragons are industrial cooperatives created to solve poverty problem in the Basque region in the first half of the twentieth century from very humble beginnings, but have over the decades flourished into a business empire comprising high-technology companies, banking services, educational, and retail network with the lowest failure rate that is unrivaled by any known model in the history of capitalism.

Even more remarkable is the resource mobilization abilities of the Mondragon system, and its equity allocation of the wealth created within the system. These industrial cooperatives are simply owned, managed and controlled by the members, which is a solution to a problem conventional corporate governance system has failed to conclusively solve. Aligning interests of corporate owners and agents (managers/controllers) is a perennial problem that will never go away in conventional corporate ownership and control structures. Yet despite the success, lowest failure rate, and perfect alignment of ownership and control, and positive income distribution qualities of the Mondragon, very scanty attention has been paid to the model in economic development and management studies. It is interesting to observe that with the global rise of social media, well-coordinated communities are fast emerging as the most sustainable value and wealth creators, and this suggests a new paradigm.

Finally, we attempt to solve the perennial problem of duality in economic structure. There are considerable costs to having an underdeveloped economy side by side with a relatively sophisticated one on the other end of shared spatial dimension such as in a city or country. We suggest five action points, which can be jointly leveraged to solve the problem of duality. We use the example of post-war Japanese economic experience to validate our suggestions. Solving duality problem requires more attention given the failure of trickle-down economics as typified by high economic growth figures for Sub-Saharan African countries, which fails to translate into rise in incomes of majority and cutting down high unemployment rates.

On environment, we recognize that poverty and environmental issues occur in space and as such are location or area-based, i.e., geographical or spatial in nature. Analysis of environmental and poverty related issues, therefore, needs to be spatially undertaken. This is informed by the fact that they occur in geographical space on the one hand, and from policy perspective, the poverty-environment nexus is relevant only if it has implications for the allocation and administration of public resources for alleviation of poverty and environmental problems on the other hand.

This book offers timely state-of-the-art geospatial solutions, e.g., remote sensing, Geographical Information System (GIS), and Gravity Recovery and Climate Experiment (GRACE) satellites mission to target the most pressing global problems. For example of GRACE satellite data are presented as efficient geospatial tools that are useful in estimating changes in stored water in the water-poverty-environment nexus. Other areas include: pollution, agriculture and disaster management, where geospatial techniques are applied under strong environmental impact assessment regulatory regimes. The geospatial solutions are quite timely and need to be urgently applied, because the poor are the most vulnerable to frequently occurring global environmental problems and disasters such as floods, droughts, polluted agricultural lands, and water shortages.

This is a TIGeR publication No. 470.

Kyoto, Japan, July 2013 — Nathaniel O. Agola
Bentley, Australia — Joseph L. Awange

Acknowledgments

Several figures in this book have been generously provided by various authors. In this regard, the second author (Joseph) would like to thank M. Motagh (GFZ) and M. Jia (Geoscience Australia). Some figures and materials also came from the work undertaken jointly with colleagues B. Heck (Karlsruhe Institute of Technology, Germany), W. Featherstone, M. Kuhn, K. Fleming and I. Anjasmara (Curtin University), M. Sharifi (Tehran University), A. Hunegnaw (University of Edin burgh), E. Forootan (Bonn University), JBK Kiema (University of Nairobi) and students B. Obera, Khandu, and T. Opande. To you all, "arigato gozaimasu" (Japanese for thank you very much). To all his Curtin University 3rd year (Satellite and Space Geodesy unit), and 2nd year (Civil Engineering) students who used materials from the draft book and provided feedback, Joseph would like to say "Danke sehr" (German for thank you very much).

Joseph also wishes to express his sincere thanks to Prof. B. Heck of the Department of Physical Geodesy (Karlsruhe Institute of Technology (KIT), Germany) for hosting him during the period of his Alexander von Humboldt Fellowship (2008–2011) when part of this book was written. In particular, his ideas, suggestions, and motivation on Chap. 13 have enriched the book considerably. Joseph is also grateful to Prof. B. Veenendaal (Head of Department, Spatial Sciences, Curtin University, Australia) for the support and motivation that enabled the preparation of this edition. He also wishes to acknowledge the support of Curtin Research Fellowship, while his stay at KIT was supported by a Alexander von Humboldt's Ludwig Leichhardts Memorial Fellowship. To all, he says, "ahsante sana" (Swahili for thank you very much). Last but not least, he wishes to thank his wife Naomi Awange and daughters Lucy and Ruth Awange for their patience and support, especially the hard times they endured when he was away in Germany.

On the other hand, the first author (Nathaniel) wishes to thank his entire family for constant support in his academic endeavors, which more often than not always robs all of us of time we could share together. To dad who pulled me into his study room filled with books, your casual instruction that when I grow up I should read all those books, gives you the place as my greatest source of inspiration in the disciplined pursuit of knowledge. Too bad your infirmity may not allow you to read this work when it comes out. Thanks to mum too for perpetual encouragement. Lastly, to Asako, Anry and Ken you kept me going in countless ways on each and every day of tireless working on this book, a big thank always!

Contents

Part I
Introduction

3

Chapter 1
Poverty: Environmental Link

The poor have traditionally taken the brunt of the blame for causing societies many problems. The most recent accussation against them is that they cause environmental degradation [17]. This is evidenced in Brundtland Commission Report [9] which states: Poverty is a major cause and effect of global environmental problems. It is therefore futile to attempt to deal with environmental problems without a broader perspective that encompasses the factors underlying world poverty and international inequality.

1.1 Understanding of the Concept of Poverty and Environment

1.1.1 Poverty

What is poverty? This is a critical factor; defining poverty is not an easy task. There will always be disagreement over what constitutes a poverty income. Some analysts define poverty in terms of the amount of income necessary to provide a family of a certain size with the minimum essentials of food, clothing, shelter, and education. This approach provides an ***absolute poverty*** standard. An absolute poverty standard establishes a specific income level for a given-size household below which the household is judged to be living in a state of poverty.

But is an absolute measure of poverty appropriate? Poverty can, after all, be relative. One's sense of poverty depends upon the incomes of others in the community. A second approach to poverty, therefore, is to measure it in relative terms. A ***relative poverty*** standard might classify a household as poor if the household's income is, say 25 % of average household's income. Thus, relative poverty standard defines poverty in terms of incomes of others.

The choice of a poverty definition will determine to a great extent the number of poor and the rate at which poverty is perceived as being eliminated. If the absolute standard is selected, rising real living standards will push more and more fami-

N. O. Agola and J. L. Awange, *Globalized Poverty and Environment*,
DOI: 10.1007/978-3-642-39733-2_1,

lies above the poverty line. According to the relative standard, only equalizing the distribution of income can eliminate poverty.

The conceptual debate on poverty is carried over to measurement. A small craft industry has emerged, especially at the international level, in measuring poverty and deprivation, often in response to the need to define targets at international conferences and measure progress against them. Different models of poverty imply different indicators. Advocates of the participatory paradigm, in particular, are wary of quantification and standardization.

How is poverty measured? Is it simply about the level of income obtained by households or individuals? Is it about lack of access to social services? Or is it more correctly understood as the inability to participate in society, economically, socially, culturally or politically? These are the questions that have occupied the attention of researchers for decades. While some researchers and development practitioners have argued that poverty can be defined only subjectively, others have maintained a contrary position. For instance, Townsend has argued that "Poverty can be defined objectively and applied consistently only in terms of the concept of relative deprivation. The term is understood objectively rather than subjectively. Individuals, families and groups in the population can be said to be in poverty when they lack the resources to obtain the types of diet, participate in the activities and have the living conditions and amenities which are customary, or are at least widely encouraged or approved, in the societies to which they belong. Their resources are so seriously below those commanded by the average individual or family that they are, in effect, excluded from ordinary living patterns, customs and activities" [49].

This 'objective' definition has implications for policy that should be recognized. Although all societies have ways of identifying and trying to deal with their problems, the social sciences are having an increasing influence upon decision-makers, both in providing information and implicitly or explicitly legitimizing action. A clear definition allows the scale and degree as well as the nature of the problem of poverty to be identified, and therefore points to the scale as well as the kind of remedial action that might be taken. Such an action may involve not just the general level of benefits, for example, but revision of relativity between benefits received by different types of families.

Poverty is a difficult concept, which has become increasingly controversial. The difficulty arises from the fact that poverty is a portmanteau term, having distinct meanings to different people. Words such as "destitution", "ill-being", "powerlessness" and "vulnerability" are so frequently used in conjunction with "poverty". Poverty may be seen in terms of people being excluded from the "living conditions and amenities which are customary, or at least widely encouraged or approved, in the societies to which they belong" [49, p. 31], or what is often referred to as a 'participation standard'. In adopting this definition, the European Community stated: "The poor shall be taken to mean persons, families and groups of persons whose resources (material, cultural, and social) are so limited as to exclude them from the minimum acceptable way of life in the state in which they live".[1] This basic problem of meaning

[1] European Community, Council Decision, 19 December 1984.

of poverty pervades the debates both on the measurement of poverty and on poverty reduction policies. Although most people have an intuitive notion of what poverty is all about, the concept has been defined in various ways and is still a subject of much controversy in academic and policy-making circles [12]. The diversity of meanings attached to poverty renders conceptualization of poverty and its operational meaning and measurement difficult as well as intractable.

While poverty can be broadly defined as an absence of well being or capacities, it is multidimensional and manifests itself in various forms. This makes the definition of poverty inadequate, if only one criterion is used. It should also be recognized that there is no single indicator that can adequately measure all dimensions of poverty. Furthermore, it should be noted that poverty and inequality, though related at some points, are different concepts. While poverty is concerned with absolute standard of living of a part of the society (the poor), inequality refers to relative living standards across the whole society. It is therefore difficult for one survey like the Welfare Monitoring Survey (WMS) series,[2] used for instance in Kenya, to capture the many dimensions of poverty. In the WMS analysis, poverty is defined in terms of total household consumption expenditure. A household is considered poor if it cannot attain some recommended food energy intake. The required level of nutrient intake is 2,250 calories per day per adult plus a minimum allowance for non-food consumption. The notion of what constitutes the poor goes beyond an attainment of a level of material well being to constitute a reasonable minimum by the standards of a given society. A household, which does not meet its calory intake requirements, but has a relatively high income, is not considered poor.

This raises the question of whether, ultimately, it is possible to take an absolutist view of poverty. Sen [45] has argued persuasively "that ultimately poverty must be seen to be primarily an absolute notion". It is of course true that there is "an irreducible absolutist core" [45] to the notion of poverty. If someone is starving, he is poor even if everyone else is starving too. But once one moves away from extreme cases it becomes much more difficult to make assertions that carry conviction or inspire confidence. For example, despite decades of work by nutritionists and physiologists, we still do not know what is the minimum calorie intake necessary for adequate nutrition [20, p. 12]. More awkward for the absolutist position, we do not even know, as some claim, whether individuals really do have fixed requirements, i.e., whether the body is able to adjust to a persistent shortfall of calories and thereby in effect avoid "undernutrition" [20, p. 12].

In the eighteenth century, Adam Smith made it clear that "necessities" were determined by "custom" and hence that poverty was relative. He said, "By necessities I understand not only the commodities which are indispensably necessary for the support of life, but whatever the custom of the country renders it indecent for creditable people, even the lowest order, to be without" [47]. Theron de Montauge bluntly

[2] Details of the Welfare Monitoring Surveys are reported in the Economic Survey 1997 and have been extensively used in this study. The Appendix of this study contains the findings of these surveys.

asserted that "poverty is measured by comparisons [14, p. 2]." It has nothing to do with physiological need for calories [20].

The absolutist view of poverty, however, does not depend on the results of physiological research. It is much more sophisticated than that. Emphasis is placed not only on the absolute need for adequate nutrition but also on the need for decent shelter, to live without shame and to have self-respect. The needs for decency, self-respect and the avoidance of shame are absolute, but the commodity requirements to fulfill these socially defined needs rise with average prosperity. In this way absolute needs or capabilities are transformed into relative requirements for commodities and income [46].

Perhaps there is enough common ground here between relativists and absolutists to make further discussion unnecessary. But once it is accepted that needs, or at least some needs, are socially defined, we are very close to accepting that poverty itself is socially defined. That is, poverty is neither a relationship between a person and a bundle of commodities nor is it a relationship between a person and a bundle of (socially defined) *needs and capabilities. It is a relationship between one person and another*. This interpretation of the nature of poverty has emerged most forcefully from the research of Marshall Sahlins and others on the economies of extremely "primitive", "backward" and "poor" people [42]. The economic anthropologists have shown that these "primitive" people do indeed have an objectively low standard of living. However, they also have much leisure time, are generous in sharing their possessions with others, are essentially egalitarian, and most important of all, do not regard themselves, individually or collectively, as poor. As Sahlins says,

> The world's most primitive people have few possessions, but they are not poor. Poverty is not a certain small amount of goods, nor is it just a relation between means and ends; above all, it is a relation between people. Poverty is a social status. As such it is the invention of civilization [43].

If civilization invented poverty and if social status defines it, presumably man in society, operating through the institutions he has created, or if necessary through new institutions, can abolish poverty [20, p. 13].

Previous operational definitions of poverty have not been expressed in thoroughgoing relativist terms, nor founded comprehensively on the key concepts of resources and styles of living. The concern has been with narrower concepts of income and the maintenance of physical efficiency. In reference to most developing economies, poverty has not been clearly defined. The issue of poverty has been confined to discussing the concepts of "relative" and "absolute" poverty, but not much else. The definition of poverty and its measurement has been influenced by the work of Rowntree, as discussed below.

Historically, the standard economic definition of poverty in terms of income and consumption date back to Booth's 1892 and Rowntree's work in Victorian England. However, this work has been challenged by modern social scientists looking for broader and more inclusive definitions of ill-being in both developed and developing countries.

Among the early studies of poverty, the work of Rowntree is most important. He had a narrow conceptualization of primary poverty in terms of nutritionally based poverty lines. In 1899, he collected detailed information about families in York. He defined families whose 'total earnings were insufficient to obtain the minimum necessaries for the maintenance of merely physical efficiency as being in primary poverty' [41]. Rowntree made use of the work of W. O. Atwater, an American nutritionist, who had experimented with the diets of prisoners in Scotland to find how nutritional intakes were related to the maintenance of body weight. Rowntree estimated the average nutritional needs of adults and children, translated these needs into quantities of different foods and hence into cash equivalent of these foods. To these costs for food he added minimum sums for clothing, fuel and household sundries according to size of family. Rent was treated as an unavoidable addition to the calculation, and was counted in full. A family was therefore regarded as being in poverty if its income minus rent fell short of the poverty line.

Nearly all subsequent studies were influenced deeply by this application of the concept of subsistence. With minor adaptations, a stream of area surveys of poverty based on Rowntree's methods was carried out in Britain, South Africa, Canada and Tanazania. The subsistence standard was used as a measuring rod, or as a basis for recommending minimum social security rates and minimum earnings in many countries [3, 4, 6, 34, 36]. In recent years, his methodology has been used to measure poverty in countries like Kenya [13, 52].

But the standards that were adopted proved difficult to defend. Rowntree's estimates of the costs of necessities other than food were based either on his own and others' opinions or, as in the case of clothing, on the actual expenditure of those among a small selection of poor families who spent the least. Does the actual expenditure of the poorest families represent what they need to spend on certain items? Neither in his studies nor in similar studies were criteria of need, independent of personal judgment, or of the minimum amounts actually spent on certain goods, put forward.

In the case of food it seemed, at first sight, that independent criteria of need had been produced. But the Rowntree's procedures have been faulted on three main grounds. First, estimates of the nutrients required were very broad averages and were not varied by age and family composition, still less by occupation and activity outside work. Secondly, the foods that were selected to meet these estimates were selected arbitrarily, with a view to securing minimally adequate nutrition at lowest cost, rather than in correspondence with diets that are conventional among the poorer working classes. Finally, the cost of food in the total cost of subsistence formed a much higher percentage than in ordinary experience. In relation to the budgets and customs of life of ordinary people, the make-up of the subsistence budget was unbalanced.

An adaptation of the Rowntree method is in use in many countries, including Kenya. Kenya's National Development Plan, 1979–1983, under the theme of meeting the "Basic Needs", used the Rowntree's nutritional method. A basic standard of nutritional adequacy was put forward by the Ministry of Economic Planning, and this standard was translated into quantities of types of food compatible with the preferences of Kenyan families. This, in turn, was then translated into the minimum costs of purchases on the market. Finally, by reference to the average sums spent per capita on food as a proportion of all income (derived from consumer expenditure surveys), it was assumed that food costs represent a fixed percentage of the total income needed by families of three or more persons.

A number of points in the argument can be examined critically. ***First***, and most important, the index is not redefined periodically to take account of changing customs and needs. In a rapidly changing society like Kenya, dietary customs and needs are liable to change equally rapidly and estimates of need must be reviewed frequently. Otherwise, there is the risk of reading the needs of the present generation as if they were those of the past. Foods are processed differently, and presented from time to time in new forms, whether in recipe or packaging. Real prices may rise without any corresponding improvement in nutritional content. No price index can cope properly with changes in ingredients, quality and availability of and 'need for' goods and services. There is no known way of how to adjust a poverty line to conform to changes in productivity.

Secondly, nutritional needs are narrowly defined. The cost of buying a minimally adequate diet for families restrict the kind and quality of their purchases and skill exercised in preparing as well as buying food, is worked out. Nothing extra is allowed for eating meals out, and the amounts are enough only for 'temporary or emergency use when funds are low.' The underlying definitions of dietary adequacy are insufficiently related to actual performance of occupational and social roles. Estimates of nutritional needs in fact include a larger element for activities that are socially and occupationally determined than for activities that are biologically and physiologically determined. Moreover, the former obviously varies widely among individuals and communities. While it may seem to be reasonable to average nutritional requirements, empirical studies of diets in relationship to incomes and activities have to be undertaken to demonstrate whether that procedure is in fact as reasonable as it purports to be.

Finally, the question of finding criteria for *needs* other than *food* is dodged by estimating food costs and then taking these as a fixed percentage of the total budget stated to be necessary. The percentage varies with households of different size and is lower for farm families than for other families. How, therefore, are the percentages chosen? This remains the nagging problem about the entire procedure. One researcher has addressed this issue as follows:

> Poverty, like beauty, lies in the eye of the beholder. Poverty is a value judgment; it is not something one can verify or demonstrate, except by inference and suggestion, even with a measure of error. To say who is poor is to use all sorts of value judgments. The concept has to be limited by the purpose, which is to be served by the definition. In the [USA] Social Security Administration, poverty was first defined in terms of the public or policy issue; to how many people, and to which ones, did we wish to direct policy concern [32, p. 37].

This may be shrewd but scarcely reassuring. Socio-economic measures cannot rest only on imaginable or even politically acceptable, but must also rest on demonstrable, definitions of social conditions. These may be difficult to apply consistently. There are bound to be difficulties and disadvantages in any approach that is well developed. In the final analysis, a definition of poverty may have to rest on value judgments.

But this does not mean that a definition cannot be objective and that it cannot be distinguished from social or individual opinion.

1.1.2 Environment

The preceding section dwelt on the concept of poverty and its definition. Poverty and the environment are interrelated as will be discussed in Sect. 1.2. In this section, similar to Sect. 1.1.1, the concept of environment and its possible definitions are presented. The word *environment* has different meaning in different jurisdictions, and is therefore widely recognized as a broad term with many interpretations and definitions. In Western Australia, for example, the *Environmental Protection Act (EPA) WA* 1986 defines environment as including water, air and land and the inter-relationship which exists among and between water, air and land, and human beings, other living creatures, plants, micro-organism and property.[3] The *Environment Protection and Biodiversity Conservation Act* 1999 (the EPBC Act) on the other hand, which is Australian-wide environmental legislation, defines environment as including[4]:

(a) ecosystems and their constituent parts, including people and communities; and
(b) natural and physical resources; and
(c) the qualities and characteristics of locations, places and areas; and
(d) heritage values of places (i.e., places included in the Register of the National Estate kept under the *Australian Heritage Council Act* 2003; and
(e) the social, economic and cultural aspects of a thing mentioned in paragraph (a), (b), (c) or (d).

In Canada, the *Canadian Environmental Assessment Act* (CEAA) 1992 define the environment as the components of the Earth, and includes

(a) land, water and air, including all layers of the atmosphere,
(b) all organic and inorganic matter and living organisms, and
(c) the interacting natural systems that include components referred to in paragraphs (a) and (b).

In general, therefore, the term '*environment*' may be used narrowly, with reference to '*green*' issues concerned with nature such as pollution control, biodiversity and climate change; or more broadly, including issues such as drinking water and sanitation provision (often known as the '*brown agenda*') [31]. For instance, Neefjes ([30], p. 2) uses the term in a broad sense, referring to the environment as a vehicle for analysing and describing relationships between people and their surroundings, now and in the future, while Bucknall ([10], p. 3) points out that the word environment generally refers to a natural resource base that provides sources and performs sink

---

[3] EPA 1986, Sect 2(a).

[4] EPCA 1999, Sect 528, Definitions.

functions, and uses a broad definition of the environment in his background paper to the World Bank's Environment Strategy [31].

Owing to the varied definitions of the term environment, certain terms and expressions that relate to it such as environmental degradation, environmental change, and environmental quality are also problematic in that they vary widely in usage within and between disciplines, and several have been used as synonyms (e.g., Johnson et al. [23]). In an attempt to correct the problem and standardize usage, Johnson et al. [23] defines or redefines 10 of the most common environmental terms, e.g., *natural environment* and *environmental change* are defined on the basis of what is meant by natural as reflected by common usage and dictionary entries while *environmental degradation, land degradation*, and *soil degradation*, are defined as any change or disturbance to the environment, land, or soil perceived to be deleterious or undesirable. In part II of the book, we will encounter some of these terms and how they are linked to poverty.

1.2 The Poverty-Environmental Link

When one speaks of both poverty and environment, two schools of thought come to mind, i.e.,

(i) poverty is a major cause of environmental degradation and that fixing it in-turn fixes environmental problems. This view is supported, e.g., by the Brundtland Commission Report ([9], p. 3), which states that poverty is a major cause and effect of global environmental problems and that addressing environmental problems will require measures that address poverty and international inequality, and
(ii) that the direct link between poverty and environmental degradation is too simplistic and that the nexus is governed by a complex web factor as illustrated by Fig. 1.1 [17].

Due to the complexity between poverty and environment expressed in (ii) above, broad interpretations of both poverty and environment requires understanding the linkages between the two. The following subsections discusses these linkages and their importance.

1.2.1 Understanding the Poverty-Environment Linkages

The simplistic view between the poverty-environment linkage is that poverty leads to environmental degradation and vice versa, and that the two are linked in a downward and mutually reinforcing cycle (cause-effect-effect-cause relationship) (e.g., Nunan et al. [31], Forsyth and Leach [18], Prakash [39]). This cyclic line of thought was informed, e.g., by the view that poverty is the primary causes of environmental destruction, the poor cannot in their present state practise sustainable development

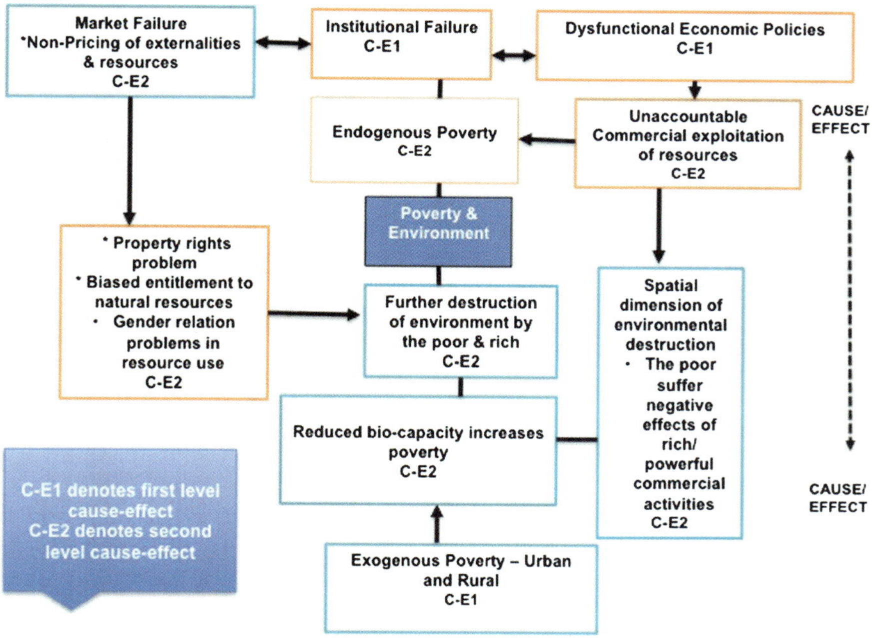

Fig. 1.1 Poverty-Environmental nexus (Intricate mix of non-linear casuality and effect)

(short-term maximisers), and if much of the environmental problem is poverty, then eliminating poverty and poor people through (economic) growth becomes key to saving the environment [8]. This is the same view propelled by the Brundtland Commission Report [9] in (ii) above.

Broad [8] rejects this circular argument and attempts to break down the conventional notions of the poverty and environment linkages by asking why are the poor poor? Recognising that some poor people act not as environmental degraders, but as environmental sustainers, and acknowledging that there are cases where the poor have become environmental activists led by three conditions based both on the people's relation to their ecosystem and on the state of the civil society. These conditions are [8]: *Environmental degradation is threatening the natural resource base off of which the poor live*; *poor people have lived in an area for some time or have some sense of permanence there*; and *civil society is politicized and organized*. Living in an area and calling it home, whereby a sense of belonging (permanence) results is expounded further in Sect. 1.2.3. Broad [8] finally concludes by stating:

> Categories for understanding poverty-environment relationship need to do better than simply rich-poor, environmentalist-non-environmentalist since the relationship between the poor and the environment depends on the "poor's historical relationship to their natural resource base, their perception of the future of that relationship, and the historical state of collective action within civil society".

Modern way of looking at the linkage between poverty and environment is moving away from the simplistic view above and adopting the more realistic complex view that incorporates *institutions* and *policy* issues. In this regard, Duraiappah [17] summarizes the factors found to be prominent in the poverty-environmental nexus as:

1. The activities by the rich and powerful are the primary contributing factors forcing groups living at the margins into poverty, i.e., a combination of *power, wealth, and greed causes environmental degradation* and *environmental degradation causes poverty* was predominant.
2. Institutional and market failures also play a prominent role in environmental degradation and subsequently poverty enhancement, i.e., a combination of *institutional and market failures being primary cause of environmental degradation* with *environmental degradation causing poverty*. In this regard, the activities of both the marginal and rich groups were influenced by these failures.
3. Endogenous poverty, i.e., poverty caused by environmental degradation was also found to be prominent. Duraiappah [17] continues "*Ninety percent of the studies show marginal groups adopting environmental-degrading activities. Of this 90 %, 10 % freely chose these activities. The remaining 90 % had no choice but to adopt unsustainable activities. The collapse or increased vulnerability of the income stream, caused in the first instance by the activities of the powerful and wealthy, left the marginal group with few options other than to adopt resource mining activities*".
4. The powerful and wealthy groups are reported not to have lost from accrued environmental degradation caused by their own activities, as well as the marginal groups. However, a rapid deterioration of natural resources implied a worsening situation for this group in the long run.

For example, the World Development Report [54] discusses the possible effects of the expected dramatic growth in the world's population, industrial output, use of energy, and demand for food [15]. For instance, the contribution of biofuel production has been associated, e.g., by Kgathi et al. [24] to influence increase in food prices. This is due to the fact that food availability is affected if food crops or productive resources such as land, labour, water are diverted from food production to that of biofuel (e.g., Kgathi et al. [24]). In addition, rise in the cost of oil also contributes to increase in food prices owing to industrial food production (e.g., farm machinery) and transportation costs on one hand and the conversion of food producing lands to biofuel production on the other hand (see, e.g., [29]). Increase in food prices leads to food insecurity with the poor being the most vulnerable. Food insecurity coupled with lack of access to safe and affordable water together with energy insecurity, to say the least, would make life for the poor, most of who depend on rainfed agriculture, unbearable. Neff et al. [29] concludes that:

> these threats—and our responses to them—will affect public health and society, not only directly through food security, but also via myriad economic, social, and environmental pathways.

[53] caution that under the existing practices, the result could be appalling environmental conditions in both urban and rural areas, and present an alternative, albeit more difficult, path—one that, if taken, would allow future generations to witness improved environmental conditions accompanied by rapid economic development and the virtual eradication of widespread poverty. Choosing this path will require that both industrial and developing countries seize the current moment of opportunity to reform policies, institutions, and aid programs [15]. Dean [15] suggest a two-fold strategy:

- ***First***, take advantage of the positive links between economic efficiency, income growth, and protection of the environment. This calls for accelerating programs for reducing poverty, removing distortions that encourage the economically inefficient and environmentally damaging use of natural resources, clarifying property rights, expanding programs for education (especially for girls), family planning services, sanitation and clean water, and agricultural extension, credit and research.
- ***Second***, break the negative links between economic activity and the environment. Certain targeted measures, described in the Report, can bring dramatic improvements in environmental quality at modest cost in investment and economic efficiency. To implement them will require overcoming the power of vested interests, building strong institutions, improving knowledge, encouraging participatory decision making, and building a partnership of cooperation between industrial and developing countries.

The complex approach view above is supported, e.g., by Prakash [39] who opine that the casual roots of environmental degradation lie in institutional and policy issues rather than in poverty, and that the relation between poverty and the environment is mediated by institutional, socio-economic and caltural factors (see e.g., Nunan et al. [31] and also Duraiappah [17]). Nunan et al. [31] propose the concept of *environmental entitlements*—defined by; government legislation, markets, common property resource-management arrangements, land tenure, customary rights, resources to 'make effective use of' for example capital and technology, and gender roles—as one approach to understanding the relationships between environment and poverty. They argue that the key issue raised by the approach is that the links between environmental change and impoverishment are not direct, but are mediated by the institutions, poor people's interactions with particular environments, and structured by macro-level processes (see e.g., Leach et al. [26], Leach and Mearns [25] and Duraiappah [17]). These sentiments are captured by Scherr [44] who put forward policies for jointly addressing poverty and environment, with the potential to generate increased livelihood security for poor farmers while also improving environmental conditions as; *local endowments*, *conditions for adoption of conservation technology* and *local institutions*. Scherr [44] then propose strategies for achieving joint reduction of poverty and improved natural resources as:

- Increasing poor people's access to natural resources essential to their livelihoods.
- Working with the poor to increase the productivity of their natural resources so they can take advantage of existing or emerging economic opportunities (by coinvesting

in on-farm natural resources of the poor, promoting agricultural technologies with environmental benefits and promoting low-risk perennial production in poor and marginal areas).

- Involving the poor in promoting good environmental management under conditions when economic incentives for doing so are not in place (by compensating the poor for conserving or managing resources important to others and by employing the poor to improve public natural resources).

1.2.2 Importance of the Poverty-Environment Linkages

It is difficult to discuss poverty without the mention of the environment. This is because the two are inter-twined such that the impact on poverty would definitely have some effect on the environment and like wise the impact on environment would have some effects on poverty. As a start, people live in the environment and rely on it for their livelihoods, health and security. For example, environmental pollution, which we discuss in Chap. 14 are likely to impact upon water resources, with a high likelihood of the poor who inhabit such environment risking health hazards. Other environmental issues pertinent to the livelihood of the poor included; soil degradation, changes in biodiversity, deforestation wood products, food and medicine, degradation of coastal areas, increasing water demands, over-abstraction and pollution, and natural disasters [16].

According to Nunan et al. [31], the environmental priorities suggested by the World Bank for the poor are *environmental health* and *natural resource management*, with the dimensions of poverty most affected by the environmental agenda being health, economic opportunity, security and empowerment [10]. Bucknall et al. [10] present environmental factors impacting on poverty as *quality of natural resource base, access to natural resources, access to water and toilets, air quality, access to environmental information*, and *ecological fragility* [31].

In the developing countries, most people who are classified as poor are found in rural areas where they earn their livelihood from toiling their farms. This is the case in Africa, Asia and parts of Eastern Europe. Scherr [44] acknowledges that agriculture remains the principle livelihood of the rural poor but adds [35]:

> Patterns of rural population growth, agricultural expansion and intensification and income growth projected for the next few decades pose serious challenges to achieving both environmental improvements and rural poverty reduction.

Besides agriculture, consequence of floods or droughts on the environment would be much felt by the poor who rely on the environment for farming on the one hand. On the other hand, poor farming methods leading to land degradation are likely to enhance the chances of environmental related disasters, e.g., floods and droughts. The situation is captured by a common farmer of Koshi River basin, Bihar (India) who in the midst of floods and drought states:

> The greatest exploiter for all of us are floods today, droughts tomorrow, earthquakes sometimes and all these multiply our trauma of deprivation, pains of poverty and hunger. These disasters take away not only our crops, shelters, lives of our families, friends, cattle but also destroy our hopes and dreams of the future. Is there any event comparable to these which causes so much human suffering and injustice? [22].

The linkage between poverty and environment, therefore, can be viewed through environmental disasters, health risks, and natural resources. Environmental disasters, whether of meteorological origin, e.g., cyclones, floods, hurricanes and droughts, or of having geological nature such as earthquakes and volcanoes impact on human life and economy, with the most vulnerable often being the poor who have little capability of coping. The gravity of the situation is captured by the 2011 drought that ravaged Somalia, parts of Kenya and Ethopia. Those mostly affected, staring death due to starvation, were the poor. Indeed, disaster trend reveal that the most vulnerable and harder hit are normally the poorest people, most of who live in developing countries. With tropical climate and unstable land forms, coupled with high population density, poverty, illiteracy and lack of infrastructure development, developing countries are more vulnerable to suffer from the damaging potential of disasters [22]. Yet the situation is not limited only to developing countries. In the big Tohoku Tsunami which struck Japan on 11th March 2011, those who were mostly affected, similar to the 2004 Tsunami of South East Asia, were the poor peasants.

From the foregoing discussion, it clearly emerges that one main link between environment and poverty revolves around *environmental disasters*, and hence the need to reduce the poor's vulnerability to environmental hazards. For the poor who rely on the environment to generate their food, environmental related disasters due to climate extremes, e.g., floods and droughts pose a great challenge. Awange et al. [2] show the relation between drought and food security situation in Lake Victoria region of East Africa, and obtain a drought cycle of 5–7 years. The Lake Victoria region becomes vulnerable as the food security is threatened, putting the poor into a more risky situation as they have little coping mechanism. Indeed, the food crisis experienced in recent times has been attributed by Chartres [11] to a growing population, changes in trade patterns, urbanization, dietary changes, biofuel production (see also Payne [33]), regional droughts, climate change, and declining availability of water that is needed to grow irrigated and rainfed crops. The link between, water, climate change, and drought on the one hand and food security on the other hand signifies the important link between poverty and environment, especially given that food is an important ingredient to define poverty as we saw in Sect. 1.1.1.

Environmental pollution as previously stated carries with it the health risk to the poor. Considering a basic necessity such as water, it is the poor who face difficulty affording clean consumable water. Unclean water brings with it risk of water-borne diseases. An example was the case of Haiti in 2011 where cholera broke out as a result of polluted water. Those who were mostly affected were the poor. Closely related to water resource is the issue of sanitation. In most areas where the poor dwell, it is normally a question of whether to focus on the stomach, i.e., food or pay attention to sanitation issues. In most cases, as we will discuss in this book, the poor cannot afford proper sanitation and as a result, they are more vulnerable to environmental health risks. As for the air pollution, though it is difficult to quantify the impact on the poor, compared to those who are not poor, it suffices to state that the poor will be hard hit to afford treatment while at the same time struggle to afford a decent livelihood.

Natural resource utilization is an area that has been at the forefront of conflict between man and the environment. An example is given by the iron ore mining near Pannawonica in the Pilbara region of Westerm Australia where in 2007, the *Environmental Protection Authority* rejected a $12 billion mining plans by Rio Tinto to protect a rare spider (troglobite) unique to the mining site.[5] This led to an uproar from the people who felt that it was unthinkable for spiders to be valued more than human beings who were going to loose their livelihood. In developing countries, though most countries now use the Environmental Impact Assessment (EIA) as a regulatory tool of protecting the environment against the exploitation of its natural resources, these EIA exist only as cosmetic and are rarely given legislation powers.

In Kenya, for example, it was in 2011 that the environment was elevated to a pillar position in the newly enacted constitution. Unlike Australia and the US, where environmental legislatives are given due consideration in resource exploitation, the Kenyan case as an example of a developing country is likely to shelve such environmental policies. This can be cited from its degradation of the Mau forest, where an attempt to conserve it generated quickly into regional and local politics. This supports the school of thought that in poverty-prone areas, environment is likely to suffer as people exploit the natural resources.

In recognition of the important link between poverty and environment, the World Bank promoted environmental improvement as a fundamental element of development and poverty reduction by stressing on three areas of intervention, which linked poverty reduction and environmental improvements. These are [7]:

(i) Enhancing the livelihoods of the poor people through natural resource management.
(ii) Preventing and reducing environmental health risks.
(iii) Reducing people's vulnerability to environmental hazards.

Nunan et al. [31] developed poverty-environment indicators by reviewing the environmental issues raised by the poor through participatory poverty assessments, which reviewed nearly 100 participatory poverty assessments (PPAs) from around the world, sources of information on environmental issues of relevance to the poor such as DFID [16], and the relevant indicator initiatives that have been or are in the process of being developed, e.g., the poverty-environment indicators developed for poverty reduction strategy programmes (PRSP) guidance by the World Bank. Environmental issues identified by Nunan et al. [31] as relevant to the poor included: environment and health (including malaria, diarrhoea and respiratory problems, particularly arising from indoor air pollution), forest cover, soil degradation, water quantity and quality, fisheries, and natural disasters. Some of these will be treated in part II of the book.

[5] The Sunday Morning Herald. http://www.smh.com.au/news/business/spiderlike-troglobite-stops-12b-mine/2007/03/29/1174761644076.html.

Indeed, that the World Bank sought to integrate environmental issues in the Poverty Reduction Strategy Programmes (PRSPs) is not surprising going by the concept of sustainability, which is discussed next.

1.2.3 The Concept of Sustainability

Sustainability has been defined as meeting the needs of current and future generations through integration of environmental protection, social advancement and economic prosperity [19]. Sustainability assessment (SA) can be performed when a proponent requests a regulator to do so (external) for the purpose of approval or internally as a mechanism for improving internal decision-making and the overall sustainability of the final proposal, see e.g., [37, 38].

Sustainable development has been defined as development which meets the needs of the present without compromising the ability of future generations to meet their own needs [9]. An alternate definition is given "as improving the quality of life while living within the carrying capacity of the supporting ecosystem" ([21], p. 23). Various interpretations have been given to the term sustainable development thereby undermining the very noble concept which underpins sustainable development. For example, Jacobs (ibid) points out the frustration and irritation common amongst participants in environmental debates to the fact that sustainability is never properly defined. Different attempts to define it can be seen, e.g., in [28]. This therefore highlights the importance of having a clear definition of sustainability development.

Lack of a clear definition can add to confusion to the political and academic debate around the term and contribute to other problems. Such problems have been identified, e.g., by [40]. Robinson (ibid) points out that unclear use of sustainable language can promote what is unsustainable leading to cosmetic environmentalism and fostering delusion. On the other hand, the [51] states that a clear definition may be unnecessary to design action programmes but basic conceptual guidelines must be drawn to set sustainability goals.

The case for a clear definition of the term sustainable development is further strengthens by [27] who provides a critical review of the definition and states that if sustainability development is to be really sustained as a development paradigm, two apparently divergent efforts are called for: Making sustainable development more precise in its conceptual underpinnings while allowing more flexibility, and diversity of approaches in developing strategies that may lead to the society living in harmony with the environment and itself.

Other advantages of a clear definition of sustainable development are [21, p. 23]:

- Sustainable development can be made operational in terms of policy if single and precise meaning can be agreed upon.
- A clear definition would enable measurable criteria upon which groups with wide differing values, political preference or assumptions about human nature could agree on whether the criteria are being met in concrete development program.

- There is a political concern amongst some environmentalist that lack of clear definition would allow masking of environmentally harmful projects as sustainable developments. In particular, the governments and private business are feared to be capable of using such "sugar coated" definitions only as "smoke screens" where in actual sense; their actions would be detrimental to the very concept of sustainable development.
- A clear definition would avoid the concept being dismissed by environmentalist as a development fad or embraced by those opposed to change in the status quo.

Clearly, the need for a precise definition of the term sustainable development can be seen from the discussion above. The hardest huddle, however, would be to agree on a single definition that will satisfy all the stakeholders who have diverse interpretation of the term. Just as the words love or democracy which does not have precise definition, perhaps the fundamentals that underpin the term sustainable (e.g., equity) need to be strengthened as there seem less acrimony over them as opposed to the definition.

Once a clear definition of sustainability has been achieved, it is worth looking at the key elements of a sustainable place, which makes people to preserve and conserve the environment. In Sect. 1.2.1, it was stated that Broad [8] listed place where people live in as a condition for becoming environmental activist. The first element of sustainable place presented by [50] is the notion of place making. Place is space given meaning and value by people and it contains social, physical and perceptual dimensions and often emphasise the importance of a sense of 'belonging' and identification with that place. Turner and Townsend [50] calls it an emotional attachment to 'place'. Instead of the term sense of belonging, Beatley and Manning [5] have used the term "sense of place" and reckon that a sense of place can be fostered if communities nurture built environment and settlement patterns that are uplifting, inspirational, and memorable. Such 'sense of place' engenders a special feeling of attachment and belonging.

A sustainable community also nurtures a sense of place by understanding and respecting its bioregional context-its topography and natural setting, its rivers, hilltops, open lands, native flora and fauna, and the many other unique elements of its natural context and respects the history and character of those existing features that nurture a sense of attachment to, and familiarity with, place. Once people are attached to a place, and feel a sense of belonging, they will be motivated to make it sustainable so as to preserve and conserve those values which they treasure.

Secondly, a sustainable place should be integrative and holistic [50]; [5, p. 33]. Rather than focusing only on one aspect, e.g., development, a sustainable place should employ strategies and solutions that will integrate the social, environment and economic aspects of a place. If one takes an example of a congested road, a sustainable place would not consider a narrow solution of expanding the congested road; rather, a broader sustainable solution of decongesting the road through examination of its land use strategies would be sought.

Thirdly, a sustainable place should strive for high quality of life, i.e., besides ecological issues, sustainable place should concern with social and human sustainability that create and support human living environment, liveable places, and communities that offer high quality of life [5, p. 30]. Turner and Townsend [50] refer to this element as the people and place dimension.

Further elements of a sustainable place are presented by [5]. According to Beatley and Manning (ibid), a sustainable place should: acknowledge fundamental ecological limits; should be restorative and regenerative; strive for high quality of life; be integrative and holistic; imply a new ethical posture; strive to be equitable and just; stress importance of community; and reflect and promote a full-cost accounting of social environmental costs of public and private decisions.

Summarising the foregoing discussion of [5] and [50], a sustainable place should therefore be a place where the cultural, social, economic and ecological aspects (layers) interact in a way that they can meet the needs of those relying on it but at the same time promotes all the sustainability layers.

1.3 Why 21st Century Solutions?

Poverty still remains one of the biggest challenges to humankind despite many cases where some nations have triumphed over the same problem. One would expect that for a problem already solved by others to a considerable extent, providing solutions in other places where there is dire need should be a matter of selection of methods and tools to deploy. Regrettably, that has not been the case with poverty challenge. Despite massive poverty reduction focus and efforts by supranational organizations, international NGOs, national governments and local non-governmental agencies, poverty still remains persistent in majority of countries.

The process of poverty reduction has also been a painfully slow one as attested by World Bank Poverty Data [48] on the percentage of those living on $1.25-a-day poverty line (2005 purchasing power parity prices). In a three-decade period starting from 1981 to 2011, out of 1.9 billion living in poverty, only 600 million people have been lifted out of poverty. This leaves about 1.3 billion people currently wallowing in abject poverty. Roughly the pace of progress in poverty reduction, going by the data here, points to a 32 % global poverty reduction rate over a three decade period. Some caveat is needed in understanding the figures given here since the measure given by the world Bank is only limited to income, and so does not measure the other dimensions of poverty.

It is also worth examining the true picture of progress in global poverty reduction efforts by casting a look at how much progress has been made on the other non-income dimensions of poverty. As such, Human Development Index (HDI) is the appropriate alternative measure upon which we can gauge progress in poverty reduction. The United Nations Development Program (UNDP), HDI Report 2013 projects a positive scenario indicating considerable improvements in HDI of the poorest countries for the period between 2000 and 2012. The report is bullish on the shifting global economic power as represented by emerging economies of Brazil, China and India, accompanying increased South–South trade and technology partnerships. The report notes that all these new developments have had positive effect on HDI in countries with the lowest indicators. The report is also cautious about declaring an all out success in tackling poverty because some countries and regions continue to lag behind, while income inequalities also still persist in many low HDI countries. In addition, UNDP's most recent program evaluation, which also included external evaluators points to a troubling poverty reduction scenario, specifically on the effectiveness of the programs, and clear returns on investments targeting poverty reduction, both directly and indirectly owing to the multi-dimensional nature of poverty. The report faults UNDP for spreading itself perhaps too thin, doing almost all things, while failing to clearly show returns on funds invested in poverty reduction.

The import of this report rests first on the idea that UNDP is supposed to be the global flagship organization on poverty reduction with best practices in tacking the challenge. As such, even the slightest doubts about the efficacy of UNDP's anti-poverty program casts a long shadow on poverty reduction efforts generally. While it may be too early to pass a verdict on UNDP's poverty reduction programs, based on one evaluation report, the World Bank statistics based on poverty head count percentage of national populations leaves us convinced beyond any reasonable doubt that global poverty reduction strategies in the extant and perhaps slowly evolving forms simply do not measure to the level of urgency within which the challenge needs to be tackled. That still large potions of many countries' population remain mired in poverty speaks volumes, to the point that no amount of defensive arguments and counter-arguments well presented can completely silence. Table 1.1 shows how large percentages of national populations are stuck in poverty despite recent sustained economic growth between 2000–2012. From the data set above, we can see that Sub-Sahara Africa and South Asia regions have not made much headway in reducing the portions of their population living in abject poverty. Madagascar, Zambia, Malawi and even Nigeria, which has experienced sustained and impressive economic growth rates stand out as the worst performers in terms of having far too many of their people living under poverty. To drive this point home vividly, lets us think that out of two Nigerians, one is living in abject poverty. This is set against the background of high average economic growth of 7.4 % over the past decade [1]. Sustained high economic growth that has failed to trickle down and uplift those living on $1.25-a-day is a shared characteristic of most of the African economies, so we can logically suggest that there is need for revision, at least, and at best adoption of new and innovative approaches to poverty reduction in the twenty first century. Yet, at the same time, we acknowledge that there are impressive cases of poverty reduction happening in tandem with sustained high economic growth. Such is the case of Vietnam, Cambodia, Brazil, Ecuador, and Ethiopia to some extent.

The critical issue here is how can we learn from these cases, and combine the lessons with innovative poverty reduction solutions given the dismal performance of approaches and solutions that have been on offer stretching from the past century into the new millennium. In essence, twenty first century solutions we strongly argue for is about being able to identify the very best methods of poverty reduction beyond the usual macroeconomic stability and openness to foreign direct investments, in combination with emerging innovative approaches to the problem. It is simply hard, if

Table 1.1 Poverty Headcount Ratio at $1.25-a-day (PPP) % of National Population. *Source* Compiled from World Bank Poverty Data [48]

Country	% of Population	Year
Nigeria	54.4	2011
Zambia	74.5	2010
Malawi	61.6	2010
Madagascar	81.3	2010
Senegal	26.6	2011
Uganda	38.0	2009
Angola	43.4	2009
Ethiopia	30.7	2011
Bangladesh	43.4	2010
India	32.7	2010
Nepal	24.8	2010
Vietnam	16.9	2008
Cambodia	18.6	2009
Georgia	18.0	2010
Philippines	18.4	2009
Bolivia	15.6	2008
Honduras	17.9	2009
Ecuador	4.6	2010
Dominican Republic	2.2	2010
Brazil	6.1	2009
Chile	1.4	2009
Uruguay	0.2	2010
Mexico	0.7	2010

not impossible to argue against the need for innovative and effective poverty reduction solutions and approaches when we know too well from the past several decades about the dismal effect anti-poverty programs have had. Finally, it is important to assert that such innovative approaches need to be multi-dimensional while still retaining focus element and measurable contributions to poverty reduction. It is within such logical orientation that this book combines the problems of poverty and environment.

1.4 Objectives and Aims of the Book

This book is intended to be of use to two main groups of readers; those who deal with poverty related issues, and those who are engaged in environmental related tasks. It is aimed at realizing two main objectives:

(1) The first task of this book is to help create an informed intellectual position about unbreakable connection between poverty and environmental issues. These problems are merely sides of one coin made of permeable material. Attributes of each side of this coin freely seep back and forth creating an inseparable mix. Simply put, any attempts to provide solutions to poverty or environmental ills cannot be separated. At the Rio+20, United Nations Conference on Sustainable Devel-

opment, the president of the African Development Bank, Dr. Donald Kaberuka drove this point home thus: "For Africa and low income countries of the world, whose essential livelihood depends on nature, poverty and environment are two sides of the same coin"—[1].

This thought paradigm lays the foundation for conception of this book, which calls for integrated solutions to poverty and environment. Practically, integrated solutions proposed in this book calls for sustainability elemement, which in the case of poor people, must address the immediate and short-term consumption needs, while also focusing on long-term solutions. Striking a balance between long-term and short-term needs of the poor calls for innovative solutions. Innovative solutions proposed in this book, e.g., geospatial-based solutions discussed in Chap. 2 justify the title reference to innovative solutions for the twenty-first Century.

(2) The second objective of this book is to draw attention to the problem of pervasively ill-conceived solutions to poverty due to lack of rigorous theoretical formulation of the problem. Proper theoretical formulation around issues beneficially leads to workable solutions. On the contrary, flawed theoretical framing of issues leads to faulty solutions. This book demonstrates the confused and flawed theoretical framework within which over the decades into present times, mainstream poverty measures and solutions continue to be prescribed both by governments and international multilateral institutions. We therefore draw attention to the problem of theoretical conception of poverty, and by logical extension bring into question often prescribed measures and solutions. Doing this is in itself very innovative. Innovative think starts by the orientation that we can question conventional practices, which more often than not are outcomes of innocent and unintentional assumptions. It is our hope that this book will spark an intellectual debate and exchange of ideas about what should be a pragmatic theoretical conception of poverty. Such debate would then lead to, not only new, effective and befitting tools for tackling poverty, but better tools and methods for arresting the problem of environmental degradation.

In Part I of this book, the concept of poverty are presented, hopefully in a manner that would be easily understood by environmentalists. Part II of the book presents the environmental link between poverty and environment.

1.5 Concluding Remarks

Two conclusions might be drawn from these brief historical attempts to define poverty. The first is that definitions, which are based on some conception of 'absolute' deprivation, disintegrate upon close and sustained examination and deserve to be abandoned. Poverty has been defined, in the words of an OECD review, "in terms of some absolute level of minimum needs, below which people are regarded as being poor, for purpose of social and government concern, and which does not change through time." The review tacitly acknowledges the intellectual weakness of this approach. In fact, people's needs, even for food, are conditioned by the society in which they live and to which they belong, and just as needs differ in different societies so they differ in different periods of the evolution of single societies. Any conception of poverty as 'absolute' is therefore inappropriate and misleading.

The second conclusion is that, though the principal definitions put forward historically have invoked some 'absolute' level of minimum needs, they have in practice represented rather narrow conceptions of relative deprivation and deserve to be clarified as such. Thus Rowntree's definition amounted in effect to a conception of nutritional deprivation relative to the level believed to be required for members of the manual working class at the turn of the century to function efficiently.

Finally, the income/consumption approach has also been criticized for paying insufficient attention to common property resources and state provided commodities as well as to vulnerability. For these reasons, both the theory and measurement of poverty have become controversial, thus leading to controversial policies to eradicate, alleviate, or reduce poverty.

Historically, the most influential definitions of poverty have been those expressed in terms of some absolute level of minimum needs, below which people are regarded as being poor, and which does not change through time. However, conceptions of poverty as 'absolute' were found to be inappropriate and misleading. People's needs, even for food, are conditioned by the society in which they live and to which they belong, and just as needs differ in different societies, so they differ in different periods of the evolution of single societies. In practice, previous definitions have represented narrow conceptions of relative deprivation sometimes associated only with what is necessary for the physical efficiency of the working classes. A fuller conception of relative deprivation needs to be adopted and spelt out.

The problem of poverty has attracted a lot of concern, and also justifiable frustration. Many of the attempts to document and explain it have been grounded in limited national and even parochial, not to say individualistic, conceptions. Until social scientists can provide the rigorous conception within which poverty can be examined, and the relationship between inequality and poverty perceived, the accumulation of data and the debates about the scale and causal antecedents of the problem will in large measure be fruitless.

References

1. African Development Bank (2012) African economic outlook- nigeria 2012. African Development Bank, Tunisia
2. Awange JL, Aluoch J, Ogallo L, Omulo M and Omondi P (2007) An assessment of frequency and severity of drought in the Lake Victoria region (Kenya) and its impact on food security. Clim Res 33:135–142
3. Batson E (1941) Social survey of cape town. Reports of the school of social science and social administration, University of Cape Town, pp 1941–1944
4. Batson E (1945) The poverty line in salisbury, Cape Town: University of Cape Town
5. Beatley T, Manning K (1997) The ecology of place: planning for environment, economy, and community. Island Press, Washington

6. Bettison DS (1960) The poverty datum line in central Africa. Rhodes Livingstone J 27:1–40
7. Bojo J, Reddy RC (2002) Mainstreaming environment poverty in reduction strategies. Environ strategy notes
8. Broad R (1994) The poor and the environment: friend or foes? World Dev 22(6):811822
9. Brundtland Commission Report (1987) Our common future. World commission on environment and development, Oxford University Press, Oxford
10. Bucknall J, Hamilton K, Kishor N, Kraus C, Pillai P (2000) Environment, in sourcebook for PRSPs (draft). World Bank, Washington
11. Chartres C (2008) Water: the forgotten crisis. Appropriate Technol 35(3):19–20
12. Collier P, Deepak L (1993) Poverty and growth in Kenya, world bank, staff working paper No. 389, Washington, DC: The World Bank; World Bank, Kenya: Employment Growth for Poverty Alleviation, Washington, DC: The World Bank
13. Crawford E, Thorbecke E (1978) Employment, income distribution, poverty alleviation and basic needs in Kenya. Report of an ILO consulting mission, Cornell University
14. De Montauge T (1977) Peasants into frenchmen: the modernization of rural france, 1870–1914. Chatto and Windus, London, p 2
15. Dean JM (1992) Trade and the environment: a survey of the literature. Background paper No. 3 for the world development report, Washington
16. DFID (2000) Achieving sustainability: poverty elimination and the environment. Department for International Development, London
17. Duraiappah A (1998) Poverty and environmental degradation: a review and analysis of the Nexus. World devel 26(12):2169–2179
18. Forsyth T, Leach M, with Scoones I (1998) Poverty and environment: priorities for research and policy: an overview study. Prepared for the united nations development programme and european commission, Institute of Development Studies, Brighton
19. Government of Western Australia (2003) Hope for the future: The Western Australian state sustainability strategy. Department of the Premier and Cabinet Perth, Western Australia. http://www.sustainability.dpc.wa.gov.au/docs/Final
20. Griffin K (1989) Alternative strategies for economic development. St. Martin's Press, Inc., New York
21. Jacobs M (1999) Sustainable development as a concept. Fairness and futurity, Oxford University Press, Oxford
22. Jayaraman V, Chandrasekhar MG, Rao UR (1997) Managing the natural disasters from space technology inputs. Elsevier Science Ltd, Great Britain
23. Johnson DL, Ambrose SH, Bassett TJ, Bowen ML, Crummey DE, Isaacson JS, Johnson DN, Lamb P, Saul M, Winter-Nelson AE (1997) Meanings of environmental terms. J Environ Qual 1997(26):581–589. doi:10.2134/jeq1997.00472425002600030002x
24. Kgathi DL, Mfundisi KB, Mmopelwa G, Mosepele K (2012) Potential impacts of biofuel development on food security in Botswana: a contribution to energy policy. Energy Policy 43: 70–79. doi:10.1016/j.enpol.2011.12.027
25. Leach M, Mearns R (1991) Poverty and the environment in developing countries. Economic and Social Research Council/Overseas Development Administration, London
26. Leach M, Mearns R, Scoones I (1997) Environmental entitlements: a framework for understanding the institutional dynamics of environmental change, IDS discussion paper No. 359, Institute of Development Studies, Brighton
27. Lele S (1991) Sustainable development: a critical review. World Dev 19:607–621
28. Mebratu D (1998) Sustainability and sustainable development: historical and conceptual review. Environ Impact Assess Rev 18:493–520
29. Neff RA, Parker CL, Kirschenmann FL, Tinch J, Lawrence RS (2011) Peak oil, food systems, and public health. Am J Public Health 101(9):1587–1597
30. Neefjes K (2000) Environments and livelihoods: strategies for sustainability. Oxfam, Oxford
31. Nunan F, Grant U, Bahiigwa G, Muramira T, Bajracharya P, Pritchard D, Vargas MJ (2002) A study of poverty-environment indicators with case studies from Nepal. Nicaragua and Uganda, Environment Policy Department Issue Paper No. 2

32. Orshansky M (1969) How poverty is measured. Monthly Labour Review, February, p 37. See Organization for Economic Cooperation and Development, Public Expenditure on Income Maintenance Programs, Studies in Resource Allocation No. 3, Paris, July 1976, pp 62–4
33. Payne WA (2010) Farming systems and food Security in sub-saharan Africa. In: Lal R, Stewart BA (eds) Food security and soil quality, Taylor and Francis, New York
34. Pillay PN (1973) A poverty datum line study among Africans in Durban, occasional paper No. 3, Department of Economics, University of Natal
35. Pinstrup-Andersen P, Pandya-Lorch R, Rosegrant M (1997) The world food situation: recent developments, emerging issues, and long-term prospects. 2020 Vision Food Policy Report, International Food Policy Research Institute, Washington
36. Poduluk JR (1967) Income distribution and poverty in Canada. Dominion Bureau of Statistics, 1968
37. Pope J (2006) What's so special about sustainability assessment? J Environ Assess Policy Manage 8: v–x. doi :10.1142/S1464333206002505
38. Pope J, Grace W (2006) Sustainability assessment in context: issues of process, policy, and governance. J Environ Assess Policy Manage 8:373–398. doi:10.1142/S1464333206002566
39. Prakash S (1997) Poverty and environment linkages in mountains and uplands: reflections on the 'Poverty Trap' Thesis, CREED working paper No. 12. International Institute for Environment and Development, London
40. Robinson J (2004) Squaring the circle. Ecol Econ 48:369–384
41. Rowntree BS (1901) Poverty: a study of town life. Macmillan, London
42. Sahlins M (1972) Stone age economics. Aldine-Atherton, Chicago
43. Sahlins M (1986) The original affluent society. Development 3:22
44. Scherr S (2000) A downward spiral? research evidence on the relationship between poverty and natural resource degradation. Food Policy 25:479–498
45. Sen AK (1983) Poor, relatively speaking. Oxf Econ Pap 35: 153
46. Sen AK (1984) Resources values and development. Blackwell, Oxford
47. Smith A (1776) The Wealth of nations, Book 5, Chapter 2 Part II, Methuen & Co., Ltd. London
48. The World Bank Poverty Data (2013) The world bank group, Washington D.C. Retrieved: http://data.worldbank.org/topic/poverty. Accessed 4 June 2013
49. Townsend P (1979) Poverty in the United Kingdom: a survey of household resources and standards of living. Penguin Books, London
50. Turner and Townsend (2004) Towards more sustainable place. Sustainable places, partnership working and urban regeneration skills. A Research Report for Turner & Townsend Group and the RICS Foundation. University of Aberdeen and Kevin Murray Associates.
51. United Nations Industrial Development Organisation (2008) Public goods for economic development. UNIDO, Vienna
52. Vandemortele JV (1982) Income distribution and poverty in Kenya: a statistical analysis. IDS, University of Nairobi, Nairobi
53. World Development Report (1992) Development and the environment. Oxford University Press
54. World Bank Development Report (1993) The world bank and the environment: the world development report 1992. World Dev 21(5):869–881

Chapter 2
Geospatial Analysis in Poverty-Environment Nexus

> Rigorous empirical studies that combine local-area environmental variables (deforestation, outdoor air quality, water quality, soil erosion, etc.) with standard household surveys are almost nonexistent. Similarly, very few local-area studies relate environmental quality to the number and characteristics of poor households. In poverty-environment analysis that is relevant for policy, the spatial dimension is critical [9].

2.1 Need for Geospatial in Poverty-Environmental Nexus

Throughout part III of the book, geospatial techniques are used to provide analysis of the poverty-environment link. Its use is informed by two main factors. First, poverty and environmental issues occur in space and as such are location or area based, i.e., geographical in nature. Dasgupta et al. [9] recommend that due to such spatial dimension of poverty-environment nexus, different environmental problems should be analyzed at different regional scales. They provide an example of pollution where they state [9]:

> In the case of pollution, for example, the theoretically appropriate scale is affected by the dispersal characteristics of the pollutant and medium: Particulate pollution from cement mills may only be dangerous in one urban region; acid rain from sulfur emissions may damage forests hundreds of miles from the source; and eutrophication from fertilizer runoff may affect ocean fisheries a thousand miles downstream from the farms that are the source of the problem. In practice, data constraints often dictate the choice of the scale.

Environmental degradation impacts on the poor, e.g., from pollution as exemplified above by [9] can take on the local (e.g., from particulate air pollution, which results in cardiorespiratory health problems), regional, or/and global (e.g., global warming).

Second, administrative economics are spatial based. To this effect, [9] writes:

> On the environment side, for example, effective regulation requires local inspection of damage sources (pollution, deforestation, etc.), as well as more centralized facilities for information collection, storage, and analysis. Environmental management is undoubtedly improved by a knowledge of local conditions, but the marginal cost of administration rises with distance

N. O. Agola and J. L. Awange, *Globalized Poverty and Environment*,
DOI: 10.1007/978-3-642-39733-2_2,

from administrative centers, because of deteriorating transport and communications quality. Generally, province- or district-level administration strikes the right balance between headquarters scale economies and the cost of dispersed monitoring and enforcement operations.

Analysis of environmental and poverty related issues, therefore, need to be spatially undertaken. This is because they occur in geographical space on the one hand, and from policy perspective, the poverty-environment nexus is relevant only if it has implications for the allocation and administration of public resources for alleviation of poverty and environmental problems on the other hand. Geospatial techniques, e.g., remote sensing, geographical information system (GIS), or Gravity Recovery and Climate Experiment (GRACE) satellites mission applied in part III of the book provide innovative solutions to the poverty-environment linkage.

In what follows (i.e., adopted from Awange and Kiema [2]), basics of geospatial, starting with the concept of space is presented in Sect. 2.2. Geodata, the basic elements of geospatial are then presented in Sect. 2.3 followed by the new paradigm of the digital Earth in Sect. 2.4. Section 2.5 then gives the fundamentals of geospatial necessary for the reader to understand its applications in part III of the book.

2.2 Dimensions of Space, Time and Scale

Understanding the characteristics of and possibilities in using geodata is premised on proper comprehension of the underlying concepts of space, time and scale, contextualized within the Earth's framework. Although these concepts are used in everyday parlance, often without much afterthought, they are not trivial at all. For instance, looking back throughout the entire history of mankind, the concepts of space and time have been the subject of animated philosophical, religious and scientific debates. In this section, we attempt to present a background of each of these dimensions of geodata, both independently and collectively, as well as highlight their relevance in influencing the character of geodata.

Space is that boundless, three-dimensional extent in which objects and events occur and have relative position and direction [5]. In analytical geometry, one examines "spaces" with different dimensionality and underlying structures. Indeed, the concept of space is considered to be of fundamental importance to an understanding of the physical universe although disagreement continues between philosophers over whether it is itself an entity, a relationship between entities, or part of a conceptual framework [36].

Philosophical debates on the nature, essence and the mode of existence of space date back to antiquity. From treatises like that championed by *Timaeus of Plato* in his reflections on what the Greeks called *khora* (i.e. space), to the physics of *Aristotle* in the definition of *topos* (i.e. place), or to even the geometrical conception of place as "*space qua extension*" by *Alhazen* [13].

Many of the classical philosophical assertions were later discussed and reformulated in the seventeenth century, particularly during the early development of classical mechanics. For example, in *Sir Isaac Newton's* view, space was absolute, in the sense that it existed permanently and independent of whether there were any matter in the space [15]. However, other philosophers like *Gottfried Leibnitz* were of the different view that space was a collection of relations between objects, given by their distance and direction from one another [36].

Up until around the eighteenth century, and within the framework of *Euclidean geometry*, space was perceived by most mathematicians to be flat. However, between the 19th and 20th centuries mathematicians began to examine *non-Euclidean geometries*, in which space was inferred to be curved, rather than flat. According to *Albert Einstein's* theory of general *relativity*, space around gravitational fields deviates from Euclidean space [6]. Furthermore, experimental tests of general relativity have confirmed that non-Euclidean space provides a better model for the shape of space as illustrated in Fig. 2.1.

Turning to the dimension of time, time is considered to be part of the measuring system used to sequence events, to compare the durations of events and the intervals between them, and to quantify rates of change such as the motions of objects [20]. The temporal position of events with respect to the transitory present is continually changing. For example, future events become present, then pass further back into the past. In the *Bible*, time is traditionally regarded as a medium for the passage of predestined events. Subsequently, there is an appointed time for everything, see e.g., *Ecclesiastes 3:1–8* [4]. Evidently, time has been a major subject in religion, philosophy, and science, but defining it in a non-controversial manner applicable to all fields of study has consistently eluded the greatest scholars [36].

Time is one of the seven fundamental physical quantities defined in the International System of (SI) Units. It is also used to define other quantities, such as velocity. An operational definition of time infers that observing a certain number of repetitions

Fig. 2.1 Concept of non-euclidean space

of one or another standard cyclical event (such as the passage of a free-swinging pendulum) constitutes one standard unit such as the second. This view is highly useful in the conduct of both advanced experiments and everyday affairs of life. However, this operational definition ignores the question whether there is something called *time*, apart from the counting activity that transits and can be measured [36].

Two contrasting assertions on time divide many prominent philosophers. The first view is that time is part of the fundamental structure of the universe, a dimension in which events occur in sequence. *Sir Isaac Newton* subscribed to this realistic view, and hence it is sometimes referred to as *Newtonian time*, see e.g., [26, 33, 34] etc. According to this view, time travel becomes a possibility as other "times" persist like frames of a film strip, spread out across the time line.

The second and opposing view contends that time does not refer to any kind of "container" that events and objects "move through", nor to any entity that "flows", but that it is instead part of a fundamental intellectual structure (together with space and number) within which humans sequence and compare events. This assertion, in the tradition of *Gottfried Leibnitz* [5] and *Immanuel Kant* (see e.g., Mattey [27], McCormick [28] etc.) holds that time is neither an event nor a thing, and thus it is not itself measurable nor can it be traveled.

Temporal measurement has occupied the minds of scientists for a long time and was the prime motivation in the disciplines of navigation and astronomy. Periodic events and periodic motion have long served as standards for units of time. Examples include the apparent motion of the sun across the sky, the phases of the moon, the swing of a pendulum, and the beat of a heart. Currently, the international unit of time, the *second*, is defined in terms of radiation emitted by cesium atoms. Time is also of significant social importance and is often viewed as having economic value as captured by the popular adage *time is money*, as well as personal value, due to an awareness of the limited and finite time in each day and in the human life span [36]. Consequently, different time scales are employed in different application domains, such as geological time [17, 18, 21], biological time [12, 19, 37] etc.

From the above discussion, regardless of the school of thought advanced, it is evident that historically, the dimensions of space and time have been closely related. As a matter of fact, it is virtually impossible to describe either of the two dimensions without inferring the other. Put together, these two dimensions represent the *space-time* concept expressed in Einstein's special relativity and general relativity theories. According to these theories, the concept of time depends on the spatial reference frame of the observer, and the human perception as well as the measurement by instruments such as clocks are different for observers in relative motion. Subsequently, the past is the set of events that can send light signals to the observer, whilst the future is the set of events to which the observer can send light signals [36].

This then brings us to the dimension of scale. The scale of a map is an important metric that defines the level of detail of geospatial information that can be extracted from such a map (see [2]). Scale also gives an indication of the resolution in the geodata. In general, a larger scale means that more geodata would be captured, including fuzzy detail that might otherwise be generalized or glossed over at smaller scales. The interpretation of scale is therefore important. For instance, by simply varying the map scale alone, the estimated distance between two points would vary. Many researches have studied the scale dimension and its perception and meaning in different applications, see e.g., [14, 22, 25, 35] etc. A review of space, time and scale from a geographer's perspective is given in Meentemeyer [29].

For many years, the dimension of scale was not explicitly integrated into data modeling. Therefore, scale was assumed to be uniform within a spatio-temporal context. This was done ostensibly to keep the whole geo-modeling problem simplified. The fact that classical maps could only be produced at one specific scale probably reiterates this. By convention, national mapping agencies had to designate certain mapping scales for different map coverages. This therefore enabled map users to identify the maps that were suitable for different applications. For example, in typical civil engineering work, whereas a scale of 1:50,000 would be appropriate at the reconnaissance or preliminary planning stage, larger scales of 1:500 – 1:2,000 would be required at the construction or maintenance phases (see, e.g., Awange [1] and [2]).

Evidently, the scale dimension has not evoked as much controversy as the twin dimensions of space and time. The issue with the scale dimension has been more to do with the scientific challenge of identifying appropriate data models and structures. Indeed, consideration of scale as an extra dimension of geographic information, fully integrated with the other dimensions, is a fairly recent proposition [31]. Whereas 3D space captures the geometrical characteristics of geodata, 4D integrates the temporal representation, with the 5D providing the scale definition. Meentemeyer [29] avers that most geographic research is now conducted with a relativistic view of space rather than a view of space as a "container". However, spatial scales for relative space are more difficult to define than those for the absolute space of cartography and remote sensing [29].

In concluding this section, it is important to recognize that the five dimensions of space, time and scale are integral to the unambiguous definition of position for they help to fully integrate 5D data modeling. Realizing this would ensure that geodata is used seamlessly with no undesirable overlaps or gaps and assuming consistency across space, time and scale dimensions. In future, probably the existence and relative importance of different classes in diverse applications could also be considered in a more integrated manner as the 6D of geodata—the *semantic* dimension [31].

2.3 Geodata

Data is simply defined as any set of raw facts or figures that have been collected, often in a systematic manner, and from which inference(s) may be drawn. Similarly, *information* is defined as any useful data that satisfies some user need(s). This is generally required to support the making of decisions. Apparently, data and information constitute the basic building blocks in the decision-making support infrastructure that also includes *evidence*, *knowledge* and *wisdom* as summarized in Table 2.1.

Table 2.1 Hierarchy of decision making support infrastructure. (Modified after Longley et al. [23])

Level of decision-making support infrastructure	Ease of sharing	Example
Wisdom ⇑	Impossible	Policies developed and accepted by stakeholders e.g., ideal use for parcel
Knowledge ⇑	Difficult (especially tacit knowledge)	Personal knowledge about places and issues e.g., adjoining parcel boundaries
Evidence ⇑	Often not easy	Results of spatial analysis of datasets or scenarios e.g., parcel area
Information ⇑	Easy	Contents of a database assembled from raw facts e.g., owner of parcel
Data	Easy	Raw facts and figures e.g., geographic coordinates

An *information system* is a combination of technical and human resources, together with a set of organizing procedures that produces information in support of decision-making usually to meet some managerial requirement. Thus an information system should be able to receive, store, process, update, output and distribute data and information. Classical information systems for general management are called *Management Information Systems* (MIS). They are distinguished from *Geographic Information Systems*, which are information systems that deal with spatially referenced data and are discussed in more detail in [2].

Data is distinguished as *geodata* (or *geospatial* data) if it can be geographically referenced in some consistent manner using for example; latitudes and longitudes, national coordinate grids, postal codes, electoral or administrative areas, watershed basins etc. As mentioned in Sect. 2.2, although geodata is normally defined in 3D in many practical applications, it needs to be redefined in 5D for geodata to be used without any restrictions in space, time or scale. The first three dimensions describe the geometric characteristics of geodata usually in 3D space. The fourth dimension provides the temporal representation that denotes how geodata has changed over time, while the scale is represented by the fifth dimension. This dimensional view of geodata is important for it ensures that there are no gaps or overlaps in the data. Furthermore, it also maintains the consistency of geodata across space, time and scale dimensions. Awange [1] and [2] review typical datums used in surveying.

Geodata may be collected by both government organizations as well as private agencies. A key characteristic of this type of data is its potential for diverse and multiple applications. Moreover, geodata can be shared and re-used by different users and applications through the *spatial data infrastructure* (SDI), see e.g., [7, 16, 24] etc. To infer the correct decision(s), it is imperative that the geodata be *accurate, complete, consistent* and *timely*. Furthermore, it is important that the required geodata be made available and that in addition, it also be allowed to flow unhindered to and between the various users and applications.

2.4 Digital Earth Concept

Digital Earth is the name given to a concept coined by former US vice president Al Gore in 1998, that describes a virtual representation of the Earth that is spatially referenced and interconnected with the world's digital knowledge archives.[1] Furthermore, the greater part of this knowledge store would be free to all via the *Internet*. However, a commercial marketplace of related products and services was envisioned to co-exist, in part in order to support the expensive infrastructure that such a system would require [36].

Clearly, many aspects of this vision have been realized, evidenced in part by the popularity of virtual globe geo-browsers such as *Google Earth*[2] for commercial, social and scientific applications as discussed in Awange and Kiema [2]. But the Gore speech outlined a truly global, collaborative linking of systems that has yet to be fully realized [36]. That vision has been continually interpreted and refined by the growing global community of interest. As technological advances have made the unlikely possible, the vision has evolved and become more concrete, and as we better understand the interdependence of the environment and social activities, there is greater recognition of the need for such a system. Digital Earth has come to stand for the large and growing set of web-based geographic computing systems worldwide. These are both useful and promising, but do not yet constitute the envisioned *global commons* [36].

The global dimension of the digital Earth concept is perhaps best captured by two excerpts from the Beijing declaration[3] on digital Earth, which state as follows [3]:

(a) Digital Earth is an integral part of other advanced technologies including: Earth observation, geo-information systems, global positioning systems, communication networks, sensor webs, electromagnetic identifiers, virtual reality, grid computation, etc. It is seen as a global strategic contributor to scientific and technological developments, and will be a catalyst in finding solutions to international scientific and societal issues;

(b) Digital Earth should play a strategic and sustainable role in addressing such challenges to human society as natural resource depletion, food and water insecurity, energy shortages, environmental degradation, natural disasters response, population explosion, and, in particular, global climate change.

A consortium of international geographic and environmental scientists from government, industry, and academia brought together by the *Vespucci Initiative for the Advancement of Geographic Information Science, and the Joint Research Center of*

---

[1] In a speech prepared for the California Science Center in Los Angeles on January 31, 1998, Gore described a digital future where school children—indeed all the world's citizens—could interact with a computer-generated three-dimensional spinning virtual globe and access vast amounts of scientific and cultural information to help them understand the Earth and its human activities.

[2] http://www.earth.google.com

[3] Ratified on September 12, 2009 at the 6th international symposium on digital earth in Beijing, Peoples Republic of China.

the European Commission published a position paper that outlined the eight key next generation digital Earth elements to include the following [8]:

(1) Not one digital Earth, but multiple connected globes/infrastructures addressing the needs of different audiences: citizens, communities, policy-makers, scientists, educationalists;
(2) Problem oriented: e.g., environment, health, societal benefit areas, and transparent on the impacts of technologies on the environment;
(3) Allowing search through time and space to find similar/analogous situations with real time data from both sensors and humans (different from what existing GIS can do, and different from adding analytical functions to a virtual globe);
(4) Asking questions about change, identification of anomalies in space in both human and environmental domains (flag things that are not consistent with their surroundings in real time);
(5) Enabling access to data, information, services, and models as well as scenarios and forecasts: from simple queries to complex analyses across the environmental and social domains;
(6) Supporting the visualization of abstract concepts and data types (e.g., low income, poor health, and semantics);
(7) Based on open access, and participation across multiple technological platforms, and media (e.g., text, voice and multi-media); and
(8) Engaging, interactive, exploratory, and a laboratory for learning and for multidisciplinary education and science.

2.5 Fundamentals of Geospatial

Having introduced the 5D datum paradigm that needs to be adequately dealt with to define geodata accurately, consistently, timely and completely so that it can be used without any restrictions in space, time or scale and further, having appreciated the truly global dimension of the digital Earth, to put everything in perspective, it is now appropriate to focus on geospatial. Like for all other disciplines elaborated in this book it is only right to begin this discussion with pertinent definitions.

Although geospatial is a fairly recent terminology, various definitions of the same have been advanced by different authors. For instance, Raju [32] describes geospatial as "the science and technology dealing with the structure and character of spatial information, its capture, its classification and qualification, its storage, processing, portrayal and dissemination, including the infrastructure necessary to secure optimal use of this information". Similarly, Ehlers [11] defines geospatial as "the art, science or technology dealing with the acquisition, storage, processing, production, presentation and dissemination of geoinformation".

The bottom line is that there is no globally accepted definition of geospatial. However, as a multidisciplinary field, geospatial has at its core different technologies that support the acquisition, analysis and visualization of geodata. The geodata is usually acquired from Earth observation sensors as remotely sensed images, analyzed by geographic information systems (GIS) and visualized on paper or on computer screens. Furthermore, it combines geospatial analysis and modeling, development of geospatial databases, information systems design, human-computer interaction and both wired and wireless networking technologies. Geospatial uses geocomputation and geovisualization for analyzing geoinformation. Typical branches of geospatial include: *cartography, geodesy, geographic information systems, global navigation satellite systems (GNSS), e.g.,* [1], *photogrammetry, remote sensing, and web mapping*. These different disciplines that have been developed over different time epochs are applied in part III of this book. By combining the ever-increasing computational power, modern telecommunications technologies, abundant and diverse geodata, and more advanced image analysis algorithms available, and integrating technologies such as remote sensing, GIS and GNSS, many opportunities for application of geospatial have been realized. Today, many applications routinely benefit from geospatial including; urban planning and land use management, in-car navigation systems, virtual globes, public health, local and national gazetteer management, environmental modeling and analysis, military, transport network planning and management, agriculture, meteorology and climate change, oceanography and coupled ocean and

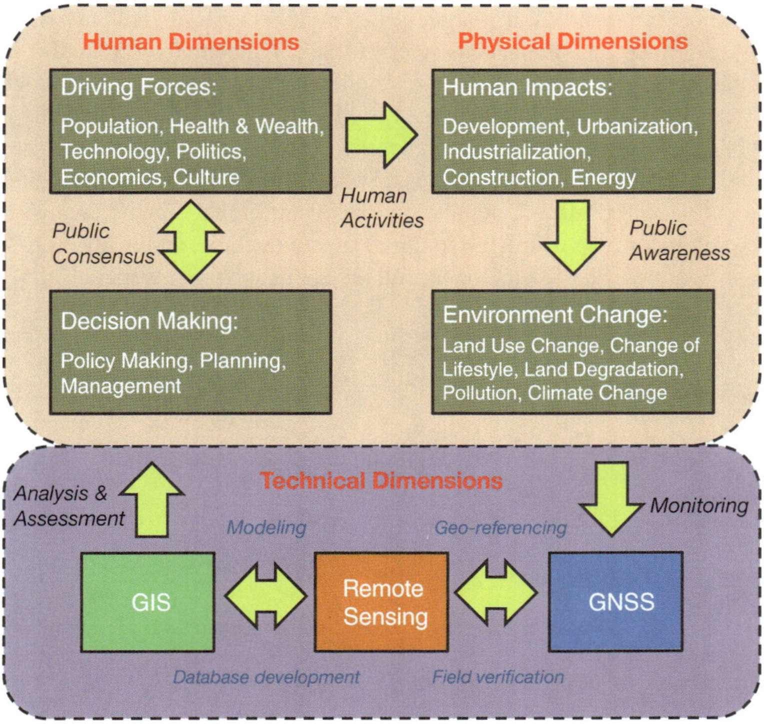

Fig. 2.2 Conceptual framework showing the role of geospatial in spatial decision support (Modified after [30])

atmosphere modeling, business location planning, architecture and archaeological reconstruction, telecommunications, criminology and crime simulation, aviation and maritime transport etc.

Consequently, geospatial has become a very important technology to decision-makers across a wide range of disciplines, industries, commercial sector, environmental agencies, local and national government, research and academia, national survey and mapping organizations, international organizations, United Nations, emergency services, public health and epidemiology, crime mapping, transportation and infrastructure, information technology industries, GIS consulting firms, environmental management agencies, tourist industry, utility companies, market analysis and e-commerce, mineral exploration etc. Increasingly, many government and non government agencies worldwide are using geodata and geospatial for managing their day to day activities. Figure 2.2 shows a conceptual framework that underlines the role of geospatial in supporting spatial decision-making.

2.6 Concluding Remarks

Although still unusual in many practical mapping constructs worldwide, a 5D coordinate reference framework is, nonetheless, desirable. This would not only ensure that geodata are defined accurately, consistently, timely and completely, but also guarantee that they are employed without any restrictions whatsoever in terms of space, time and/or scale. There is no doubt that, perhaps more that ever before, humanity faces a myriad of complex and demanding challenges today. These include natural resource depletion, food and water insecurity, energy shortages, environmental degradation, intermittent natural disasters, population explosion, global climate change etc. To develop pragmatic and sustainable solutions to address these and many other similar challenges requires the use of geodata and the application of geospatial.

References

1. Awange JL (2012) Environmental monitoring using GNSS. Global navigation satellite system, Springer, Berlin, New York
2. Awange JL, Kiema JBK (2013) Environmental geoinformatics. Monitoring and management, Springer, Berlin, New York
3. Beijing (2009) Proceedings of the 6th international symposium on digital earth (ISDE6), 9–12th Sept Beijing
4. Bible (2011) King James Version. www.kingjamesbibleonline.org. Accessed on 4/05/2011
5. Burnham D (2006) Gottfried Wilhelm Leibniz (16461716) Metaphysics 7. Space, time, and indiscernibles. The internet encyclopedia of philosophy
6. Carnap R (1995) An introduction to the philosophy of science. Courier Dover Publications, New York, p 300
7. Cookbook SDI (2004) Deeveloping spatial data infrastructures: The SDI cookbook. Global spatial data infrastructur. In: Nebert DD (ed) Technical working group chair, GSDI

8. Craglia M, Goodchild MF, Annoni A, Camara G, Gould M, Kuhn W, Mark D, Masser I, Maguire D, Liang S, Parsons E (2008) Next-generation digital earth. Int J Spat Data Infrastruct Res 3:146–167
9. Dasgupta S, Deichmann U, Meisner C, Wheeler D (2005) Where is the poverty environment nexus? Evidence from Cambodia, Lao PDR, and Vietnam. World Dev 33(4):617638. doi:10.1016/j.worlddev.2004.10.003
10. Duraiappah A (1998) Poverty and environmental degradation: a review and analysis of the Nexus. World Dev 26(12):2169–2179
11. Ehlers M (2003) Geoinformatics and digital earth initiatives: a German perspective. Int J Digital Earth 1(1):17–30
12. Enright JT (1965) The search for rhythmicity in biological time-series. J Theor Biol 8(3):426–468
13. El-Bizri Nader (2007) In defence of the sovereignty of philosophy: al-Baghdadi's critique of Ibn al-Haytham's geometrisation of place. Arabic Sci Philos Hist J 17:57–80 Cambridge University Press
14. Fisher P, Wood J, Cheng, (2004) Where is Helvellyn? Fuzziness of multi-scale landscape morphometry. Trans Inst Brit Geogr 29:106–128
15. French AP, Ebison M (2007) Introduction to classical mechanics. Chapman and Hall, London p 341
16. Groot R, McLaughlin J (eds) (2000) Geospatial data infrastructure: concepts, cases and good practice. Oxford University Press, New York
17. Haq BU (2006) The geological time table Wallchart, Elsevier, 6th ed
18. Harland WB, Armstrong, RL, Craig, LE, Smith, AG, Smith, DG (1989) A geologic time scale. Cambridge university press, Cambridge
19. Hochachka PW, Guppy M (1987) Metabolic arrest and the control of biological time. ISBN 0-674-56976-8
20. Internet Encyclopedia of Philosophy (2011) http://www.iep.utm.edu/. Accessed on 13/03/2011
21. Kulp JL (1961) Geologic time scale. Science 133(3459):1105–1114. doi:10.1126/science.133.3459.1105
22. Levin SA (1992) The problem of pattern and scale in ecology. Ecology 73(6):1943–1967
23. Longley PA, Goodchild MF, Maguire DJ, Rhind DW (2005) Geographic information systems and science. John Wiley, West Sussex, England
24. Maguire DJ, Longley PA (2005) The emergence of geoportals and their role in spatial data infrastructures. Comput Environ Urban Syst 29(1):3–14
25. Mandelbrot B (1967) How long Is the coast of Britain? In: statistical self-similarity and fractional dimension. Science 156(3775):636–638
26. Markosian N. (2002) Time. In: Zalta EN (ed)The stanford encyclopedia of philosophy. The Metaphysics Research Lab, Stanford University, Stanford
27. Mattey GJ (1997) Critique of pure reason, lecture notes: philosophy 175 UC Davis. http://hume.ucdavis.edu/mattey/phi175/practicallechead.html. Accessed 30 Aug 2013
28. McCormick M (2006) Immanuel Kant (17241804) Metaphysics: 4. Kant's Transcendental Idealism, The Internet encyclopedia of philosophy
29. Meentemeyer V (1989) Geographical perspectives of space, time, and scale. Landscape Ecol 3(3–4):163–173. doi:10.1007/BF00131535,Springer
30. Murai S (1999) GIS Work Book: Fundamental and Technical courses. Vols.1 & 2. National Space Development Agency of Japan (NASDA) / Remote Sensing Technology Center of Japan (RESTEC). Japan Association of surveyors.
31. Oosterom P, Stoter J (2010) 5D Data Modelling: full integration of 2D/3D Space, time and scale dimensions. Fabrikant SI et al (eds) GIScience 2010, LNCS 6292. Springer, Heidelberg, p 310324
32. Raju PLN (2003) Fundamentals of geographic information systems. In: Sivakumar MVK, Roy PS, Harmsen K, Saha SK (eds) Workshop: satellite remote sensing and GIS applications in agricultural meteorology, India

33. Rynasiewicza R (1995a) By their properties, causes and effects: Newton's Scholium on time, space, place and motion. Part I: the text. Stud Hist Philos Sci 26:133–153
34. Rynasiewicza R (1995b) By their properties, causes and effects: Newton's Scholium on time, space, place and motion. Part II: the context. Stud Hist Philos Sci 26:295–321
35. Tate N, Wood J (2001) Fractals and scale dependencies in topography. In: Tate N, Atkinson P (eds) Modelling scale in geographical information science, Wiley, Chichester, p 3551
36. Wikipedia (2011) http://en.wikipedia.org/wiki/. Accessed on 28/08/2011
37. Winfree AT (2001) The geometry of biological time. Springer, [ISBN 0387989927], p 777

Part II
Poverty

3

Chapter 3
Theoretical Framework for Analyzing Poverty

3.1 Introduction: Analysis of the Concept of Poverty

The concepts of poverty have developed rapidly over the last thirty years, and international attention is now focused more sharply on poverty reduction than it was twenty years ago. The Development Committee of the OECD countries has proposed reduction by half by 2015 the proportion of people living in extreme poverty. This proposal has been widely adopted by the international community. But exactly what this target might mean is obscured by the bewildering ambiguity with which the term is used, and by the many different indicators proposed to monitor poverty. Is the interest on income poverty or human development? Or, is it on sustainable livelihood or social inclusion? Or is it on current consumption or future security? Different concepts imply different interventions.

The current debate is inconclusive on a number of issues. For instance, there are debates on the importance of monetary variables, an objective or subjective measures, and on the link between material income and the wider 'functioning' in society. There is general agreement that money income (or consumption) on its own is an imperfect measure of welfare, and also the need to take into account variability of measures over time. The idea of relative deprivation is also widely accepted. However, there are different views about the relative importance of non-monetary variables, like self-esteem, and about the weight that should be given to the views expressed by poor people themselves.

These conceptual problems have arisen because poverty is a difficult and complex concept. There are, at least, four different concepts and indices of poverty: Income Poverty Index, Human Poverty Index, Human Development Index, and the Quality of Life Index. The last of these are of recent origin. National Poverty Eradication Plan (NPEP) in various countries presents an elaborate structure of how poverty is to be eradicated. However, the various NPEP fails to discuss the concept of poverty to be eradicated. In fact, NPEP does not address the last three concepts of poverty. In failing to do so, it becomes difficult to ascertain clearly what it is that is being eradicated. Indeed, it is difficult to distinguish between the type of poverty conceived

N. O. Agola and J. L. Awange, *Globalized Poverty and Environment*,
DOI: 10.1007/978-3-642-39733-2_3, © Springer-Verlag Berlin Heidelberg 2014

and the policy promulgated to deal with it. This chapter discusses the various types of poverty as well as the problems inherent in the conception of poverty.

3.2 Adjustment Poverty and Structural Poverty

In a country undertaking economic reforms, there are, broadly speaking, two types of poverty: adjustment poverty and structural poverty. Since the early 1980s, economists have paid a lot of attention to adjustment poverty and what compensatory measure governments can take. It is important to distinguish between the two kinds of poverty that exist in various countries. The first is the "adjustment poverty" that is, the poverty created during the structural adjustment process. The second is the "structural poverty" which is the poverty that existed before governments embarked on structural adjustment programs. It encompasses the vast majority of the poor. Adjustment poverty is largely an urban phenomenon; structural poverty exists in both cities and rural areas. Both kinds of poverty are addressed in this study.

3.3 Towards an Understanding of the Concept of Poverty

What is poverty? This is a critical factor; defining poverty is not an easy task. There will always be disagreement over what constitutes a poverty income. Some analysts define poverty in terms of the amount of income necessary to provide a family of a certain size with the minimum essentials of food, clothing, shelter, and education. This approach provides an absolute poverty standard. An absolute poverty standard establishes a specific income level for a given-size household below which the household is judged to be living in a state of poverty.

But is an absolute measure of poverty appropriate? Poverty can, after all, be relative. One's sense of poverty depends upon the incomes of others in the community. A second approach to poverty, therefore, is to measure it in relative terms. A relative poverty standard might classify a household as poor if the household's income is, say 25 % of average household's income. Thus, relative poverty standard defines poverty in terms of incomes of others.

The choice of a poverty definition will determine to a great extent the number of poor and the rate at which poverty is perceived as being eliminated. If the absolute standard is selected, rising real living standards will push more and more families above the poverty line. According to the relative standard, only equalizing the distribution of income can eliminate poverty.

The conceptual debate on poverty is carried over to measurement. A small craft industry has emerged, especially at the international level, in measuring poverty and deprivation, often in response to the need to define targets at international conferences and measure progress against them. Different models of poverty imply different indicators. Advocates of the participatory paradigm, in particular, are wary of quantification and standardization.

How is poverty measured? Is it simply about the level of income obtained by households or individuals? Is it about lack of access to social services? Or is it more correctly understood as the inability to participate in society, economically, socially, culturally or politically? These are the questions that have occupied the attention of researchers for decades. While some researchers and development practitioners have argued that poverty can be defined only subjectively, others have maintained a contrary position. For instance, Townsend has argued that "Poverty can be defined objectively and applied consistently only in terms of the concept of relative deprivation. The term is understood objectively rather than subjectively. Individuals, families and groups in the population can be said to be in poverty when they lack the resources to obtain the types of diet basic requirements, participate in the activities and have the living conditions and amenities which are customary, or are at least widely encouraged or approved, in the societies to which they belong. Their resources are so seriously below those commanded by the average individual or family, that they are in effect excluded from ordinary living patterns, customs and activities" [17].

This 'objective' definition has implications for policy that should be recognized. Although all societies have ways of identifying and trying to deal with their problems, the social sciences are having an increasing influence upon decision-makers, both in providing information and implicitly or explicitly legitimizing action. A clear definition allows the scale and degree as well as the nature of the problem of poverty to be identified, and therefore points to the scale as well as the kind of remedial action that might be taken. Such an action may involve not just the general level of benefits, for example, but revision of relativity between benefits received by different types of families.

Poverty is a difficult concept, which has become increasingly controversial. The difficulty arises from the fact that poverty is a portmanteau term, having distinct meanings to different people. Words such as "destitution", "ill-being", "powerlessness" and "vulnerability" are so frequently used in conjunction with "poverty". Poverty may be seen in terms of people being excluded from the "living conditions and amenities which are customary, or at least widely encouraged or approved, in the societies to which they belong" [17, p. 31], or what is often referred to as a 'participation standard'. In adopting this definition, the European Community stated: "The poor shall be taken to mean persons, families and groups of persons whose resources (material, cultural, and social) are so limited as to exclude them from the minimum acceptable way of life in the state in which they live".[1] This basic problem of meaning of poverty pervades the debates both on the measurement of poverty and on poverty reduction policies. Although most people have an intuitive notion of what poverty is all about, the concept has been defined in various ways and is still a subject of much controversy in academic and policy-making circles [4]. The diversity of meanings attached to poverty renders conceptualization of poverty and its operational meaning and measurement difficult as well as intractable.

---

[1] European Community, Council Decision, 19 December 1984.

While poverty can be broadly defined as an absence of well being or capacities, it is multidimensional and manifests itself in various forms. This makes the definition of poverty inadequate, if only one criterion is used. It should also be recognized that there is no single indicator that can adequately measure all dimensions of poverty. Furthermore, it should be noted that poverty and inequality, though related at some points, are different concepts. While poverty is concerned with absolute standard of living of a part of the society (the poor), inequality refers to relative living standards across the whole society. It is therefore difficult for one survey like the Welfare Monitoring Survey (WMS) series,[2] used for instance in Kenya in recent years, to capture the many dimensions of poverty. In the WMS analysis, poverty is defined in terms of total household consumption expenditure. A household is considered poor if it cannot attain some recommended food energy intake. The required level of nutrient intake is 2,250 calories per day per adult plus a minimum allowance for non-food consumption. The notion of what constitutes the poor goes beyond an attainment of a level of material well being to constitute a reasonable minimum by the standards of a given society. A household, which does not meet its calorie intake requirements, but has a relatively high income, is not considered poor.

This raises the question of whether, ultimately, it is possible to take an absolutist view of poverty. Sen [14] has argued persuasively "that ultimately poverty must be seen to be primarily an absolute notion". It is of course true that there is "an irreducible absolutist core" [14] to the notion of poverty. If someone is starving, he is poor even if everyone else is starving too. But once one moves away from extreme cases it becomes much more difficult to make assertions that carry conviction or inspire confidence. For example, despite decades of work by nutritionists and physiologists we still do not know what is the minimum calorie intake necessary for adequate nutrition [7, p. 12]. More awkward for the absolutist position, we do not even know, as some claim, whether individuals really do have fixed requirements, i.e., whether the body is able to adjust to a persistent shortfall of calories and thereby in effect avoid "undernutrition" [7, p. 12].

In the 18th century, Adam Smith made it clear that "necessities" were determined by "custom" and hence that poverty was relative. He said, "By necessities I understand not only the commodities which are indispensably necessary for the support of life, but whatever the custom of the country renders it indecent for creditable people, even the lowest order, to be without" [16]. Theron de Montauge bluntly asserted that "poverty is measured by comparisons [6, p. 2]." It has nothing to do with physiological need for calories [7].

The absolutist view of poverty, however, does not depend on the results of physiological research. It is much more sophisticated than that. Emphasis is placed not only on the absolute need for adequate nutrition but also on the need for decent shelter, to live without shame and to have self-respect. The needs for decency, self-respect and the avoidance of shame are absolute, but the commodity requirements to fulfill these socially defined needs rise with average prosperity. In this way absolute

[2] Details of the Welfare Monitoring Surveys are reported in the Economic Survey 1997 and have been extensively used in this study. The Append of this study contains the findings of these surveys.

needs or capabilities are transformed into relative requirements for commodities and income [15].

Perhaps there is enough common ground here between relativists and absolutists to make further discussion unnecessary. But once it is accepted that needs, or at least some needs, are socially defined, we are very close to accepting that poverty itself is socially defined. That is, poverty is neither a relationship between a person and a bundle of commodities nor is it a relationship between a person and a bundle of (socially defined) *needs and capabilities. It is a relationship between one person and another.* This interpretation of the nature of poverty has emerged most forcefully from the research of Marshall Sahlins and others on the economies of extremely "primitive", "backward" and "poor" people such as the Bushmen of the Kalahari Desert, hunter-gatherers in the forests of South America and the aboriginal Australians [12]. The economic anthropologists have shown that these "primitive" people do indeed have an objectively low standard of living. However, they also have much leisure time, are generous in sharing their possessions with others, are essentially egalitarian, and most important of all, do not regard themselves, individually or collectively, as poor. As Sahlins (1972) says,

> The world's most primitive people have few possessions, but they are not poor. Poverty is not a certain small amount of goods, nor is it just a relation between means and ends; above all, it is a relation between people. Poverty is a social status. As such it is the invention of civilization [13].

If civilization invented poverty and if social status defines it, presumably man in society, operating through the institutions he has created, or if necessary through new institutions, can abolish poverty [7, p. 13].

Previous operational definitions of poverty have not been expressed in thorough-going relativist terms, nor founded comprehensively on the key concepts of resources and styles of living. The concern has been with narrower concepts of income and the maintenance of physical efficiency. In reference to most developing economies, poverty has not been clearly defined. The issue of poverty has been confined to discussing the concepts of "relative" and "absolute" poverty, but not much else. The definition of poverty and its measurement has been influenced by the work of Rowntree, as discussed below.

Historically, the standard economic definition of poverty in terms of income and consumption date back to Booth's 1892 and Rowntree's work in Victorian England. However, this work has been challenged by modern social scientists looking for broader and more inclusive definitions of ill-being in both developed and developing countries.

Among the early studies of poverty, the work of Rowntree is most important. He had a narrow conceptualization of primary poverty in terms of nutritionally based poverty lines. In 1899, he collected detailed information about families in York. He defined families whose 'total earnings were insufficient to obtain the minimum necessaries for the maintenance of merely physical efficiency as being in primary poverty' [11]. Rowntree made use of the work of W. O. Atwater, an American nutritionist, who had experimented with the diets of prisoners in Scotland to find how nutritional intakes were related to the maintenance of body weight. Rowntree estimated the average nutritional needs of adults and children, translated these needs into quantities of different foods and hence into cash equivalent of these foods. To these costs for food he added minimum sums for clothing, fuel and household sundries according to size of family. Rent was treated as an unavoidable addition to the calculation, and was counted in full. A family was therefore regarded as being in poverty if its income minus rent fell short of the poverty line.

Nearly all subsequent studies were influenced deeply by this application of the concept of subsistence. With minor adaptations, a stream of area surveys of poverty based on Rowntree's methods was carried out in Britain, South Africa, Canada and Tanazania. The subsistence standard was used as a measuring rod, or as a basis for recommending minimum social security rates and minimum earnings in many countries [1–3, 9, 10]. In recent years, this methodology has been used to measure poverty in countries like Kenya [5, 18].

But the standards that were adopted proved difficult to defend. Rowntree's estimates of the costs of necessities other than food were based either on his own and others' opinions or, as in the case of clothing, on the actual expenditure of those among a small selection of poor families who spent the least. Does the actual expenditure of the poorest families represent what they need to spend on certain items? Neither in his studies nor in similar studies were criteria of need, independent of personal judgment, or of the minimum amounts actually spent on certain goods, put forward.

In the case of food it seemed, at first sight, that independent criteria of need had been produced. But the Rowntree's procedures have been faulted on three main grounds. First, estimates of the nutrients required were very broad averages and were not varied by age and family composition, still less by occupation and activity outside work. Secondly, the foods that were selected to meet these estimates were selected arbitrarily, with a view to securing minimally adequate nutrition at lowest cost, rather than in correspondence with diets that are conventional among the poorer working classes. Finally, the cost of food in the total cost of subsistence formed a much higher percentage than in ordinary experience. In relation to the budgets and customs of life of ordinary people, the make-up of the subsistence budget was unbalanced.

An adaptation of the Rowntree method is in use in many developing countries. Kenya's National Development Plan, 1979–1983, for instance, under the theme of meeting the "Basic Needs", used the Rowntree's nutritional method. A basic standard of nutritional adequacy was put forward by the Ministry of Economic Planning, and this standard was translated into quantities of types of food compatible with the preferences of Kenyan families. This, in turn, was then translated into the minimum costs of purchases on the market. Finally, by reference to the average sums spent per capita on food as a proportion of all income (derived from consumer expenditure surveys), it was assumed that food costs represent a fixed percentage of the total income needed by families of three or more persons.

A number of points in the argument can be examined critically. First, and most important, the index is not redefined periodically to take account of changing customs and needs. In a rapidly changing society like Kenya, dietary customs and needs are liable to change equally rapidly and estimates of need must be reviewed frequently. Otherwise, there is the risk of reading the needs of the present generation as if they were those of the past. Foods are processed differently, and presented from time to time in new forms, whether in recipe or packaging. Real prices may rise without any corresponding improvement in nutritional content. No price index can cope properly with changes in ingredients, quality and availability of and 'need for' goods and services. There is no known way of how to adjust a poverty line to conform to changes in productivity.

Secondly, nutritional needs are narrowly defined. The cost of buying a minimally adequate diet for families restrict the kind and quality of their purchases and skill exercised in preparing as well as buying food, is worked out. Nothing extra is allowed for eating meals out, and the amounts are enough only for 'temporary or emergency use when funds are low.' The underlying definitions of dietary adequacy are insufficiently related to actual performance of occupational and social roles. Estimates of nutritional needs in fact include a larger element for activities that are socially and occupationally determined than for activities that are biologically and physiologically determined. Moreover, the former obviously varies widely among individuals and communities. While it may seem to be reasonable to average nutritional requirements, empirical studies of diets in relationship to incomes and activities have to be undertaken to demonstrate whether that procedure is in fact as reasonable as it purports to be.

Finally, the question of finding criteria for *needs* other than *food* is dodged by estimating food costs and then taking these as a fixed percentage of the total budget stated to be necessary. The percentage varies with households of different size and is lower for farm families than for other families. How, therefore, are the percentages chosen? This remains the nagging problem about the entire procedure. One researcher has addressed this issue as follows:

> Poverty, like beauty, lies in the eye of the beholder. Poverty is a value judgment; it is not something one can verify or demonstrate, except by inference and suggestion, even with a measure of error. To say who is poor is to use all sorts of value judgments. The concept has to be limited by the purpose, which is to be served by the definition? In the [USA] Social Security Administration, poverty was first defined in terms of the public or policy issue; to how many people, and to which ones, did we wish to direct policy concern [8, p. 37].

This may be shrewd but scarcely reassuring. Socio-economic measures cannot rest only on imaginable or even politically acceptable, but must also rest on demonstrable, definitions of social conditions. These may be difficult to apply consistently. There are bound to be difficulties and disadvantages in any approach that is well developed. In the final analysis, a definition of poverty may have to rest on value judgments. But this does not mean that a definition cannot be objective and that it cannot be distinguished from social or individual opinion.

3.4 Concluding Remarks

Two conclusions might be drawn from these brief historical attempts to define poverty. The first is that definitions, which are based on some conception of 'absolute' deprivation, disintegrate upon close and sustained examination and deserve to be abandoned. Poverty has been defined, in the words of an OECD review, "in terms of some absolute level of minimum needs, below which people are regarded as being poor, for purpose of social and government concern, and which does not change through time." The review tacitly acknowledges the intellectual weakness of this approach. In fact, people's needs, even for food, are conditioned by the society in which they live and to which they belong, and just as needs differ in different societies so they differ in different periods of the evolution of single societies. Any conception of poverty as 'absolute' is therefore inappropriate and misleading.

The second conclusion is that, though the principal definitions put forward historically have invoked some 'absolute' level of minimum needs, they have in practice represented rather narrow conceptions of relative deprivation and deserve to be clarified as such. Thus Rowntree's definition amounted in effect to a conception of nutritional deprivation relative to the level believed to be required for members of the manual working class at the turn of the century to function efficiently.

Finally, the income/consumption approach has also been criticized for paying insufficient attention to common property resources and state provided commodities as well as to vulnerability. For these reasons, both the theory and measurement of poverty have become controversial, thus leading to controversial policies to eradicate, alleviate, or reduce poverty.

Historically, the most influential definitions of poverty have been those expressed in terms of some absolute level of minimum needs, below which people are regarded as being poor, and which does not change through time. However, conceptions of poverty as 'absolute' were found to be inappropriate and misleading. People's needs, even for food, are conditioned by the society in which they live and to which they belong, and just as needs differ in different societies, so they differ in different periods of the evolution of single societies. In practice, previous definitions have represented narrow conceptions of relative deprivation sometimes associated only with what is necessary for the physical efficiency of the working classes. A fuller conception of relative deprivation needs to be adopted and spelt out.

The problem of poverty has attracted a lot of concern, and also justifiable frustration. Many of the attempts to document and explain it have been grounded in limited national and even parochial, not to say individualistic, conceptions. Until social scientists can provide the rigorous conception within which poverty can be examined, and the relationship between inequality and poverty perceived, the accumulation of data and the debates about the scale and causal antecedents of the problem will in large measure be fruitless.

References

1. Batson E (1941) Social survey of Cape Town, reports of the school of social science and social administration. University of Cape Town, New York
2. Batson E (1945) The poverty line in salisbury. University of Cape Town, New York
3. Bettison DS (1960) The poverty datum line in central Africa. Rhodes Livingstone J 21(27): 32–63
4. Collier P, Deepak L (1993) Poverty and Growth in Kenya, World Bank, Staff Working Paper No. 389, Washington, DC: The World Bank; World Bank, Kenya: employment growth for poverty alleviation, Washington, DC: The World Bank
5. Crawford E, Thorbecke E (1978) Employment, income distribution poverty alleviation and basic needs in Kenya, report of an ILO consulting mission. Cornell University, Ithaca
6. De Montauge T (1977) Peasants into Frenchmen: the modernization of rural France, 1870–1914. Chatto and Windus, London, p 2
7. Griffin K (1989) Alternative strategies for economic development. St. Martin's Press, New York
8. Orshansky M (1969) How poverty is measured. Monthly labour review, February, p 37. See Organization for Economic Cooperation and Development, Public expenditure on income maintenance programs, studies in resource allocation No. 3, Paris, July 1976, pp 62–64
9. Pillay PN (1973) A poverty datum line study among Africans in Durban, Occasional Paper No. 3. Department of Economics, University of Natal
10. Poduluk JR (1967) Income distribution and poverty in Canada. Dominion Bureau of Statistics, 1968
11. Rowntree BS (1901) Poverty: a study of town life. Macmillan, London
12. Sahlins M (1972) Stone age economics. Aldine-Atherton, Chicago
13. Sahlins M (1968) Notes on the Original Affluent Society. In: Lee RB, DeVore I (eds) Man the Hunter, New York: Aldine Publishing Company, pp. 85–89
14. Sen AK (1983) Poor, relatively speaking. Oxf Econ Pap 35:153–169
15. Sen AK (1984) Resources, values and development. Oxford, Blackwell
16. Smith A (1776) The Wealth of Nations, Book 5, Chapter 2 Part II, Methuen & Co., Ltd. London
17. Townsend P (1979) Poverty in the United Kingdom: a survey of household resources and standards of living. Penguin Books, London
18. Vandemortele JV (1982) Income distribution and poverty in Kenya: a statistical analysis. University of Nairobi, Nairobi

Chapter 4
Economic Theories of Poverty and Poverty Eradication

"I believe that we can create a poverty-free world because poverty is not created by poor people. It has been created and sustained by the economic and social systems that we have designed for ourselves; the institutions and concepts that make up that system; the policies that we pursue."

Muhammad Yunus

4.1 Rigorous Theoretical Grounding

The role of economic theory in policy formulation cannot be overemphasized. Theory provides the foundation upon which policy decisions are conceived, deliberated and eventually made. Ideally, the adoption of economic policy should depend heavily on economic theory, which should provide the necessary framework for policy formulation and decision making, as well as for predicting the likely outcomes of policy prescriptions. It is important therefore that decision-makers should have a clear understanding of the theoretical underpinnings of policies to be adopted. Therefore it is pertinent that we address the question of whether there an economic theory of poverty. How relevant are the prevailing economic theories in analyzing and adopting policies for global poverty eradication? What role do economic doctrines received from 'above' play in the formulation of anti-poverty policies? It goes without questioning that should fault lie in theoretical framework upon which poverty eradication policies and measures are formulated, then such poverty eradication measure are from the onset destined to fail.

Economists prefer to provide a theoretical framework for discussing policy. Theory provides parameters within which to analyze and measure a phenomenon at hand. Yet, economic theory is seriously lacking in the discussion presented in majority of National Poverty Eradication Plans (NPEP) as detailed in the Poverty Reduction Strategy Papers (PRSPs). The absence of a theoretical framework creates a

N. O. Agola and J. L. Awange, *Globalized Poverty and Environment*,
DOI: 10.1007/978-3-642-39733-2_4, © Springer-Verlag Berlin Heidelberg 2014

situation in which there is policy formulation without theoretical basis. For example, the Plans make predictions about what is expected to happen at each of the implementation stages, but without buttressing it with theoretical assumptions underpinning the expected results. Indeed, it is difficult to see how policy-makers reached conclusions on the likely outcomes of policy prescriptions for each stage without resorting to theoretical framework to provide guidance in drawing those conclusions.

The above statement does not presuppose that there exist viable economic theories of poverty. It should be noted that the received economic doctrines from above as embodied in PRSPs provide weak theoretical structures that do not lend themselves readily to credible analysis and policy prescriptions concerning poverty alleviation. The doctrines behind the PRSPs originate from the two controlling institutions, the World Bank and International Monetary Fund, and include the Neoclassical, dual labour market, the radical, the new political theories of poverty, and Sen's theory of poverty. These are discussed below. Each one of these theories leaves a lot to be desired. Indeed, given the dearth of poverty, it may well be that some of the most innovative ways of tackling poverty may well come from the developing countries, rather than from the doctrines from the two controlling institutions, their main funding sources and the accompanying business interests.

4.2 The Neoclassical Economic Theory of Poverty

Economic development has traditionally focused on the gap between the rich and the poor countries and on ways the process of growth, especially in poor countries, could be accelerated. Only very recently has the gap between the rich and the poor people in both the developed and developing countries been noticed. This belated concern with absolute and relative poverty as well as with income distribution has presumably been promoted by the realization that, even though development occurs and per capita incomes grow the numbers of poor people also increases.

The legacy of the neoclassical paradigm of development is probably as much responsible as any other single factor for the blindness of development economics with respect to income distribution. Three components can be distinguished in the traditional dogma on income distribution. First, income inequality is necessary for growth and efficiency. In the neoclassical paradigm, specialization and exchange are predicated on the existence of differing initial endowments. Differences in incomes merely reflect the underlying differences in physical and human assets. Moreover, income inequality is likely to increase the savings rate and thereby raise investment and the rate of growth.

Secondly, the neoclassical economists argue that, in the presence of exchange and growth, income inequality is bound to decrease. Indeed, in the harmonious world of automatic and equilibrating adjustment mechanisms of the neoclassical paradigm, income trickles down and spreads. Therefore, should we be able to take care of the rate of growth in GNP, poverty would be taken care of. Economic development reality glaringly defy this theoretical framework.

Thirdly, proponents of the neoclassical paradigm maintain that, should unwanted income differences still exist after the operation of the "trickling-down" mechanisms, the Keynesian economics offer another line of defense: the reliance on taxes and subsidies for redistributing income and reducing or even eliminating poverty. Such a scheme is especially attractive when it refers to redistribution of the increment of GNP, of the national dividend growth, as it were, and therefore does not impair future growth possibilities.

The proponents of the traditional neoclassical approach have, however, overlooked three critical points. The first is that the differences in personal incomes seem much too vast to be explained by differences in factor endowments alone. Secondly, the magnitude of the transfers necessary in order to reduce absolute poverty, or at least to decrease relative income inequality, has been seriously underestimated. Thirdly, the traditional approach has failed to recognize that the institutions responsible for growth are not neutral to income distribution. Indeed, it is almost inevitable that the largest portion of the increment in GNP will go to the wealthiest few. Moreover, it is difficult and virtually impossible to devise and implement tax reforms that would redistribute the national dividend to any significant extent.

That income distribution has long been a neglected subject in economics is also evidenced by the relative scarcity of analytical and empirical work on the subject. Despite conceptual shortcomings and paucity of data, the available estimates of income inequality almost invariably suggest that incomes are significantly more unequally distributed in the developing than in the developed countries. Moreover, the relatively high level of income inequality in developing countries persists and in many cases increases, at least in the early stages of development [1, 2, 8].

This evidence suggests that economic dualism of the "have" and "have-nots" exists and persists over time. In our opinion, this is explained by the persistence of market imperfections that originally gave rise to dualism. These imperfections usually result from the efforts of individual groups to establish rent-maximizing positions. Motivations to enhance one's absolute as well as relative position are, obviously, neither temporary nor transitory phenomena. Moreover, once these positions are established, they tend to be reinforced by devoting the rents at least partly to attempts to further reduce effective competition. Public officials and civil servants can also benefit from market imperfections, forming a powerful political coalition for the preservation of distortions and for the unequal distribution of the rents that arise from them. In this way, dualism and the unequal distribution of income are intimately interconnected and both are perpetuated.

This view of dualism is contrary to the neoclassical theory of dualistic development, which considers dualism a temporary aberration, bound to disappear as development takes hold. The approach to the subject adopted in this book shares at least a common tone with that of the sociologists and anthropologists, who have considered dualism as a permanent phenomenon to be reconciled with, as well as with that of the radical economists, who have seen "separatism" or "ghetto economic development" as the most promising means of achieving growth with redistribution. In our view, dualism can solved as presented in our solutions for the twenty-first

Century Chapter, with an elucidation of how Japan solved the problem of duality in the course of their economic development.

For over a century, measurement of poverty relied exclusively on the concept of the poverty line. This device has constituted the standard economic definitions of poverty in terms of income and consumption and date back to Booth's 1892 and Rowntree's work in Victorian England [3]. Rowntree's narrow conceptualization of 'primary' poverty in terms of nutritionally based poverty lines has been challenged by modern social scientists looking for broader and more inclusive definitions of ill-being in both developed and developing countries. The income/consumption approach has also been criticized for paying insufficient attention to common property resources and state-provided commodities as well as to vulnerability. For these reasons, both the traditional theory of poverty line and its measurement have become controversial. Consequently, there have been controversies over policies to eradicate, alleviate, or reduce poverty.

The Neoclassical economic theory of poverty is based on the concept of marginal productivity of labour. Although there are significant variations among theorists, the core of the orthodox economic theory, as it seems to have been applied to income distribution and poverty, might be said to consist of the certain ingredients. Amongst these are the assumptions of perfect competition and market equilibrium—believed to be sufficiently borne out by the market processes of advanced capitalist economies to demonstrate a strong relationship between wages and marginal productivity. According to the marginal productivity theory, the owners of the factors of production receive a price that equals the factor's marginal revenue product (MRP). Under this system, those who are more productive (those who have high MRPs because of special skills or hard effort) will receive more than those with low MRPs. Lester Thurow has explained that, 'If an individual's income is too low, his productivity is too low. The individual's income can be increased only if his productivity can be raised' [10, p. 26]. This approach provided the initial theoretical justification for the subsequent examination of productivity components, like education, skill or ability and experience, in explaining variations in wages. It also fitted into conventional theories of demand and supply by permitting a fairly sophisticated elaboration of the productivity characteristics of the labour that was supplied. As a result, human capital theory evolved. Gordon summed it up this way: 'Employers demand what workers supply stocks of "human capital" embodied in individuals' [7, p. 31].

The main problem about this theory is that different factors are invoked to explain different aspects of income distribution and a consistently interrelated set of concepts is not presented. The concepts of 'managerial responsibility' and 'hierarchic organization' are believed to explain high incomes, but it could be asked why these should apply only to high incomes, and at what point they cease to apply and whether there is any evidence that they apply to lower incomes.

4.3 Dual Labour Market Theory of Poverty

A large number of economists have begun to adopt approaches that depart from the neoclassical orthodox assumptions. Some of them direct attention to the nature of the demand for labour and to forces other than individual characteristics that determine wage levels. Others direct attention to 'aspects of the labour market such as trade unions, employers' monopoly power and government intervention, which mean that there is no longer perfect competition'. Studies of local conditions have favoured adoption of a concept of a segmented labour market. Accumulated evidence is in favor of there being an internal labour market within an established firm or plant, which appeared in many respects to be insulated from the outside labour market. This observation had further led to the conception of a dual labour market. On the one hand, attention was called to a 'primary' sector in which employment was stable, where pay was good and where there were strong trade unions. On the other hand, attention was also called to a 'secondary' sector in which employment was unstable, where pay tended to be low, prospects of promotion poor and unions of small importance. In explaining poverty, then, emphasis was placed as much on the disadvantageous characteristics of the secondary labour market as upon the characteristics of the individual holding such jobs. But the explanation offered by the dual labour market theory was inadequate to explain in full the phenomenon of poverty. A new theory therefore emerged to explain the poverty problem.

4.4 Radical Economic Theory of Poverty

The 'radical' economic theory of poverty draws heavily on the Marxist tradition, 'but it has molded and recast Classical Marxism in response to modern social and historical developments' [4]. According to such theory, the market price of a product affects the value of an individual's marginal product, just as it does according to orthodox theory. Supply and demand, reinforced by competition, affect an individual's productivity. But the radical theory 'also postulates that the class division in society and the relative distribution among classes will affect the distribution of individual income as well. An individual's class will, ultimately, affect both his productivity, through the allocation of social resources to invest in the workers of his class and through the differential access of different classes to different kinds of complementary capital, and his relative share of final product' [7, p. 53].

Some economists think, "The radical theorists' criticism of the orthodox approach is more effective than their expositions of alternatives. Their challenge is simply that they do not offer a theory of the labour market that can replace neoclassical theory" [5]. The problem is that there is disagreement on the criteria of what makes for good theory. Those who assume a competitive and perfectly functioning labour market preclude serious discussion of such factors as trade unions, employers' monopoly power and state intervention. None the less, so-called radical theory remains to be developed. There are those who have argued that non-participants in the labour force, including old people and the handicapped, are in poverty as a result of their past labour force status. A comprehensive summary of the radical approach is Watchel [11] but little attempt has been made to analyze the low income status of those not in labour force, to widen the discussion of inequality of distribution from earnings to other resources, further to widen the discussion of income recipients from individuals to income units and households, and, finally, to analyze the contemporary class structure in any depth.

4.5 The New Political Economy

The central paradigm of the new political economy is simple: scholars see states as mounted, as it were, upon markets; they then analyze forms of governmental intervention that take place, infer or impute the objectives that lead to them, and evaluate their effects on efficiency and distribution. The new political economy assumes the existence of perfectly competitive markets and builds its analysis upon the market distortions introduced by governments. These distortions are used to measure the power of private interests and political forces.

An important contribution of this approach has been to document the high costs of forms of market intervention commonly found in the developing countries and to show not only the losses in terms of efficiency and growth but also the social costs of patterns of inequality. Indeed, the new political economy has led to a comprehensive reevaluation of the relative merits of governments and markets in the developing countries. Traditionally, development economists stressed the **failure of markets** and advocated an active role for governments. The new political economy instead emphasizes the failure of governments and advocates a wider role for markets. Given the ideological divisions that traditionally center upon the relative roles of governments and markets, highlighting the positive role of markets has made the new political economy controversial and important.

The new political economy draws almost exclusively upon market economics. However, not all economics is market economics. Indeed, contemporary frontiers in microeconomics focus on environments where the conditions for markets to operate do not hold. A major lesson in much of this literature is that non-market forms of organization enable welfare-enhancing exchanges to take place—exchanges that could not take place in market environments. The presumption of market-oriented economists is that non-market forms of intervention impose economic costs; the presumption of these alternative approaches is that they promote gains in welfare. It is difficult for the new political economists to deal with the literature on non-market economics. The market imperfection and market failure paradigms provided much of the justification for the activist prescriptions they seek to overturn.

The new political economy postulates that governments often do end up with policies that constitute an ordering of choices for their societies. These policies, however, do not represent the preferences of some single actors. They represent the outcome of a political struggle, in which competing interests with rival visions of the social good seek the power to impose policies that are consonant with their preferences upon the collectivity. To account for government policy choices, then, we must study how competition among these preferences is structured and thereby leads to choices by governments.

4.6 Functional Theory of Poverty

The functional theory points out that in all societies there are different social positions or statuses. These vary in pleasantness and difficulty and 'functional importance' for society. In order to guarantee that all positions are filled, certain rewards have to be associated with them. Hence, inequality is necessary so that the positions are filled. Functions of Poverty Gans [6] a sociologist, argues that society is so preoccupied outwardly with the 'costs' of poverty that it fails to identify the corresponding benefits, or rather, the groups or values who benefit. He describes fifteen sets of functions, as follows:

1. Poverty helps to ensure that dirty, dangerous, menial and undignified work gets done.
2. The poor subsidize the affluent by saving them money (for example, domestic servants, medical guinea pigs, and the poor paying regressive taxes).
3. Poverty creates jobs in a number of professions (e.g. drug peddlers, prostitutes, pawnshops, army, and police).
4. The poor buy shoddy, stale and damaged goods (e.g. several days-old bread, vegetables, and second-hand clothes) which prolongs their economic usefulness, and similarly use poorly trained and incompetent professional people, such as doctors and teachers.
5. The poor help to uphold the legitimacy of dominant norms by providing examples of deviance (e.g. the lazy, spendthrift, dishonest, and promiscuous).
6. The poor help to provide emotional satisfaction, evoking compassion, pity and charity, so that the affluent may feel righteous.
7. The poor offer affluent people vicarious participation in sexual, alcoholic and narcotic behaviour.
8. Poverty helps guarantee the status of the non-poor.
9. The poor assist in the upward mobility of the non-poor. (By being denied educational opportunities or being stereotyped as stupid or unteacheable, the poor enable others to obtain the better jobs).
10. The poor add to the social viability of non-economic groups (e.g. fund-raising, running settlements, other philanthropic activities).
11. The poor perform cultural functions, like providing labour for Egyptian pyramids, Greek temples, and medieval churches.
12. The poor provide 'low' culture, which is often adopted by the more affluent (e.g. jazz, blues, spirituals, and country music).

13. The poor serve as symbolic constituencies and opponents for several political groups (being seen either as the depressed or as 'welfare chiselers').
14. The poor can absorb economic and political costs of change and growth in societies (e.g. reconstruction of city centres, industrialization).
15. The poor play a relatively small part in the political process and indirectly allow the interests of others to become dominant and distort the system.

Gans concludes that 'phenomena like poverty can be eliminated only when it either becomes sufficiently dysfunctional for the affluent or when the poor can obtain enough power to change the system of social stratification' [6].

4.7 Sen's Approach to Poverty Analysis

Amartya K. Sen contends that traditional welfare economics, which stresses the revealed preferences or desire-based utilities of individuals in their acts of choice, lacks enough information about people's preferences to assess the social good. Accordingly, as an alternative, Sen's welfare theory relies not on individuals' attainments (for example, of basic needs) but individuals' capabilities, an approach he believes can draw on a richer information base. For Sen, living consists of a vector of interrelated functioning (beings and doings), such as being adequately nourished, avoiding premature mortality, appearing in public without shame, being happy, and being free. Yet Sen does not assign particular weights to these functionalities, as well, being is a "broad and partly opaque concept," which is intrinsically ambiguous.

Sen focuses on a small number of basic functionalities central to well-being: to be adequately fed and sheltered and in good health. GNP per capita, which measures capabilities, does not correlate closely with these functions, which indicate attainments as Sen contends. One example is life expectancy, a proxy for health, which, at 75 years, is almost as high for Costa Rica as for the US (76 years), which has an income per head 12 times as high.

For Sen, poverty is not low well being but the inability to pursue well being because of the lack of economic means. This lack may not always result from a deficiency of capabilities. Poverty is the failure of basic capabilities to reach minimally acceptable levels [9].

4.8 Concluding Remarks

In conclusion, it can be readily seen that economic theories from 'above' have not succeeded in effectively explaining the phenomenon of poverty. Each one of the theories discussed above has something to say about poverty, but none has succeeded in presenting a conclusive analysis from which anti-poverty policies could be drawn. This means that those involved in poverty alleviation or eradication must venture into activities, in which there are no clearly articulated theories to go by. This conclusion does extend to the activities of the World Bank and IMF, two institutions behind the incidental global convergence in least developed countries' national governments' economic development policy and implementation approach to poverty reduction. In the absence of a credible economic theory on which poverty reduction needs to be based, the PRSPs might prove to be another embarrassingly costly economic experiment ending in lamentable failure just as was the case of the infamous structural adjustment policies (SAPs).

References

1. Adelman I, Morris CT (1973) Economic growth and social equity in developing countries. Stanford University Press, Stanford
2. Ahluwalia MS, Carter NG, Chenery HB (1974) Growth and poverty in developing countries. J Dev Econ 6:229–431
3. Booth C (1892) Life and labour of the people of London. Macmillan, London
4. Bosanquet N, Doeringer P (1973) Is there a dual labour market in Britain? Econ J 83:421–435
5. Cain GG (1975) The challenge of dual and radical theories of the labour market to orthodox theory. In: Proceedings of the American economic association, May
6. Gans H (1972) The positive functions of poverty in society. Am J Sociol 78(2):1972–1973
7. Gordon DM (1972) Theories of poverty and underemployment. Lexington Books, Lexington
8. Kuznets S (1955) Economic growth and income inequality. Am Econ Rev 45:1–28
9. Sen AK (1992) Inequality reexamined. Harvard University Press, Cambridge
10. Thurow LC (1969) Poverty and discrimination. Brookings Institution, Washington
11. Watchel HM (1971) Looking at poverty from a radical perspective, review of radical political economics, summer 1971. In: Roby P (ed) The poverty establishment, Prentice-Hall, Englewood Cliffs, 1974

Chapter 5
Traditional Measures of Poverty Lines

5.1 Introductory Remarks

The development of poverty measures has taken three historical stages. The first stage, beginning from 1892 and lasting for more than half a century, was the head-count ratio of poverty measurement spearheaded by Booth [2] and popularized by Rowntree [7]. It is the ratio of the number of poor individuals to the total number of individuals in the population surveyed. The head count ratio, however, does not tell us how far below the poverty line the poor are. Because of this deficiency, another group of measures spearheaded by Sen superseded the headcount ratio. Despite this development in measurement of poverty, many countries often use the headcount ratio in measuring poverty.

The second stage of poverty measurement consists of a class of measures that elaborates in detail, on the depth of poverty. One poverty measure that has been found manageable in presenting information on the poor in an operationally convenient manner is the FGT measure, developed by Foster et al. [4]. This measure decomposes a class of poor people. It takes into account the first simple measure of poverty, which is the headcount ratio or the incidence of poverty, which is labeled $P_\alpha = 0$. When it takes into account the depth of poverty, it is the poverty gap ratio denoted by $P_\alpha = 1$. This gives the proportion of the average poor from the poverty line and can be used to estimate the amount of resources required to bring the expenditure of every poor person up to the poverty line, thereby eliminating absolute poverty. However, the measure is insensitive to redistribution among the poor.

A third measure, $P_\alpha = 2$ gives the severity of poverty and produces the coefficient of variation of expenditure distribution of the poor, which reflects the degree of inequality among the poor. However, its monetary values are difficult to interpret.

N. O. Agola and J. L. Awange, *Globalized Poverty and Environment*,
DOI: 10.1007/978-3-642-39733-2_5,

5.2 Determination of the Poverty Line

Definition of poverty in absolute terms attempts to specify the levels of absolute deprivation on the basis of widely accepted norms, which identify the minimum requirements in terms of food, and non-food expenditures needed to satisfy the minimum basic needs. The minimum value of food basket is derived from the basic minimum nutrient requirements (calories) for a healthy growth and maintenance of human body. This necessitates assumption of a similar basket of food items for the whole region and/or country for ease of comparability. The FAO/WHO minimum recommended daily energy allowance (RDA) of 2,250 calories per adult equivalent is used to derive the food poverty line. In other words, the food poverty line presented of about US$10, is the minimum monthly consumption expenditure required to meet the recommended daily energy intake (of 2,250 calories) from the chosen basket of food items. A household with monthly food expenditure of less than the above amount per equivalent adult is therefore deemed to be food poor.

The overall poverty, which encompasses the food and non-food basic requirements, is derived by adding to the food poverty line the mean of the non-food expenditure items, such as education and health, transport, clothing and footwear, and so on, incurred by households falling in the band of minus 20 % and plus 10 % of the food poverty line. Individuals and/or households falling around the poverty line choose this band, since the interest is to establish the minimum expenditure of goods and services. On the other, the hard-core poor are defined as those who would not meet their minimum calorie requirements, even if they concentrated all their spending on food. In this case, these are individuals and/or households whose total monthly expenditures are less than US$10 per adult equivalent.

The purpose of the foregoing analysis is to develop rational approaches for focusing policy attention to a country's vulnerable regions and other deprived socio-economic groups. For instance, the extent of poverty in a developing country like Kenya, by region and between urban and rural sectors, has been presented in the country's Economic Survey 2005.[1] The poverty calculations are obtained on the basis of two welfare lines, the 'extreme poverty line' and the 'poverty line'. Three different poverty measures are usually applied; the poverty incidence (corresponding to a 'count' of the poor below the respective poverty lines), poverty depth (which takes into account the distance from the poverty line), and poverty severity (which attaches greater weight, the greater the distance below the poverty line). Each one of these poverty measures will be discussed below.

This study measures the well being of individuals by total consumption expenditures, not by total income. The problem of the measurement of poverty has two aspects. First, it is necessary to specify a poverty line, the threshold level of income below which an individual is considered poor, and which may reflect the socially accepted minimal standard of living. Second, the intensity of poverty suffered by those below the threshold income must be measured once the poverty line is specified.

---

[1] Republic of Kenya, Economic Survey 1997.

Most of the literature on poverty concerns the number of people below the poverty line. But this factor, as such, does not reflect the intensity of poverty suffered by the poor. The problem is how poor are the poor. They may have incomes that approximate the threshold level, or they may have no incomes at all.

5.2.1 Approaches to Defining a Poverty Line

The most common approach to defining a poverty line is to estimate the cost of a bundle of goods and services, which would ensure that the so-called 'minimum basic needs' of individuals are just met. To estimate such costs, the first step involves estimating the minimum money cost for food, which would satisfy the average nutritional needs of families of different sizes. To these costs, one must add the rent paid and certain minimum amounts for clothing, fuel and sundries to arrive at a poverty line for a family of a given size. This approach to specify the poverty line, pioneered by Rowntree [7], may be called the physical subsistence approach.

The physical subsistence approach has been used since the early 1900s, especially for national economic planning purposes. The subsistence minimum consumption basket is then used to target households with low incomes. The cost of this basket consists of food and non-food costs and is estimated as follows.

1. Food products
2. Non-food products including:
    (a) Clothes, underwear and footwear;
    (b) Hygiene and sanitary items, medicine
    (c) Household and everyday use items
3. Services
4. Taxes and payments having decided upon the poverty line, we next compute poverty indices that would measure the intensity of poverty.

Comparisons of the magnitude and severity of poverty can provide direct evidence of an economy's progress in raising living standards of the poor and throw light on how specific macroeconomic changes and public policies affect the poor. Several difficult methodological issues cloud such comparisons, however. The chosen indicator of a household's economic well being must be readily quantified, it must reflect the range of factors that contribute to well-being, and it must be comparable across sectors, regions, and periods. Having chosen an indicator of individual well being, an equally contentious issue is the selection of a minimum acceptable level of that indicator beyond which a person is not deemed to be poor: the poverty line. Finally, there is the choice of a summary statistic with which to aggregate information on poverty across individuals or households.

Several measures have been proposed. This section presents a brief review of various poverty measures used in the literature. The empirical results based on some of these measures are presented in the subsequent sections. The poverty measures reviewed here are the following.

5.2.2 Head-Count Ratio

The most commonly used, the head-count index, measures poverty simply as the fraction of the population, which is poor. It is the simplest measure of poverty.

$$H = \frac{q}{n} \tag{5.1}$$

where (q = Number of poor or the number of persons below the poverty line; and n = Total number of persons). The head-count ratio or the incidence of poverty is the ratio of the number of poor individuals to the total number of individuals in the population surveyed. The head count ratio, however, does not tell us how far below the poverty line the poor are. While H is extremely easy to compute, and it gives 'incidence of poverty', it does not tell the amount by which the income of the poor falls short of the poverty line. In other words, head-count ratio does not tell us about the extent or the severity of poverty of a person or group of persons. This crude poverty index does not take account of the income-gap among the poor. This indicator of poverty, however, has been widely used in the empirical literature on poverty for various countries.

5.2.3 Poverty Gap and the Income Gap

The second poverty measure reported and which addresses the depth of poverty is the poverty gap ratio. This gives the proportion of the average poor from the poverty line and can be used to estimate the amount of resources required to bring about the expenditure of every poor up to the poverty line, thereby eliminating absolute poverty. However, the measure is insensitive to redistribution among the poor.

If the degree of misery suffered by an individual is proportional to the income shortfall of that individual from the poverty line, then the sum total of these shortfalls may be considered an adequate measure of poverty. Such a measure is called the poverty gap ratio and can be written as:

$$g = \int_0^z g(x)f(x)dx = \frac{H(z - \mu_p^*)}{z} \tag{5.2}$$

where

$$g(x) = \frac{(z - x)}{z}, \tag{5.3}$$

z being the poverty line, $f(x)$ is the density function of income x, H is the head-count ratio and p is the mean income of the poor.

The measure g will provide adequate information about the intensity of poverty if all the poor are assumed to have exactly the same income, which is less than the poverty line. In practice, of course, income among the poor is unequally distributed and, therefore, g will not be sensitive to income inequality among the poor. Another version of measuring the poverty gap is:

$$g_i = Z - Y_i(Z > 0). \tag{5.4}$$

Z is the poverty line; $Y = (y_1, y_2, \ldots, y_n)$ is the vector of the individual incomes arranged in descending order; g_i is the income shortfall of the ith individual, i.e. the amount by which the ith individual's income is less than the poverty line.

If one were to compare two or more populations, it would be desirable to normalize the absolute poverty gap (or income shortfall) by the number of poor and the poverty line itself. Such a normalization procedure yields the measure I, income gap-ratio:

$$I = \sum_{i=1}^{q} \frac{g_i}{q} Z \tag{5.5}$$

We give the poverty gap per poor person in relation to the poverty line. This measure also gives an idea of the amount that would be required to alleviate poverty. The same incidence of poverty can be consistent with two income distributions with different poverty gaps.

The main criticism against the two poverty measures reported above is that they are insensitive to income distribution among the poor. Neither of these measures incorporates the effect of transfer of income from the poor to some one else who is poorer.

Several poverty measures have been proposed in the literature, which are sensitive to income inequality among the poor. Some of the most important of these are discussed below in Sen [8], Kakwani [5], Foster et al. [4].

5.2.4 Sen's Measure of Poverty

The third measure gives the severity of poverty and reduces the co-efficient of variation of expenditure distribution of the poor, which reflects the degree of inequality among the poor. However, its monetary values are difficult to interpret. Sen [8] has given an index, P, which incorporates all the three factors, viz,

(a) Number of poor
(b) Poverty-gap
(c) Transfer of income from not so poor to a poorer person.

P is defined as:

$$P = \frac{2}{(q+1)n.Z} \sum_{i=1}^{q} (Z - Y_i)(q + 1 - i), \tag{5.6}$$

where, $(q + 1 - i)$ is the rank order weight attached to the income gap, $(Z - Y_i)$, of the ith poor individual. The income of the poor are arranged in descending order, i.e. $y_q, y_{q-1}, \ldots, y_1$, such that $y_q > y_{q-1}$. The richest among the poor (with income y_q) gets weight 1 and the poorest of the poor with income y_1 gets weight q. Sen derives his index in an axiomatic fashion. His poverty measure is defined as the "normalized weighted sum of the income gaps of the poor". The two axioms required to develop his index are:

(a) Income weighting scheme
(b) Normalization procedure

According to the first axiom, the weight on the income of the poor is his own rank in the income ordering below the poverty line. As per the normalization procedure when all the poor have the same income, the index assumes value equal to the proportion of people below the poverty line multiplied by the proportionate average income gap. Sen's index, P, lies between 0 and 1. It assumes the value zero when everybody's income is above the poverty line (i.e. when $q = 0$), and value 1 when every poor's income is zero. If q is large, P can be expressed as

$$P = \frac{q}{n.Z}[(Z - y_p)(1 - G_p)], \tag{5.7}$$

where, G_p is the Gini Coefficient of incomes of the poor and y_p is the mean income of the poor. The effect of the weighting scheme is "to augment the average poverty gap by the Gini Coefficient times mean income of the poor".

It may be noted that in Sen's measure the rank-order welfare function is rather special. The relative weight on a person's income depends entirely on the rank of the person in income-ordering and not on the amount of his income per se. Consequences of this kind of ordering is, as pointed out by Anand [1, p. 121], is that there is a trade-off between the mean income of the poor (y_p) and the equality $(1 - G_p)$ in their income distribution. The trade-off is indicated by the factor $y_p(1 - G_p)$. It is thus possible for P to decrease when the poor have become poorer in absolute term (i.e., their absolute poverty has increased as y_p has decreased so long as equality in their distribution $(1 - G_p)$ has increased more than proportionately).

5.2.5 Fishlow's Measure of Poverty Alleviation Through Redistribution

A variant of Fishlow's index **xxii** expresses the poverty gap as a function of the income of the non-poor required to eliminate poverty and is expressed as below:

$$F = \frac{q}{ny - qy_p}[(Z - y_p)(1 - G_p)], \tag{5.8}$$

y is the overall mean per capita income. The relationship between F and Sen's index (P) is given by

$$F = \frac{Z}{y - \frac{q}{ny_p}P}, \tag{5.9}$$

F indicates the burden on the non-poor in terms of reduction in their income if the poverty gap is to be closed through redistribution of income.

It may be noted that if the poverty gap is to be closed through redistribution of income alone, the income of the non-poor should be large enough to avoid dragging the non-poor into the state of poverty through transfer of income. Anand [1, p. 121] has shown that this would suggest a different sort of normalization of Sen's index in which poverty gap is expressed as a fraction of the income of the non-poor over and above the poverty level. This measure, $F*$, can be expressed as

$$F* = \frac{z}{y - \frac{q}{n}y_p - Z(1 - \frac{q}{n})}P, \tag{5.10}$$

A few comments on measure F are in order:

(a) $F*$ may fall even without any decrease in number or increase in income of the poor, if the incomes of the non-poor increase. This leads to an anomalous situation as F falls without reduction in poverty.
(b) $F*$ is not necessarily bound by unity; it can exceed unity if the augmented poverty gap could exceed the income of the non-poor. It is, therefore, important that the mean income of the society exceeds poverty line (i.e. $y > Z$), if the redistribution of income has to be a meaningful instrument of poverty alleviation.
(c) An implicit assumption behind the redistribution of income being an important factor of poverty alleviation is that the size of income of the society remains unchanged as a result of redistribution. This ignores disincentive effect on work (labour supply functions), distribution of skills, etc., arising from raising of marginal tax rate on the income of the non-poor.

The above comments suggest that the problem of measurement of poverty needs to be conceptualized separately from the possibilities for its alleviation.

5.2.6 Ray's Measure

Avariant of $P\alpha$ (FGT measure) is given by Ray [6], which combines additive decomposability with poverty aversion. The combination of FGT and Clark et al. [3] criteria is expressed as Additive Decomposability and Aversion Monotonicity Axiom.

$$P * \alpha = \sum_{k=1}^{m} W_k(\alpha).P_k(\alpha) \tag{5.11}$$

where $P_k(\alpha)$ is the sub-group poverty index and it increases with α. $W_k(\alpha)$ corresponding weight which increases with a for sub-group with "above average" poverty and decreases for those with "below average" poverty. $P * \alpha$ also increases with α. Ray's Poverty Measure is given by

$$P * \alpha = \frac{g}{nz} \sum_{i=1}^{q} \frac{g_i^{\infty}}{g} \tag{5.12}$$

where

$$g = \frac{\sum_{i=1}^{q} g_i}{g} \tag{5.13}$$

is the 'mean poverty gap'. Ray shows that the above measure satisfies not only the criteria proposed by Sen. $P * \alpha$ can be made to satisfy the Transfer Sensitivity Axiom by choosing the value of α sufficiently larger than unity, depending on the magnitude of g_i's. Ray shows that the overall poverty measure can be written as a linear combination of the sub-group poverty indices as

$$P * \alpha = A_1 P_{\alpha 1} + A_2 P_{\alpha 2} \tag{5.14}$$

where $P_{\alpha 1}$ and $P_{\alpha 2}$ are the poverty indices for group 1 and group 2 and A_1 and A_2 are the respective weights, which are given by

$$\begin{aligned} A_1 &= \left[\frac{n_1}{n_1 + n_2}\right]\left[\frac{g_1}{W_1 g_1 + W_2 g_2}\right]^{\infty - 1} \\ A_2 &= \left[\frac{n_2}{n_1 + n_2}\right]\left[\frac{g_2}{W_1 g_1 + W_2 g_2}\right]^{\infty - 1} \quad \textit{and} \\ W_i &= \frac{g_i}{g_i + g_2} (i = 1, 2) \end{aligned} \tag{5.15}$$

represents ith group's share of the poor. The index can be generalized for m groups.

$$P * \alpha = \sum_{j=1}^{m} A_j P_{\infty} \tag{5.16}$$

where A_j is the 'mean poverty gap' of jth group and $g = \alpha W_j g_j$ is the mean poverty gap for the aggregate population.

5.2.7 Decomposability and the Poverty Measures

One of the desirable characteristics of poverty measure is its decomposability, i.e. the overall poverty measure should be capable of being decomposed into sub-group poverty. This would enable us to quantify the contribution of a sub-group (say, occupation or any socio-economic category) to the overall poverty. Sen's measure, which uses rank-order weighting, does not satisfy the criterion that if a sub-group's poverty increases, it should increase the overall poverty also.

An important aspect to be considered in the context of poverty measure is the Aversion Monotonicity Axiom due to Clark et al. [3] according to which the more 'averse' one is to poverty, the greater does he perceive the level of poverty. The decomposability criterion (sub-group Monotonicity Axiom) requires that when incomes in a given sub-group change without changing that of the remaining sub-groups, the sub-group and total poverty must move in the same direction. One poverty measure that has been found manageable in presenting information on the poor in an operationally convenient manner is the measure developed by Foster et al. [4] (FGT measure hereafter). It satisfies the transfer axiom as well as decomposability property without violating the basic criteria proposed by Sen. In short, FGT measure has the following features:

(a) It is additively decomposable (weighted average of sub-group poverty levels) in which population share is the weight.
(b) It satisfies the basic properties proposed by Sen.
(c) It incorporates the concept of relative deprivation of poverty.

The FGT measure is expressed as a combination of the following:

(a) Inequality (in this case it is the squared coefficient of variation).
(b) Head-Count Ratio
(c) Income-gap Ratio.

The concept of 'aversion to poverty' is incorporated by generalizing the index in terms of a parametric family of measures in which a particular parameter (∞) can be treated as an indicator of aversion to poverty. Such poverty measure is expressed as:

$$P\alpha(y; Z) = \frac{1}{n}\sum_{i=1}^{q}\frac{g_i^{\infty}}{Z} \tag{5.17}$$

$P\alpha$ is equal to:

- $P_0 = H$, Head-Count Ratio for $\alpha = 0$
- $P_1 = H.1$, (renormalized income gap) for $\alpha = 1$

$$P_2 = \frac{1}{NZ^2}\sum_{i=1}^{q} g_i^2 \ for \ \alpha = 2 \tag{5.18}$$

'α' is interpreted as the poverty aversion parameter; a large 'α' gives greater emphasis to the poorest of the poor.[2] It is shown that $P\alpha$ satisfies the transfer sensitivity axiom for $\alpha > 2$. There are variants of the FGT, the most important of which are the discrete and continuous versions:

(i) **Discrete Version**: Once an appropriate poverty line has been determined, such as per capita estimates of individual consumption or income may then be reduced to a single summary statistic using the class of decomposable $P\alpha$ measures proposed by Foster et al. [4].

$$P\alpha = \frac{1}{N}\sum_{i=1}^{N}[Max((z - y_i)/z, 0)]^{\infty} \tag{5.19}$$

where y_i = real income (or consumption) per capita, z = the poverty line, N = the number of individuals in the sample population $P\alpha$ = measure of inequality aversion. When $P\alpha = 0$, the Head Count Ratio (which measures the proportion of the population below the poverty line) is produced, showing the percentage of people in poverty. When $P\alpha = 1$, then the index gives the poverty gap: the mean shortfall of consumption below the poverty line as a proportion of the poverty line, showing the depth of poverty; When $P\alpha = 2$, the weight given to shortfalls from the poverty line increases with the shortfall—indicating the severity of poverty index.

(ii) **Continuous Version**: If the sample population is continuous rather than discrete, the Foster-Greer-Thorbecke $P\alpha$ measures may be calculated using the formula:

$$P\alpha = \int_0^Z ((z - y_i)/z)^{\infty} f(y)dy \tag{5.20}$$

where $f(y)$ is the income distribution function.

(iii) **Another Version of the FGT**

$$P\alpha = \frac{1}{n}\sum_{i=1}^{n}\{max[PL - C_i]/PL\}^{\infty} \tag{5.21}$$

where PL = the poverty line, n = population size, C_i = real consumption per capita of person i, $P\alpha$ = measures inequality aversion, and $\alpha = 0$, the index gives a head-count: the percentage of people in poverty. $\alpha = 1$, then the index gives the poverty gap: the mean shortfall of consumption below the poverty line as a proportion of the poverty line. $\alpha = 2$, the weight given to shortfalls from the poverty line increases with the shortfall.

---

[2] As a becomes large, it considers the position of the poorest household. The above measure also satisfies the Transfer Sensitivity Axiom due to [5] according to which if a transfer $t > 0$ of income takes place from a poor person with income y_i to another poor person with income $y_i + d(d > 0)$, then for a given d, the magnitude of the increase in poverty must decrease as y_i increases.

5.3 Concluding Remarks

Measurement of poverty is quite an important exercise, given that a reliable measurement tool provides a metric against which progress, the lack of progress or a worsening of poverty condition in a country or region can be measured. However, as can be gleaned from the developments of measurement criteria presented in this chapter, the problem of arriving at a good measurement methodology is very much linked to the problems of definition of poverty and as well as the varied approaches to poverty alleviation. It is notable that poverty alleviation approaches tend to put emphasis on different dimensions of poverty itself as would be discussed in the later chapters.

References

1. Anand S (1983) Inequality and poverty in Malaysia measurement and decomposition
2. Booth C (1892) Life and labour of the people of London. Macmillan, London
3. Clark S, Hemming R, Ulph D (1981) On indices for the measurement of poverty. Econ J 91: 515–526
4. Foster J, Greer J, Thorbecke E (1984) A class of decomposable poverty measures. Econometrica 52:761–766
5. Kakwani N (1980) On a class of poverty measures. Econometrica 48:437–446
6. Ray R (1989) A new class of decomposable poverty measures. Indian Econ J 36(4):30–38
7. Rowntree BS (1901) Poverty: a study of town life. Macmillan, London
8. Sen AK (1976) Poverty: an ordinal approach to measurement. Econometrica 44:219–231

Chapter 6
Recent Developments

6.1 Introductory Remarks

From the 1990s, there have been new and powerful concepts applied to the analysis of poverty and for formulation of anti-poverty policies. These new concepts and approaches have not been addressed though in most National Poverty Eradication Plans (NPEP). Most countries NPEP discusses the concept of poverty only in terms of the traditional consumption/expenditure approach. From this, the Plans draw conclusions about the poverty line, the depth of poverty, upon which formulation of policies to eradicate poverty are then based. The emphasis of the NPEPs is on expenditure on food and non-food items. However, recent developments in poverty analysis and measurement have gone far beyond the simplistic approach adopted in most countries' NPEP. Three main strands of thought have recently come to dominate the discussion of poverty: Sens approach to poverty measurement, Human Poverty index, Human Development Index, and the Physical Quality of Life Index (PQLI). We argue that these new indices are much more relevant than the traditional measures contained in the NPEPs. The omission of discussion of these new measures does not augur well for the recent thinking and policy formulation that keep pace with new theories and practices regarding poverty issues. For that reason, it is necessary to discuss these new approaches in detail.

6.2 Three Perspectives on Poverty

There are three main perspectives on poverty. The first one is income poverty. According to this perspective, a person is poor if, and only if, her income level is below the defined poverty line. Many countries have adopted income poverty line to monitor progress in reducing poverty incidence. Often the cut-off poverty line is defined in terms of having enough income for a specified amount of food. This approach dominates most countries perspective.

N. O. Agola and J. L. Awange, *Globalized Poverty and Environment*,
DOI: 10.1007/978-3-642-39733-2_6, © Springer-Verlag Berlin Heidelberg 2014

The second approach is the "Basic Needs perspective. According to this perspective, poverty is deprivation of material requirements for minimally acceptable fulfillment of human needs, including" food. This concept of deprivation goes well beyond the lack of private income: it includes the need for basic health and education and essential services that have to be provided by the community to prevent people from falling into poverty. It also recognizes the need for employment and participation.

The third approach is the "capability" perspective, spearheaded by Amartya Sen. Under this perspective, poverty represents the absence of some basic capabilities to function a person lacking the opportunity to achieve some minimally acceptable levels of these functioning. The functioning relevant to this analysis can vary from such physical ones as being well nourished, being adequately clothed and sheltered and avoiding preventable morbidity, to more complex social achievements such as partaking in the life of the community. The capability approach reconciles the notions of absolute and relative poverty, since relative deprivation in incomes and commodities can lead to an absolute deprivation in minimum capabilities. This approach has not been effectively adopted in most countries- with regard to real application reflected in anti-poverty measures.

Human Poverty Index (HPI) is a relatively new subject that is not addressed in the National Poverty Eradication Plan. We feel that the HPI is such an important consideration that should be referred to when discussing poverty in most if not all countries. The significance of HPI lies in the range of issues it covers.

However, given the pervasiveness of poverty in poor countries, the HPI developed in one country may not necessarily be the same as the one developed for another country. The index used is dependent on the context and the variables chosen to reflect that. The nature of poverty in rich countries deserves a specialized study—and a more specialized index—focusing on those deprivations particularly relevant for those countries. Thus, the HPI developed for Mexico must necessarily be unique to the country, and the same applies to Sub-Sahara African country like Zimbabwe.

The Human Poverty Index (HPI) concentrates on deprivation in three essential elements of human life longevity, knowledge and a decent standard of living. The first deprivation relates to survival the vulnerability to death at a relatively early age. The second deprivation, relates to knowledge being excluded from the world of reading and communication. The third relates to a decent living standard in terms of overall economic provisioning. HPI has weaknesses—in data and as a concept. Like all measures it cannot capture the totality of human poverty. But by combining in a single poverty index the concerns that often get pushed aside when focus is on income alone, the HPI makes a useful addition to the measures of poverty.

Where is human poverty index most pervasive? The HPI exceeds 50 % in six African countries—Niger, Sierra Leone, Burkina Faso, Ethiopia, Mali, Mozambique—implying that more than half their people suffer several forms of human poverty. Altogether, 35 of the 78 developing countries for which the HPI was calculated have HPI exceeding 33 %. Comparing HPI with income measures of poverty based on a US$1-a-day poverty line reveals interesting contrasts:

(a) Both income poverty and human poverty are pervasive, affecting a quarter to a third of the people in the developing world.
(b) Sub-Saharan Africa and South Asia have the highest incidence of both income and human poverty at about 40 %.
(c) Most of the Arab States have made remarkable progress in reducing income poverty, now a merely 4 %, but face a large backlog of human poverty (32).
(d) Latin America and the Caribbean, with an HPI of 15 %, has reduced human poverty in many countries, but income poverty is still 24 %.
(e) In Egypt, Guinea, Morocco, Pakistan and 10 other countries the proportion of people in human poverty exceeds the proportion in income poverty.
(f) In Guinea-Bissau, Honduras, Kenya, Peru and Zimbabwe the proportion of people in income poverty exceeds the proportion in human poverty.

6.2.1 Measurement of Poverty and the Human Poverty Index

Can the concept of human poverty be targeted and monitored? Can an overall measure of poverty be developed that can inform as well as be used for policy? Can an internationally comparable measure be defined?

We therefore discuss the human poverty index (HPI) in an attempt to bring together in a composite index the different features of deprivation in the quality of life to arrive at an aggregate judgment on the extent of poverty in a community. The Human Development Report 1996 and later Reports attempted this through a particular version of the "capability poverty measure". The HPI pursues the same approach, focusing on a broader and more representative set of variables, in a consistent relationship to the human development index (HDI).

Like many other concepts, human poverty is larger than any particular measure, including the HPI. As a concept, human poverty includes many aspects that cannot be measured—or are not being measured. It is difficult to reflect them in a composite measure of human poverty. Critical dimensions of human poverty excluded from the HPI for these reasons are lack of political freedom, inability to participate in decision-making, lack of personal security, inability to participate in the life of a community and threats to sustainability and inter- generational equity.

6.2.2 Sens Approach to Poverty Analysis

Amartya K. Sen contends that traditional welfare economics, which stresses the revealed preferences or desire-based utilities of individuals in their acts of choice, lacks enough information about people's preferences to assess the social good. Accordingly, as an alternative, Sens welfare theory relies not on individuals attainments (for example, of basic needs) but individuals capabilities, an approach he believes can draw on a richer information base. For Sen, living consists of a vector of interrelated functioning (beings and doings), such as being adequately nourished, avoiding premature mortality, appearing in public without shame, being happy, and being free. Yet Sen does not assign particular weights to these functioning, as well being is a "broad and partly opaque concept", which is intrinsically ambiguous.

Sen focuses on a small number of basic functioning central to well being: to be adequately fed and sheltered and in good health. Sen contends that GNP per capita, which measures capabilities, does not correlate closely with these functioning, which indicate attainments. One example is life expectancy, a proxy for health, which, at 75 years, is almost as high for Costa Rica as for the US (76 years), which has an income per head 12 times as high.

For Sen, *poverty is not low well-being but the inability to pursue well-being because of the lack of economic means*. This lack may not always result from a deficiency of capabilities. Poverty is the failure of basic capabilities to reach minimally acceptable levels [2].

6.3 Poverty Depends on the Context

The nature of the main deprivations varies with the social and economic conditions of the community in question. The choice of indicators in the HPI cannot but be sensitive to the social context of a country. For example, an index that concentrates on illiteracy and premature mortality may be able to discriminate between Pakistan and Sri Lanka more easily than it can between say France and Germany.

Issues of poverty in the developing countries involve many factors. The most important of these are hunger, illiteracy, epidemics and the lack of services or safe water. Such issues may not be so central in the more developed countries, where hunger is rare, literacy is close to universal, most epidemics are well controlled, health services are typically widespread and safe water is easy to tap. Not surprisingly, studies of poverty in the more affluent countries concentrate on such variables as "social exclusion." These can be forceful deprivations and very hard to eliminate in all countries. But they take on relatively greater prominence in the affluent ones. There is no real possibility of constructing an index of human poverty that would be equally relevant in different types of countries.

Given the pervasiveness of poverty in poor countries, the HPI developed is aimed at the context and the variables chosen reflect that. The nature of poverty in rich countries deserves a specialized study—and a more specialized index—focusing on those deprivations particularly relevant for those countries. This is even more so given the relative rise in numbers of poor people in developed countries like Japan, a country that has long distinguished itself for equitable distribution of wealth and income amongst its population.

6.4 The Three Indicators of the Human Poverty Index

As already discussed above, the HPI presented in this book concentrates on the deprivation in three essential elements of human life already reflected in the HDI—longevity, knowledge and a decent living standard. The first deprivation relates to **survival**—the vulnerability to death at a relatively early age—and is represented in the HPI by the percentage of people expected to die before age 40. The second dimension relates to **knowledge**—being excluded from the world of reading and communication—and is measured by the percentage of adults who are illiterate. The third aspect relates to a decent standard of living, in particular, overall economic provisioning. A composite of three variables—the percentage of people with access to health services and to safe water and the percentage of malnourished children under five represent this.

A few observations must be made about this last variable and about why income does not feature in the HPI. The logic underlying the construction of the economic provisioning variable is that the GNP included in the HDI is actually an amalgam of private and public facilities, since public services are paid out of aggregate national income.

Private income could not be an adequate indicator of an individual's economic facilities, which also include crucial public services (such as health care arrangement and a safe water supply). But why is private income not chosen to supplement the information on public facilities?

One of the problems in assessing the prevalence of income poverty is that the use of the same poverty line in different countries can be very misleading because of the variation in "necessary" commodities. Depending on the prevailing patterns of consumption—clothing, accommodation and such tools of communication and interaction as radios and telephones—many provisions are taken to be essential for social participation in one community without being treated as such in another. As a result, the minimum income needed to escape social estrangement can be quite different between communities.

Given the social pressure, these felt "needs" might compete—for relatively poorer people in rich countries—even with the provision of resources for food, nutrition and health care. This can explain the prevalence of some hunger and malnutrition, especially among children, even in the Untied States, where incomes are high but inequalities generate a heavy burden of "necessity" in the direction of socially obligated consumption, often to the detriment of health and nutritional spending. So, the assessment of poverty on the basis of a low minimum cut-off income used for poor countries fails to show any poverty in generally affluent societies, even when the relatively poor in those societies may lack social participation and may even suffer from hunger and malnutrition.

An alternative is to use different poverty lines in different countries. But it is not easy to decide what the appropriate variations would be and how the respective poverty lines could be estimated. The official national lines cannot serve this purpose, since they reflect other influences, especially political ones, and cannot be used for international comparisons. The general need for a variable cut-off line of poverty is easier to appreciate than it is to find adequate values for variable poverty lines in different communities.

A more practical possibility is to be less ambitious and focus on material deprivation in hunger and malnutrition, not on income. A very high proportion of personal income goes to food and nourishment, especially for poor people in poor countries. For this we can use information on food intake, which relates to personal incomes. Alternatively, there are estimates of malnutrition, but these are influenced by a number of variables, such as metabolic rate, climatic conditions, activity patterns and epidemiology. Since our concern is with the lives that people can lead, there is a case for going straight to the prevalence of malnutrition. This is what is done in HPI. It concentrates specifically on the malnutrition of children, which is relatively easier to measure and for which usable data are more uniformly available.

For public provisions, access to health services and to safe water was chosen. Combining this two-access variable with the prevalence of malnutrition gives a fairly broad picture of economic provisioning—private and public—to supplement the information on survival and literacy. These are the basic informational ingredients of the HPI. It must be emphasized that there is some inescapable arbitrariness in any such choice. The choice was made on the basis of balancing considerations of relevance on the one hand, and the availability and quality of data on the other. There are inevitable compromises made, and it would be highly insensitive to pretend that even the variables that have been included stem from high-quality data for every country. There has been an attempt, in these selections, to strike a balance between the demands of relevance and the need for tolerably usable data, and these choices would certainly remain open to criticism and public scrutiny. Such criticism and scrutiny is necessary for meaningful progress and realization of any new innovations both in conception and development of poverty measurements.

6.5 Uses of the Human Poverty Index

The human poverty index can be used in at least three ways.

1. It can be used as a tool for advocacy. If poverty is to be eradicated, public opinion and support needs to be mobilized to the cause. The HPI can help summarize the extent of poverty along several dimensions, the distance to go, and the progress made. Income poverty also needs to be measured but income alone is too narrow a measure.
2. As a planning tool for identifying areas of concentrated poverty within a country. The HPI has been used in many countries to rank districts or counties as a guide to identifying those most severely disadvantaged in terms of human development. It can thus be used as a planning tool by identifying those most seriously affected by human poverty. Though ranking by any one index alone would be possible say, by illiteracy rate, lack of access to health services or the percentage in income

poverty the HPI makes possible a ranking in relation to a combination of basic deprivations, not one alone. Based on HPI, countries can now create poverty maps, which are useful tools not only in planning, but also as communication tools in advocacy campaigns.

3. As a research tool. The HPI has been used especially when a researcher wants a composite measure of development. For such uses, other indicators have sometimes been added to the HPI. The HPI could be similarly used and enriched especially if other measures of poverty and human deprivation were added, such as employment.

The HPI provides an alternative GNP, for assessing a countrys standing in basic human development or its progress in human development over time. It does not displace economic measures but can serve as a simple composite complement to other measures like GNP.

The HPI can similarly serve as a useful complement to income measures of poverty. It will serve as a strong reminder that eradicating poverty will always require more than increasing the incomes of the poorest.

Construction of the Human Poverty Index

The procedures for constructing the HPI, including weighing and aggregation are presented as follows:

The process of aggregation can be sensitive to the overlaps in the three dimensions of the HPI. For example, consider a case in which in each of the three categories of deprivation, 30 % of people fail to meet the minimum requirement. This can be so because the same 30 % of people may fail in all three fields. But it can also be that a different 30 % fail in each category. Or we may have some combination of the two extremes. In the first extreme case only 30 % are affected by poverty, but they are deprived on all three fronts. In contrast, in the second extreme case as many as 90 % of the population are deprived altogether, but each group has inadequacy in merely one field. Even though information on overlaps (or covariance) is not easy to obtain, (since data regarding the different variables come from different sources), these distinctions can be important in describing poverty. They can also be crucial for causal analysis, since deprivation of one kind often feeds others.

However, when it comes to constructing an index, it is not easy to decide whether 30 % of people with inadequacies of all three types represent larger social poverty than 90 % of people having one deficiency each. It is a matter of the importance to be given to depth vis–vis breadth. For the purpose of the HPI, the two cases have been treated as equivalent, so that in some sense depth and breadth have been equally considered.

There is a further issue to be addressed in deriving an aggregate index, namely that of substitutability between the three components of the HPI. This is done through an explicit procedure of using an additional weight (x). When x is taken to be 1, perfect substitutability is presumed, and the aggregate is obtained by simply averaging the three deprivations. The opposite case of no substitutability corresponds to x being taken to be infinity. In that case the largest of the percentage shortfalls rules the roost.

For example, if 30 % fail in field one, 50 % in field two and 45 % in field three, then the overall extent of poverty, in this case, is simply 50 %.

Perfect substitutability is too extreme an assumption and goes against the sensible requirement that as the deprivation in some field becomes relatively more acute; the weight placed on removing deprivation in that field should increase. Nor is the other extreme, zero substitutability, very easy to support, since it implies that any increase in deprivation in any category other than the one with the highest rate of deprivation must leave the aggregate poverty measure completely unchanged. Both extremes are avoided by choosing an intermediate value of x.

6.6 Human Development Index

The United Nations Development Program has formulated another indicator of development, shifting the emphasis to human development [1]. Since its launch in 1990 the **Human Development Report** has defined human development as the process of enlarging people's choices. The most critical ones are to lead a long and healthy life, to be educated and to enjoy a decent standard of living. Additional choices include political freedom as well as other guaranteed human rights and various ingredients of self-respect—including what Adam Smith called the ability to mix with others without being "ashamed to appear in public". These are among the essential choices, the absence of which can block many other opportunities. Human development is thus a process of widening people's choices as well as raising the level of well being achieved.

If human development is about enlarging choices, poverty means that opportunities and choices most basic to human development are denied—to lead a long, healthy, creative life and to enjoy a decent standard of living, freedom, dignity, self-respect and the respect of others.

Longevity of life as an indicator of development is based on three considerations: the intrinsic value of longevity, its value in helping people pursue various goals, and its association with other characteristics (such as good health and nutrition). Life expectancy at birth is an indicator of longevity. For the second key component of knowledge, literacy figures are a crude reflection of access to education. To derive the indicator of command over resources needed for decent living, GDP per capita is the best available indicator, despite some limitations. People do not isolate different aspects of their lives. Instead, they have an overall sense of well being. Therefore, it was advocated that a composite **Human Development Index** (HDI), integrating the three components of longevity, literacy and income, be constructed. It is notable that in 2010, HDI has been improved to reflect inequality in each of the three components (health, education and income), and so Inequality-adjusted HDI (IHDI) is used. In a situation of perfect equality, the values of IHDI and HDI are equal, but as inequality rises, the IHDI falls below the HDI. Inequality is a major attribute of poverty stricken countries so the average values for health, education, and income are discounted according to each components? level of inequality. Therefore, IHDI is computed as the geometric mean of the values for health, education, and income adjusted for inequality.

6.6.1 Contrast Between Human Development and Human Poverty

The contrast between human development and human poverty reflects two different ways of evaluating development. One way, the "conglomerative perspective", focuses on the advances made by all groups in each community, from the rich to the poor. This contrasts with an alternative viewpoint, the "deprivational perspective", in which development is judged by the way poor and the deprived fare in each community. Lack of progress in reducing the disadvantages of the deprived cannot be "washed away" by large advances—no matter how large the progress- made by the better-off people.

Interest in the process of development concerns both perspectives. At a very basic level, the lives and successes of everyone should count, and it would be a mistake to make our understanding of the process of development completely insensitive to the gains and losses of those who happen to fare better than others. It would go against the rights of each citizen not to be counted, and also clash with the comprehensive concerns of universalistic ethics. Yet a part—a big part- of the general interest in the progress of a nation concentrates specifically on the state of the disadvantaged.

6.6.2 Construction of the Human Development Index

The HDI is constructed in three steps. First, step involves defining the measure of deprivation that a country suffers in each of the three basic variables life expectancy, literacy, and log of real GDP per capita. A maximum and minimum value is determined for each of the three variables, given the actual values. The deprivation measure then places a country in the range of 0–1, as defined by the difference between the maximum and the minimum.

The second step is to define an average deprivation indicator. Taking a simple average of the three indicators does this. The third step is to measure the HDI as 1 minus the average deprivation [3]. To illustrate the application of this formula, take the specific data for a country like Kenya:

1. Life expectancy deprivation = (78.4 − 59.4)/(78.4 − 41.8) = 0.519, where 78.4 and 41.8 are the maximum and minimum life expectancy, respectively, and 59.4 is life expectancy at birth in Kenya.
2. Literacy deprivation = (100.0 − 60.0)/(100.0 12.3) = 0.456, where 100.0 and 12.3 are the maximum and minimum literacy rate in Kenya.
3. GDP deprivation = 3.68 2.90/3.68 2.34 = 0.582, where 3.68 and 2.34 are the log of maximum and minimum per capita GDP, and 2.90 the log of Kenyas GDP.

4. Kenyas average deprivation $= (0.519 + 0.456 + 0.582)/3 = 0.519$. Kenyas HDI $=$ $10.519 = 0.481$.

6.7 Concluding Remarks

What emerges from the above discussions of poverty measurement tools is that the National Poverty Eradication (NEP) programs have left out some critical aspects of the scourge, which by all means need to be incorporated in both the monitoring and metrics for determining both the scale and intensity of poverty. It also emerges that there are clear differences in the type of poverty afflicting people in different economies and regions. Precisely, poverty in the developed world has its own unique characteristics as compared to poverty in most of the developing world. Consequently, no universal measurement tool can be used to measure global poverty. The use of a universal measure of poverty (or some universal measures as is the case presently), tends to disguise or hide poverty in the developed economies. A full fledged attention to poverty in the developed countries matched with different and befitting measurement and monitoring tools would be timely and appropriate in the face of rise in poverty incidence in the developed countries.

References

1. ILO/UNDP (1972) Employment, Incomes and Equality: A strategy for increasing productive employment in Kenya, Geneva
2. Sen AK (1992) Inequality reexamined. Harvard University Press, Cambridge, Mass
3. UNDP (1990) Human development reports. Oxford University Press, New York

Chapter 7
The Human Development Perspective

7.1 Many Dimensions of Poverty

Concerns with identifying people affected by poverty and the desire to measure it have at times obscured the fact that poverty is too complex to be reduced to a single dimension of human life. It has become common for countries to establish an income-based or consumption-based poverty line. Although income focuses on an important dimension of poverty, it gives only a partial picture of the many ways human lives can be blighted. Someone can enjoy good health and live quite long but be illiterate and thus cut off from learning, from communication and from interactions with others. Another person may be literate and quite well educated but prone to premature death because of epidemiological characteristics or physical disposition. Yet a third may be excluded from participating in the important decision making processes affecting her life. In all these cases, the deprivations cannot be fully captured by the level of each individual's income. Also, people perceive deprivation in different ways—and each person and community defines whatever deprivation faced and disadvantages that affect their lives in varied ways such that any concurrence is a coincidental matter.

Poverty of lives and opportunities—or human poverty—is multidimensional in character and diverse rather than uniform in content. In the 1970s the concept of social exclusion came into the literature to analyze the condition of those who are not necessarily incomepoor though many are that too—but who are kept out of the mainstream of society even if not income-poor. The inadequacy of traditional definitions of poverty, based on incomes and consumption, was widely acknowledged to explain these new concerns.

It is in the deprivation of the lives that people can lead that poverty manifests itself. Poverty can involve not only the lack of the necessities of material well being, but the denial of opportunities for living a tolerable life. Life can be prematurely shortened. It can be made difficult, painful or hazardous. It can be deprived of knowledge and communication. And it can be robbed of dignity, confidence and self-respect—as well as the respect of others. All are aspects of poverty that limit and blight the lives of many millions in the world today.

N. O. Agola and J. L. Awange, *Globalized Poverty and Environment*,
DOI: 10.1007/978-3-642-39733-2_7, © Springer-Verlag Berlin Heidelberg 2014

7.1.1 Physical Quality of Life Index

Social indicators are hard to combine into an overall measure of a country's social development. In 1979, an effort to do so was made by the Washington-based Overseas Development Council (ODC). The ODC's physical quality index (PQLI) was based on three widely available indicators of the basic-human-needs variety: life expectancy at age one, infant mortality rate, and literacy rate. This index was put forward by Morris D. Morris.

According to Morris, even in a monetized economy, GNP may increase over time without improving well being. The poorest groups of society may not benefit much, or not at all, from rising national income alone. On the other hand, a country can be made better off in terms of social security, social harmony, longevity of life, and mortality rates through appropriate programs, without any increase in GNP. Morris, therefore, emphasized that the objective of development should be the satisfaction of essential human needs, along with increasing the output of goods and services. This equires a mix of policies that would have positive synergy with one another. He adopted life expectancy, the infant mortality rate and the literacy rate as the three most universal and important components of development and, based on them, developed a composite indicator of development called the Physical Quality of Life Index (PQLI). He believed that, by this approach, it was possible to get a better picture of the quality of life than was possible with per capita income or GNP alone.

7.1.2 Construction of the Physical Quality of Life Index

To construct the Physical Quality of Life Index, the performance of an individual country is placed on a scale of 0 to 100, where 0 represents an explicitly defined worst performance and 100 represents an explicit best performance. Once performance for each indicator is scaled to this common measure, a composite index can be calculated by averaging the three indicators, giving equal weight to each of them. In the case of basic literacy, as data are reported as a percentage of the population of fifteen years and above who are literate, the percentage figures correspond to the index numbers. In the case of infant mortality, the known range is between 2.29 and 7 per thousand, so that a 2.22 point change in the infant mortality rate shows a 1 point change in the infant mortality index. In the case of life expectancy, the range is between 38 and 77 years, so that a change in life expectancy of 0.39 years results in a one-point change in the index.

7.1.3 Human Development Index

The United Nations Development Program has formulated another indicator of development, spawning a major shift and emphasis to human development. Since its launch in 1990 the Human Development Report has defined human development as the process of enlarging people's choices. The most critical ones are to lead a long and healthy life, to be educated and to enjoy a decent standard of living. Additional choices include political freedom; other guaranteed human rights and various ingredients of self-respect—including what Adam Smith called the ability to mix with others without being "ashamed to appear in public". These are among the essential choices, the absence of which can block many other opportunities. Human development is thus a process of widening people's choices as well as raising the level of well being achieved.

If human development is about enlarging choices, poverty means that opportunities and choices most basic to human development are denied—to lead a long, healthy, creative life and to enjoy a decent standard of living, freedom, dignity, self-respect and the respect of others. Longevity of life as an indicator of development is based on three considerations: the intrinsic value of longevity, its value in helping people to pursue various goals, and its association with other characteristics (such as good health and nutrition). Life expectancy at birth is an indicator of longevity. For the second key component of knowledge, literacy figures are a crude reflection of access to education. To derive the indicator of command over resources needed for decent living, GDP per capita is the best available indicator, despite some limitations. People do not isolate different aspects of their lives. Instead, they have an overall sense of well being. Therefore, it was advocated that a composite Index of Human Development (HDI), integrating the three components of longevity, literacy and income, be constructed.

7.1.4 Construction of the Human Development Index

The Human Development Index (HDI) is constructed in three steps. First, define the measure of deprivation that a country suffers in each of the three basic variables life expectancy, literacy, and log of real GDP per capita. A maximum and minimum value is determined for each of the three variables, given the actual values. The deprivation measure then places a country in the range of 0–1, as defined by the difference between the maximum and the minimum. The lowest HDI ranges from 0.00 to 0.499, while medium score ranges from 0.500 to 0.799, and the high score ranging between 0.800 and 1.

The second step is to define an average deprivation indicator. Taking a simple average of the three indicators does this. The third step is to measure the HDI as 1 minus the average deprivation. To illustrate the application of this formula to a country such as Kenya: **iii**

(1) Life expectancy deprivation = (78.4 – 59.4) / (78.4 – 41.8) = 0.519, where 78.4 and 41.8 are the maximum and minimum life expectancy, respectively, and 59.4 is life expectancy at birth in Kenya.
(2) Literacy deprivation = (100.0 – 60.0) / (100.0 – 12.3) = 0.456, where 100.0 and 12.3 are the maximum and minimum literacy rate in Kenya.

(3) GDP deprivation = 3.68 – 2.90 / 3.68 – 2.34 = 0.582, where 3.68 and 2.34 are the log of maximum and minimum per capita GDP, and 2.90 the log of Kenya's GDP.
(4) Kenya's average deprivation = (0.519 + 0.456 + 0.582) / 3 = 0.519.
(5) Kenya's HDI = 1 – 0.519 = 0.481.

7.1.5 Contrast between Human Development and Human Poverty

The contrast between human development and human poverty reflects two different ways of evaluating development. One way, the "**conglomerative perspective**", focuses on the advances made by all groups in each community, from the rich to the poor. This contrasts with an alternative viewpoint, the "**deprivational perspective**", in which development is judged by the way poor and the deprived fare in each community. Lack of progress in reducing the disadvantages of the deprived cannot be "washed away" by large advances—no matter how large—made by the better-off people.

Interest in the process of development concerns both perspectives. At a very basic level, the lives and successes of everyone should count, and it would be a mistake to make our understanding of the process of development completely insensitive to the gains and losses of those who happen to fare better than others. It would go against the rights of each citizen to be counted, and also clash with the comprehensive concerns of universalistic ethics. Yet a part—a big part—of the general interest in the progress of a nation concentrates specifically on the state of the disadvantaged.

7.2 How Does Human Poverty Relate to Other Approaches?

Over the years, the concept of poverty has been defined in different ways. Poverty in the human development approach draws on each of these perspectives, but draws particularly on the capability perspective. In the capability concept the poverty of a life lies not merely in the impoverished state in which the person actually lives, but also in the lack of real actual lives, but also in the lack of real opportunity—due to social constraints as well as personal circumstances—to lead valuable and valued lives.

In the capability concept the focus is on the functioning that a person can or cannot achieve, given the opportunities she has. Functioning refers to the various valuable things a person can do, or be, such as living long, being healthy, being well nourished, mixing well with others in the community, and so on.

The ***capability approach*** concentrates on functioning information, supplemented by considering, where possible, the options a person had but did not choose to use. For example, a rich and healthy person who becomes ill nourished through fasting can be distinguished from a person who is forced into malnutrition through a lack of means or as a result of suffering from a parasitic disease. In practice such discrimination is difficult when dealing with aggregate statistics (as opposed to detailed micro studies of individuals), and the practical uses of the capability concept in poverty analysis have been mainly with simple functioning data. The Human Development Report too presents information that is essentially about living conditions and functioning.

In choosing particular aspects of living for special investigation in a poverty study, there is need for public discussion. There is an inescapable element of judgment in any such selection. In constructing any index of poverty (such as the human poverty index), the selections and the weights have to be explicitly stated and clarified so that public scrutiny can occur. It is very important that the standards to be used are not determined on a topdown basis, but are open to—if possible, emerge from—a participatory, democratic process. One of the purposes of Human Development Report has been precisely to facilitate such a process, and this applies to poverty analysis as well.

The application of "sustainable livelihood approach" to the study of poverty has particularly emphasized the need for local participation. In this approach each community can define criteria of well being and the key elements of deprivation as they appear in the local context. This process brings out the concerns and worries of vulnerable people that are persistently neglected in national statistics and in many studies of poverty.

7.3 The Human Development Index and the Human Poverty Index

While human development focuses on progress in a community as whole, human poverty focuses on the situation and progress of the most deprived people in the community. The distinction between the two is analogous to the distinction between GNP and the income-based poverty index. In the income-based perspective, poverty incidence is needed to monitor progress in eliminating poverty. In the same way, the HPI is needed to judge the extent of human poverty in a country and to monitor its progress.

The growth rate of GNP per person gives an account of progress seen in the conglomerative/aggregate perspective—everyone's income counts in the GNP total. In contrast, the reduction of an income-based poverty index—such as the decline in the proportion of people below the poverty—line income—uses the deprivation perspective, concentrating only on the incomes of the poor. In this income—based perspective, it would make little sense to argue that since GNP is already based on income information, any income based poverty measurement be a substitute for GNP. Nor would it be sensible to suggest that the availability of GNP as an indictor makes it redundant to seek a measure of income poverty measures use the income information in different perspectives—with GNP taking a conglomerative view and the income poverty measures focusing specifically on people poor in income.

The relationship between HDI and the HPI has to be seen in a similar way. Both have to use the rich categories of information associated with human development—characteristics of human lives and quality of living that go far beyond what income information can provide. But while the HDI uses these characteristics in the conglomerative perspective, the HPI must use them in the deprivation perspective. The availability of GNP measures does not obviate the need for an income-based poverty indicator, nor does the HDI measure eliminate the need for an HPI.

7.3.1 Values and Ranking of the Human Poverty Index

Estimates of the HPI have been prepared for 78 developing countries having adequate data. The procedure for computing the index and the full results are presented in technical note 2. The HPI value indicates the proportion of the population affected by the three key deprivations in their lives—showing how widespread human poverty is. At the top of the rankings are Trinidad and Tobago, Cuba, Chile, Singapore, Costa Rica—these countries have reduced human poverty to an HPI value of less than 10 %. In other words, these countries have reduced human poverty to the point at which it affects less than 10 % of the population.

At the bottom are seven countries whose HPI exceeds 50 %—Niger, Sierra Leone, Burkina Faso, Ethiopia, Mali, Cambodia and Mozambique. And in almost half the 78 countries covered, the HPI exceeds 34 %, implying that about a third of their people suffer human poverty.

7.4 How Does the HPI Compare with Income-based Measures of Poverty?

Some countries have done better in reducing income poverty than human poverty. In Cote d'Ivoire and Egypt less than 20 % of the people are income—poor, but 35 % or more are affected by human poverty. These countries could pay more attention to reducing basic deprivations in choices and opportunities, especially by extending access to basic education and health services. Other countries have done better in reducing human poverty than income poverty—China, Costa Rica, Kenya, Peru, the Philippines, Zimbabwe, These countries have invested heavily in reducing deprivations in basic human capabilities. Progress in reducing poverty in income and progress in reducing poverty in human choices and opportunities do not always move together. Regression analysis indicates a weak relationship between the head-count index of income poverty and HPI. So in monitoring progress, the focus should not be on income poverty alone, but on indicators of human poverty as well.

Comparing the HPI with the HDI reveals stark contrasts in some countries. These differences can alert policy-makers to the need to make human development better distributed, more pro-poor. The HDI measures the overall progress of a country in human development. It can mask unequal distribution of that progress and the widespread human poverty that remains. The countries at the bottom of the HPI rankings also rank near the bottom in the HDI. In these countries the overall progress in human development has been too low to raise the majority of their people from poverty.

HPI estimates for regions show that:

- Human poverty affects a quarter of the developing world's population, while income poverty affects a third.
- Human poverty is most widespread in South Asia, affecting about 40 % of the people.
- Progress in reducing human poverty and income poverty does not always go together. The contrasts are most stark in the Arab States, where income poverty was reduced to 4 % by 1993 but human poverty was still 32 %, and in Latin America and the Caribbean, where human poverty has been reduced to 15 % but income poverty is still 24 %.
- The trends in human poverty in developing countries with available data show that although all were able to reduce the incidence of human poverty during the past two decades, the extent and pattern of reduction differed.
- While Mexico and Thailand were able to reduce the incidence of human poverty by two-thirds, Peru, starting from a similar base, reduced it by less than a fifth. A similar comparison can be made for Costa Rica and Panama.
- Estimating separate HPIs for groups or regions reveals disparities and contrasts within countries, and pinpoints concentrations of poverty.

7.5 New Conceptual Issues in the Measurement of Poverty

The determination of the number of poor people in any society depends crucially on the poverty measures used. Different measures will give different figures. The most commonly used figure is the "poverty line", which itself is a difficult and controversial concept. Concerns with identifying people affected by poverty and the desire to measure it have at times obscured the fact that poverty is too complex to be reduced to a single dimension of human life. It has become common for countries to establish an incomebased or consumption-based poverty line. Although income focuses on an important dimension of poverty, it gives only a partial picture of the many ways human lives can be blighted. Someone can enjoy good health and live quite long but be illiterate and thus cut off from learning, from communication and from interactions with others. Another person may be literate and quite well educated but prone to premature death because of epidemiological characteristics or physical disposition. Yet a third may be excluded from participating in the important decision making processes affecting her life. The deprivation in all these cases cannot be fully captured by the level of income.

Also, people perceive deprivation in different ways—and each person and community defines the deprivation and disadvantages that affect their lives in different ways.

Most National Poverty Eradication Programs use the income/expenditure method to determine the poverty line.

Major Conceptual Issues in Poverty Measurement
There are four major conceptual issues in poverty measurement: the choice of poverty indicator, the determination of the poverty line, the unit of analysis, and the choice of equivalent scale. These four issues are frequently discussed in the poverty and development economics literature. We shall review each one of these.

7.5.1 Choice of Poverty Index

Poverty may be defined in terms of a single broad indicator of economic resources. Such a definition may be restrictive. However, if the concern about poverty takes the form of concern about basic needs, such as food, housing, and clothing, the focus should be on individual items of consumption, and poverty would need to be measured in a multidimensional way, rather than in terms of a single indicator. The same consideration may apply if we are concerned with deprivation of commodities. Concern about poverty may relate to social or demographic indicators, such as life expectancy or literacy.

If the measure of disadvantage is limited to a single index of economic resources, then a natural choice may appear to be total consumption or expenditure plus home produced goods and services. A household is then said to be poor if its total consumption is below a specified amount. But most studies of poverty in advanced countries record poverty on the basis of total income rather than consumption.

In most developing economies, however, the well being of individuals is measured by total consumption expenditures, not by total income. This is for a number of reasons. The most important of these is that consumption tends to fluctuate much less during the course of a month or a year than income. Moreover, experience has also shown that people tend to provide more accurate information about their consumption behaviour than about their income sources. Furthermore, if the expenditure data can be used for welfare analysis, this has the compelling advantage that the poverty lines can be derived from the data itself and need not be adopted from other surveys.

A number of steps are required to arrive at usable consumption figures for households in many developing economies where poverty is rampant. First, information on household purchases of food items is converted into monetary aggregates based on households' reported quantities and prices paid. Second, the calorie equivalent of the observed food consumption per household is calculated. Third, non-food expenditures are evaluated and priced, paying particular attention to the valuation of water and the 'consumption' of durable goods such as refrigerators, houses, houses, or cars. Finally, nominal expenditures are adjusted for all households due to the variation in prices among different areas and regions within most countries.

7.5.2 *Poverty Lines*

A poverty datum line is used to draw a distinction between the poor and the non-poor in society. The World Development Report 1990 used an upper line of US$370 (in purchasing power parity dollars) per capita for the poor as a cut-off point for absolute poverty, and a lower poverty line of US$275 for those in abject poverty. The poverty line is used to calculate the proportion of the national population falling below it. Two types of poverty are usually analyzed: absolute poverty and relative poverty.

7.5.2.1 Absolute Poverty

There are many types of poverty lines. In the analysis of WMS surveys, poverty lines are expressed in terms of equivalent adult consumption expenditure calculated using two popular money-metric concepts of poverty, namely, absolute and relative poverty. Three poverty lines are presented in this work: food, overall, (food and non-food) and hard-core. A poverty line is one that is fixed over time in terms of the living standards indicator for the entire country. It does not take into consideration time or place. Proponents of absolute poverty concept argue that there are some basic elements of welfare that every human being has a right to and their attainment is not dependent on scarcity of local resources but is inspired by the universalistic valuation of human dignity.

Absolute poverty is below the income that secures the bare essentials of food, clothing, and shelter. Other essentials may be added. Thus, determining this level is a matter of judgment, so that it is difficult to make comparisons between countries. Moreover, what is considered poverty varies according to the living standards of the time and region. World Bank economists Martin Ravallion, Garav Datt, and Dominique van de Walle show that national poverty lines increase with mean consumption, although poverty lines are below the mean in all cases (iv). Accordingly, many people classified as poor by their government may be materially better off than many in the past decades or others today who are not considered poor.

Recognizing that the perception of poverty has evolved historically and varies tremendously across cultures, Ravallion, Datt, and van de Walle set an extreme poverty line and a poverty line (v). The lower line, the extreme poverty line, recognized as the absolute minimum by international standards, is based on a standard set in India, the country with the most extensive literature on the subject, and close to the poverty line of perhaps the poorest country, Somalia.

The definition of poverty used by the World Bank is based on previous work by its economists, particularly that of Montek S. Ahluwalia, Nicholas G. Carter, and Hollis B. Chenery. These economists, used Indian data (hence the use of I$) as their yardstick and assumed a population with a "normal" distribution by age and gender, define the extreme poverty line as the income needed to attain basic nutritional needs, that is, a daily supply of 2,250 calories per person. A figure of $275 purchasing power adjusted dollars or I$275 per capita in 1985 (vi). The 2,250 calories would be met by the following diet: 5 g of leafy vegetables, 110 g of other vegetables (potatoes, root vegetables, gourds, and so on), 90 g of milk, 35 g of oil, 35 g of sugar, 10 g of flesh foods (fish and meats), 45 g of pulses (peas or other legumes), and 395 g of cereals (rice, corn, millet, or wheat). To illustrate, the 395 g of cereals might consist of about 2 cups of hot prepared rice, equivalent in weight to 54 % of the total diet (vii).

The World Bank's upper poverty line, below which persons are designated as poor, was 370 purchasing-power adjusted dollars or I$370 per capita in 1985. This poverty line provides for consumption in excess of the bare physical minimum. However, it varies from country to country, reflecting the cost of participating in the everyday life in society. Given information on income distribution, poverty is determined by finding the percentage of the population with income of less than I$370. The extreme poverty is found by finding the share of the population with an income of less than I$275. The assumption is that two persons with the same purchasing power adjusted income (not including non-income factors, such as access to public services) living in different countries will have the same measured poverty.

7.5.2.2 Accompaniments of Absolute Poverty

Nearly half the world population has entered the twenty-first century living in absolute poverty, meaning that they suffer the following deprivations: (1) Three- to four-fifths of their income is spent on food; the diet is monotonous, limited to cereals, yams, or cassava, a few vegetables, and in some regions, a little fish or meat. (2) About 60 % are undernourished and millions are severely malnourished. Energy and motivation are reduced; performance in school and at work is undermined; resistance to illness is low; and the physical and mental development of children is often impaired. (3) One of every 10 children born die within the first year; another dies before the age of 5; and only five reach the age of 45. (4) Beginning in 1975, the World Health Organization and UNICEF expanded immunization against the major diseases affecting the children in most poor countries. Immunization rates increased rapidly, and deaths from these diseases fell substantially in many poor countries from the 1980s to the 1990s. Still fewer than 50 % of the children in absolute poverty are vaccinated against measles, diptheria, and whooping cough, which have been virtually eliminated in rich countries. (5) Majority of the poor lack access to clean and plentiful water and even a larger proportion lack an adequate system for disposing of their wastes. Lack of sanitation contributes to diseases such as diarrhea diseases yearly. (6) Average life expectancy in most poor countries is tragically low, compared to 77 years in developed countries. (7) Only about one-third to two-fifths of the adults are literate in most poor countries. (8) Only about four of every 10 children complete more than 4 years of primary school. (9) The poor are more likely to be concentrated in environmentally marginal and vulnerable areas. Therefore, they face higher rates of unemployment and underemployment, and have higher fertility rates than those who are not poor, perhaps a natural survival response to the high infant mortality rate (viii).

7.5.2.3 Relative Poverty

Definition of poverty in relative terms relates to the type of poverty analysis that endeavors to take into account the actual deprivation with respect to the average levels of satisfaction of needs in that society. Relative income/expenditure measures such as income share of the poorest 40 %, or the per cent of the population below a line defined in reference to the mean or median, e.g. at 2/3 of the mean income/expenditure are effective measures of inequality given that poverty need not fall as either aggregate income/expenditure or the income/expenditure of the poor or both rises. By contrast, with an absolute poverty line, poverty must fall if the income/expenditures of all those who are poor increase. Given the arbitrariness of the relative poverty approach, one can easily over—or under-estimate the population above or below the poverty line. Secondly, the method cannot give an indicative comparison among regions. For this reason, relative poverty measures are less relevant to developing countries in terms of policy interventions targeting the poor. The application of absolute poverty concept is therefore preferred in developing countries where the majority of the people live below the subsistence level, since in such situations one would be interested in knowing the size and distribution of those in absolute deprivation rather than relative deprivation. This is the generally expressed view by many governments and donors involved in poverty reduction programs rather than advocating more egalitarian strategies.

7.6 Measuring National Food Poverty Lines

Many poor countries also use the food poverty method to determine the poverty lines. These lines indicate the insufficiency of economic resources to meet basic minimum needs in food. There are three approaches to measuring food poverty:

7.6.1 Cost-of-Basic Needs Method

This approach sets the poverty line at the cost of a basic diet for the main age, gender and activity groups, plus a few essential non-food items. A survey then establishes the proportion of people living in households with consumption (or sometimes income) below this line. The basic diet may consist of the least expensive foods needed to meet basic nutritional requirements, the typical adult diet in the lowest consumption quintile or the investigator's notion of a minimal but decent diet. The choice of both the food and the non-food components included is necessarily arbitrary.

7.6.2 Food Energy Method

This method focuses on the consumption expenditure at which a person's typical food energy intake is just sufficient to meet a predetermined food energy requirement. Dietary energy intake, as the dependent variable, is regressed against household consumption per adult equivalent. The poverty line is then set at the level of total consumption per person at which the statistical expectation of dietary energy intake exactly meets average dietary energy requirements. The problem with this method is the caviar caveat: groups that choose a costly bundle of foods are rewarded with a higher poverty line than that for more frugal eaters.

7.6.3 Food Share Method

This method derives the cost of a consumption plan to acquire just sufficient nutrients. If the cost of basic nutrients is a third of total consumption, the poverty line is fixed at three times that cost. All three approaches are sensitive to the price used to determine the cost of the bundle. And all three concentrate mainly on calories or dietary energy, because protein deficiency due to inadequate economic resources is perceived to be rare in most societies.

7.6.4 Poverty Lines for International Comparisons

A poverty line set at US$1 (1985 PPP$) a day per person is used by the World Bank for international comparison. This poverty line is based on consumption. A poverty line of US$2 (PPP$) a day is suggested for Latin America and the Caribbean. For Eastern Europe and the ICS countries, a poverty line of US$4 (1990s) has been used. For comparison among industrial countries, a poverty line corresponding to the US poverty line of US$14.40 (1985 PPP$) a day per person has been used.

7.6.5 Equivalence Scales

In comparing household budgets, it is important to recognize that households differ in consumption and size. Thus, a simple comparison of aggregate household consumption can be misleading about the well being of individual members in a given household. Some analysts therefore use some form of normalization such as "consumption per adult equivalent". The key question is, "equivalent" in what sense? Earlier studies on poverty in countries such as Kenya used equivalence scales that covered age groups of 0–4 (weighted by a factor of 0.24), 5–14 (weighted by a factor of 0.65), and ages 15 and above (weighted by a factor of 1.0). The general practice in developing equivalence scales is to undertake a comprehensive and intensive household consumption survey to monitor and measure actual quantities consumed by each household member of various commodities produced or purchased by the household. This practice recognizes the fact that both gender and cultural influences determine, to a large extent, individual entitlements at household level.

7.7 Concluding Remarks

7.8 Footnotes

1 Morris D. Morris, Measuring the Conditions of the World's Poor: The Physical Quality of Life Index (New York: Pergamon Press 1979) 2 UNDP, Human Development Report, 1990 (iii) For details, see UNDP, Human Development Report. (New York: Oxford University Press 1990) 3 UNDP, Human Development Report, 2006 (iv). Martin Ravallion, Garav Datt, and Dominique van de Walle, "Quantifying Absolute Poverty in the Developing World," Review of Income and Wealth 37 (December 1991): 345–61. (v) Martin Ravallion, Garav Datt, and Dominique van de Walle, "Quantifying Absolute Poverty in the Developing World," Review of Income and Wealth 37 (December 1991): 345–61. (vi) Partha Dasgupta, An Inquiry into Well-Being and Destitution (Oxford: Clarendon Press, 1993), p. 404. (vii) For constructing a least—cost diet, see Patrick J. Gormely, "Are High-Protein Foods Economically Efficient?" Food Policy 3 (November 1978): 280–88. (viii) World Bank, World Development Report, 1993 (New York: Oxford University Press, 1993); World Bank, World Development Report, 1992 (New York: Oxford University Press, 1992), p. 5; World Bank, World Development Report, 1990 (New York: Oxford University Press, 1990).

Chapter 8
Emerging Conceptual Issues and Rural Urban Dichotomy

The determination of the number of poor people in any society depends crucially on the poverty measures used. Different measures will give different figures. The most commonly used figure is the "poverty line", which itself is a difficult and controversial concept. Concerns with identifying people affected by poverty and the desire to measure it have at times obscured the fact that poverty is too complex to be reduced to a single dimension of human life. It has become common for countries to establish an income-based or consumption-based poverty line. Although income focuses on an important dimension of poverty, it gives only a partial picture of the many ways human lives can be blighted. Someone can enjoy good health and live quite long but be illiterate and thus cut off from learning, from communication and from interactions with others. Another person may be literate and quite well educated but prone to premature death because of epidemiological characteristics or physical disposition. Yet a third may be excluded from participating in the important decision making processes affecting her life. The deprivation of none of them can be fully captured by the level of their income.

Also, people perceive deprivation in different ways—and each person and community defines the deprivation and disadvantages that affect their lives. As such, careful consideration needs to be given even when analyzing urban and rural poverty. There are obviously clear differences and similarities in the forms of deprivations faced by urban and rural poor. A good analysis of rural and urban poverty should reflect the diversity and complexity identifiable in the dichotomy. Such analysis hence leads to appropriate conceptualization of the deprivations, and consequentially, fitting remedial policies and action programs. A quick look at the use of poverty line as a poverty measure instrument of both urban and rural poverty fails to capture the underlying economic arrangements in the livelihoods of urban and rural poor. This is simply because the urban poor rely more on income earned from elsewhere as compared to rural poor whose economic livelihood relies considerably on non-market transactions that are hardly captured in the poverty line measure instrument. In addition, it is worth noting that even within the same country, the cost of living is much higher in urban areas than in rural areas. Also with the urban-rural poverty diversity, certain

N. O. Agola and J. L. Awange, *Globalized Poverty and Environment*,
DOI: 10.1007/978-3-642-39733-2_8,

deprivations tend to be more pronounced as in the case insecurity in slum dwellings of the urban poor. Another front on which an attack can be waged against the poverty line measure instrument is its failure to capture the systemic nature and interaction of factors that feed into and sustains the ruthless poverty machine. Insecurity, for instance, feeds into economic relations of slum dwellers and shapes the very nature, cost and details of simple but vital transactions. Below is a true depiction of how pervasive rent seeking happens in one of the biggest urban slums in the world (and perhaps the same situation prevails in all slums around the major world urban centers in Latin America, parts of Asia and other African countries). The depiction is based on the personal observations and interviews of the authors in the course of the research for this book.

Kibera slum in Nairobi, Kenya, presents a case in point of the pervasive rent-seeking behavior in slum dwellings by persons who wriggle themselves into various positions of influence, with the result of considerably driving up the cost of most services. One of the instances we observed was the case of putting up a small house/shelter by a slum dweller. Because land rights and property ownership laws and security situation are weak in the slum, as soon as work begins to put up the shelter, a few extortionists will turn up to stake their claim on the parcel of land on which construction has just begun, and forthwith demand small payment to let the construction proceed. But, as if that is not enough, another version of extortionist will again turn up offering protection by proxy, by way of close ties to the local administration official. It follows that this other version of extortionist goes on to demand his take for introduction to the administration official and on the other hand also demands the administration official's take. In what turns out like the pouring of an acoustic agent into a fresh wound, the slum dweller we interviewed narrated how he must also pay a guard at the gate to the construction site he visits daily to seek work on a daily basis. The experiences of our interviewee are but just the tip of the iceberg in what can be considered an extortionist-chain of rent-seekers in slum dwellings. Such experiences only serve to drive prices up for most if not all of the essential services and goods the urban poor need for their livelihood. However, such a scenario is quite a contrast to the experiences of the rural poor, thus underscoring the need for different conceptual analysis and remedial intervention. Life is relatively more orderly in rural areas where customary rules and family ties serve to reign in any deviants.

Consequently, this chapter will deal with the new conceptual issues in the first part, and then proceed to discuss the dichotomy of urban-rural poverty and its implications for conceptual analysis, policy formulation and possible program actions.

8.1 Major Conceptual Issues in Poverty Measurement

There are four major conceptual issues in poverty measurement: the choice of poverty indicator, the determination of the poverty line, the unit of analysis, and the choice of equivalent scale. These four issues are frequently discussed in the literature. We shall review each one of these.

Choice of Poverty Index

Poverty may be defined in terms of a single broad indicator of economic resources. Such a definition may be restrictive. However, if the concern about poverty takes the form of concern about basic needs, such as food, housing, and clothing, the focus should be on individual items of consumption, and poverty would need to be measured in a multidimensional way, rather than in terms of a single indicator. The same consideration may apply if we are concerned deprivation of commodities. Concern about poverty may relate to social or demographic indicators, such as life expectancy or literacy.

If the measure of disadvantage is limited to a single index of economic resources, then a natural choice may appear to be total consumption or expenditure plus home produced goods and services. A household is then said to be poor if its total consumption is below a specified amount. But most studies of poverty in advanced countries record poverty on the basis of total income rather than consumption.

In most poverty stricken countries, however, the well being of individuals is measured by total consumption expenditures, not by total income [1]. This is for a number of reasons. The most important of these is that consumption tends to fluctuate much less during the course of a month or a year than income. Moreover, experience has also shown that people tend to provide more accurate information about their consumption behaviour than about their income sources. Furthermore, if the expenditure data can be used for welfare analysis, this has the compelling advantage that the poverty lines can be derived from the data itself and need not be adopted from other surveys.

A number of steps are required to arrive at usable consumption figures for households in countries with poverty incidence. First, information on household purchases of food items is converted into monetary aggregates based on households' reported quantities and prices paid. Second, the calorie equivalent of the observed food consumption per household is calculated. Third, non-food expenditures are evaluated and priced, paying particular attention to the valuation of water and the 'consumption' of durable goods such as refrigerators, houses, or cars. Finally, nominal expenditures are adjusted of all households for the variation in prices among different areas and regions within the country.

Poverty Lines

A poverty datum line is used to draw a distinction between the poor and the non-poor in society. The World Development Report 1990 [8] used an upper line of US$370 (in purchasing power parity dollars) per capita for the poor as a cut-off point for absolute poverty, and a lower poverty line of US$275 for those in abject poverty. The poverty line is used to calculate the proportion of the national population falling below it. Two types of poverty are usually analyzed: absolute poverty and relative poverty.

Absolute Poverty

There are many types of poverty lines. In the analysis of WMS surveys, poverty lines are expressed in terms of equivalent adult consumption expenditure calculated using two popular money-metric concepts of poverty, namely, absolute and relative poverty. Three poverty lines are presented in this paper: food, overall, (food and non-food) and hard-core. A poverty line is one that is fixed over time in terms of the living standards indicator for the entire country. It does not take into consideration time or place. Proponents of absolute poverty concept argue that there are some basic elements of welfare that every human being has a right to and their attainment is not dependent on scarcity of local resources but is inspired by the universalistic valuation of human dignity.

Absolute poverty is below the income that secures the bare essentials of food, clothing, and shelter. Other essentials may be added. Thus, determining this level is a matter of judgment, so that it is difficult to make comparisons between countries. Moreover, what is considered poverty varies according to the living standards of the time and region. World Bank economists [4] show that national poverty lines increase with mean consumption, although poverty lines are below the mean in all cases. (i) Accordingly, many people classified as poor by their government may be materially better off than many in the past decades or others today who are not considered poor.

Recognizing that the perception of poverty has evolved historically and varies tremendously across cultures, Ravallion, Datt, and van de Walle set an extreme poverty line and a poverty line. (ii) The lower line, the extreme poverty line, recognized as the absolute minimum by international standards, is based on a standard set in India, the country with the most extensive literature on the subject, and close to the poverty line of perhaps the poorest country, Somalia.

The definition of poverty used by the World Bank is based on previous work by its economists, particularly that of Montek S. Ahluwalia, Nicholas G. Carter, and Hollis B. Chenery. These economists, used Indian data (hence the use of 1$ (One Dollar)) as their yardstick and assumed a population with a "normal" distribution by age and gender, define the extreme poverty line as the income needed to attain basic nutritional needs, that is, a daily supply of 2,250 calories per person. A figure of $275 purchasing power adjusted dollars or 1$ (One Dollar) 275 per capita in 1985. (iii) The 2,250 calories would be met by the following diet: 5 g of leafy vegetables, 110 g of other vegetables (potatoes, root vegetables, gourds, and so on), 90 g of milk, 35 g of oil, 35 g of sugar, 10 g of flesh foods (fish and meats), 45 g of pulses (peas or other legumes), and 395 g of cereals (rice, corn, millet, or wheat). To illustrate, the 395 g of cereals might consist of about 2 cups of hot prepared rice, equivalent in weight to 54 % of the total diet (iv).

The World Bank's upper poverty line, below which persons are designated as poor, was $370 purchasing power adjusted dollars or $370 per capita in 1985. This poverty line provides for consumption in excess of the bare physical minimum. However, it varies from country to country, reflecting the cost of participating in the everyday life in society.

Given information on income distribution, poverty is determined by finding the percentage of the population with income of less than 1$ (One dollar) 370. The extreme poverty is found by finding the share of the population with an income of less than 1$ (One dollar) 275. The assumption is that two persons with the same purchasing power adjusted income (not including non-income factors, such as access to public services) living in different countries will have the same measured poverty.

8.2 Accompaniments of Absolute Poverty

The more than 1 billion people living in absolute poverty suffer the following deprivations: (1). Three- to four-fifths of their income is spent on food; the diet is monotonous, limited to cereals, rice, bread, cornmeal, yams, or cassava, a few vegetables, and in some regions, a little fish or meat and (2). About 60 % are undernourished and millions are severely malnourished. Energy and motivation are reduced; performance in school and at work is undermined; resistance to illness is low; and the physical and mental development of children is often impaired and (3). One of every 10 children born die within the first year; another dies before the age of 5; and only five reach the age of 45 and (4). Beginning in 1975, the World Health Organization and UNCEF expanded immunization against the major diseases affecting the children around the globe. Immunization rates increased rapidly, and deaths from these diseases fell substantially from the 1980s to the 1990s. Still fewer than 50 % of the children in absolute poverty are vaccinated against measles, diptheria, and whooping cough, which have been virtually eliminated in rich countries and (5). Two-thirds of the global poor lack access to clean and plentiful water and even a larger proportion lack an adequate system for disposing of their wastes. Lack of sanitation contributes to diarrhea diseases yearly and (6). Average life expectancy is also about 50 years, compared to 77 years in developed countries and (7). Only about one-third to two-fifths of the adults are literate and (8). Only about four of every 10 children complete more than 4 years of primary school and (9). The global poor are more likely to be concentrated in environmentally marginal and vulnerable areas, face higher rates of unemployment and underemployment, and have higher fertility rates than those who are not poor.

Relative Poverty

Definition of poverty in relative terms relates to the type of poverty analysis that endeavors to take into account the actual deprivation with respect to the average levels of satisfaction of needs in that society. Relative income/expenditure measures such as income share of the poorest 40 %, or the percent of the population below a line defined in reference to the mean or median, e.g. at 2/3 of the mean income/expenditure are effective measures of inequality given that poverty need not fall as either aggregate income/expenditure or the income/expenditure of the poor or both rises. By contrast, with an absolute poverty line, poverty must fall if the income/expenditures of all those who are poor increase. Given the arbitrariness of the relative poverty approach, one can easily over- or under-estimate the population above or below the poverty line. Secondly, the method cannot give an indicative comparison among regions. For this reason, relative poverty measures are less relevant to developing countries in terms of policy interventions targeting the poor. The application of absolute poverty concept is therefore preferred in developing countries where the majority of the people live below the subsistence level, since in such situations one would be interested in knowing the size and distribution of those in absolute deprivation rather than relative deprivation. This is the generally expressed view by many governments and donors involved in poverty reduction programs rather than advocating more egalitarian strategies.

8.3 Measuring National Food Poverty Lines

Currently, many countries use the food poverty method to determine the poverty lines. These lines indicate the insufficiency of economic resources to meet basic minimum needs in food. There are three approaches to measuring food poverty:

Cost-of-Basic Needs Method

This approach sets the poverty line at the cost of a basic diet for the main age, gender and activity groups, plus a few essential non-food items. A survey then establishes the proportion of people living in households with consumption (or sometimes income) below this line. The basic diet may consist of the least expensive foods needed to meet basic nutritional requirements, the typical adult diet in the lowest consumption quintile or the investigator's notion of a minimal but decent diet. The choice of both the food and the non-food components included is necessarily arbitrary.

Food Energy Method

This method focuses on the consumption expenditure at which a person's typical food energy intake is just sufficient to meet a predetermined food energy requirement. Dietary energy intake, as the dependent variable, is regressed against household consumption per adult equivalent [2]. The poverty line is then set at the level of total consumption per person at which the statistical expectation of dietary energy intake exactly meets average dietary energy requirements. The problem with this method is the caviar caveat: groups that choose a costly bundle of foods are rewarded with a higher poverty line than that for more frugal eaters.

Food Share Method

This method derives the cost of a consumption plan to acquire just sufficient nutrients. If the cost of basic nutrients is a third of total consumption, the poverty line is fixed at three times that cost.

All three approaches are sensitive to the price used to determine the cost of the bundle. And all three concentrate mainly on calories or dietary energy, because protein deficiency due to inadequate economic resources is perceived to be rare in most societies.

Poverty Lines for International Comparisons

A poverty line set at US$1 (1985 PPP$) a day per person is used by the World Bank for international comparison. This poverty line is based on consumption. A poverty line of US$2 (PPP$) a day is suggested for Latin America and the Caribbean. For Eastern Europe and the ICS countries, a poverty line of US$4 (1990s) has been used [5, 9, 10]. For comparison among industrial countries, a poverty line corresponding to the US poverty line of US$14.40 (1985 PPP$) a day per person has been used.

Equivalence Scales

In comparing household budgets, it is important to recognize that households differ in consumption and size. Thus, a simple comparison of aggregate household consumption can be misleading about the well being of individual members in a given household. Some analysts therefore use some form of normalization such as "consumption per adult equivalent". The key question is, "equivalent" in what sense? Earlier studies on poverty in most developing countries used equivalence scales that covered age groups of 0–4 (weighted by a factor of 0.24), 5–14 (weighted by a factor of 0.65), and ages 15 and above (weighted by a factor of 1.0). The general practice in developing equivalence scales is to undertake a comprehensive and intensive household consumption survey to monitor and measure actual quantities consumed by each household member of various commodities produced or purchased by the household. This practice recognizes the fact that both gender and cultural influences determine, to a large extent, individual entitlements at household level.

8.4 Global Urban Poverty

Urban poverty is one of the biggest challenges facing the world in the twenty-first century. The figures speak for themselves. According to UN projections, in 2025, the world's total urban population will rise to 4.54 billion, an increase of 1.69 billion people. 94 % of this increase will occur in developing countries. And the projections are that urbanization will increase at a great speed, as compared with the industrialized world [8].

In the industrialized world, it took more than a generation to urbanize. However, urbanization is happening with a great speed in Africa and Asia. Africa, which is currently the least urbanized has the highest rate of urban growth—4 % compared to 2 % for the rest of the world. In terms of sheer numbers, urban populations are greatest in Asia, where more than 1.3 billion people, half the world's total urban population, reside in urban areas. Klaus Toepfer, the Acting Executive Director of the United Nations Centre for Human Settlements (Habitat) has observed that: "Not only are we living in an urbanizing world, we are also experiencing an unprecedented urbanization of poverty" [6].

In many cities of the developing world, up to a half of the urban population live in slums or squatter settlements. Even though the twenty-first century has been dubbed the 'Urban Millennium', and there may be the assumption that urban dwellers have better living conditions and as such are better off as compared to rural dwellers, UN-Habitat [7] report shows that in the actual sense and by most measures, the urban dwellers suffer from a set of conditions, called 'urban penalty', which makes them worse off than their rural counterparts. Despite the promise of increased opportunities and a better quality of life, city life for many urban residents is a negative and degrading experience. Slum dwellers often do not have access to the most basic amenities such as water and sanitation and many are faced with constant threat of eviction. Cities, both in the developing world and in industrialized countries, are becoming breeding grounds for crime, congestion and pollution. These social ills have been brought about by urban poverty caused by poor planning and poor governance, mismanagement and corruption.

Poor Governance as a Cause of Urban Poverty in Asia and Africa

What is the solution? The answer is that there is need for better urban governance. For many years, governments have blamed urban ills, including poverty, on lack of financial resources to implement development programs. Recently, the blame has shifted to a global economy that favors the industrialized countries, and mounting debt payments to international lending institutions. But, as a recent United Nations report reveals (UN HABITAT's State of the World's Cities Report 2006/7) [7], it is not so much the amount of resources that determine the level of poverty in a country or city, but how the resources are used. If anything, poor governance is the leading cause of poverty.

Embracing democracy is often not enough, and certainly not a panacea. In fact, representative democracy will have little or no impact on poverty, if it is not supported by mechanisms that enhance transparency and promote people's participation. The bottom line requirement for good governance is accountability in the use of public funds and the shifting of decision-making powers to those of most affected by them—local governments, civil society organizations and the people themselves. This does not mean that local governments and civil society should take over the legitimate functions of government. But when the distance between government and the governed is too wide, abuse of power and resources is more likely. This breeds poverty.

In an effort to address these urban poverty issues, Habitat is now focusing on strengthening the role of local governments to facilitate transparency and accountability in urban governance structures and institutions. As part of its Global Campaign on Urban Governance, the UN Habitat, in cooperation with local government representatives from around the world, is currently engaged in formulating a World Charter for Local Self- Government which, if adopted, will increase local autonomy and promote decentralization of policies around the world. The ultimate aim is to get the World Charter adopted by the UN General Assembly. If the Charter eventually becomes a UN Convention, it will serve as a "Magna Carta of responsiveness to the needs of citizens".

Habitat is also working closely with the World Bank on an initiative that aims at improving the lives of 100 million slum dwellers by 2020, by building on successful community-based upgrading programs and by addressing the broader policy and institutional issues that often make such efforts unsustainable.

The call for decentralization and devolution of authority to the local level is based on the premise that when people are involved in decisions that affect their lives, they are more likely to support and sustain those decisions. They also have a greater stake in ensuring that resources are used more efficiently to meet actual needs. Besides, it is harder to mismanage funds or misuse authority in an environment where one's actions are more detectable.

Participatory democracy and transparent local government structures have a great potential for improving the lives of millions of people around the world. Through a series of regional, sub-regional and thematic meetings, citizens scrutinize the past year's expenditure, agree upon current priorities, and allocate resources for new projects.

Now that the urban millennium is upon us, the promotion of good urban governance will be at the forefront of all development efforts. How well or how badly cities perform will determine how well or badly nations perform. As countries develop, cities will account for a growing share of the national income. Already cities in both developing and industrialized countries account for between 55 and 85 % of Gross National Products.

Example of Initiative to deal with the urban poverty: Britain launches an Urban Project

A new initiative by the British Government seeks to improve lives for the poor in Kenya's urban centers, and reform most settlements currently marked by inferior housing, overpopulation and lack of the most basic amenities. The Nairobi Urban Poverty Partnership Project was launched in the city in the year 2000 with the British Government providing Sh325 million ($2.8 million) to start it off [3]. The British government agency recently committed the same sum to a similar project in Mombasa that is currently getting underway.

Through Britain's Department for International Development (DFID), which is responsible for administering the Sh2.8 billion annual development program in Kenya, the British Government plans to provide more funds to strengthen the Nairobi Informal Settlements Co-ordination Committee (NISCC)—the body to oversee the new project.

The British Government has laid out elaborate plans to get the project started in the city and has made it clear that it will in the coming few months increase its "close and constructive" work with the government and other stake holders to come up with effective strategies to reverse these trends throughout the country. The project is a partnership between many different stake holders and its main arm is the Nairobi Informal Settlements Co-ordination Committee, which brings together Provincial Administration, the Nairobi City Council, non-governmental organizations, private sector and donor representatives. The plan is expected to be replicated in various urban centers once it proves its efficacy to address the growing challenge of urban poverty in most towns in the country.

The involvement of several organizations—public, private and non-governmental organizations underscores the gravity of the urban poverty problem. Similar initiatives are underway in most cities in Asia and Latin America, but this is only the beginning. It is still premature measures such as the million-acre settlement scheme, and the special to evaluate the efficacy of the initiative, though we suspect that like in other poverty reduction measures that have been undertaken for decades, it might be also a case of negligible response a bit too late.

Other responses to urban poverty are such like that of the International Labour Organization proposal of strategy for increasing productive employment and Social Dimensions of Development Program. Despite this, poverty has increased over the years, particularly in the 1990s, as the economies of Least Developed Economies (LDC) continued to perform dismally. Both the real economic growth rate and real per capita income have continued to decline, leading to deterioration in access to basic needs such as food, clean water, clothing, shelter, health services and education. In majority of the LDCs, majority of the populations still live in rural areas, though urban poverty has increased significantly with recent estimates suggesting that 49 % of the urban populations are now below the poverty line as compared with 29 % in 1994. The majority of LDCs urban poor live in informal settlements, accounting for approximately 50 % of the urban population—millions people crowded on just 5 % of the total urban residential areas.

Most of these settlements are however characterized by poor housing conditions, high densities and a lack of the most basic urban services with residents exposed to severe health risks. Morbidity and mortality rates, especially among children, are also significantly higher than in other areas of the City.

Other notable ills in these settlements are that these are temporary shelters and other housing structures that are at risk of fire and flood, the majority of residents are tenants, living in single rooms, the land tenure system is haphazard and prone to abuse and vulnerability, insecurity, drug abuse and crime.

It is notable that efforts aimed at tackling urban poverty around the world are based on partnerships of a wide range of organizations, while recognizing the importance of local participation and ownership in the intervention measures and action plans. This is a sharp contrast to the general poverty reduction measures in poor countries that for several decades were conceived and run by institutions in far-flung cities such Washington D.C.

8.5 Global Rural Poverty

At the turn of the twenty-first century, still the largest number of poor people in Asia, Africa and Latin America live in rural areas. It is notable that the incidence of poverty even in countries that have boasted the highest rates of sustained economic growth and development like China, still has majority of its poor in the rural areas. In addition, poverty as measured by low income tends to be at its worst in rural areas, even when we discount the often-substantial differences in the cost of living between cities and countryside. The problem of malnutrition, lack of education, low life expectancy, and substandard housing are also, as a rule, more severe in rural areas. The extent of poverty varies greatly among rural areas in Asia, Africa and Latin America the country [11].

Many of the rural poor are located in and confined to regions where arable land is scarce, agricultural productivity is low, and drought, floods, and environmental degradation are common. Such areas are often isolated in every sense. Opportunities for non-farm employment are few, and the demand for labour tends to be highly seasonal. Others among the rural poor live in regions that have a more promising endowment of natural resources but lack access to social services (education and health) and infrastructure (irrigation, information and technical assistance, transport, and market centers).

Although urban incomes are generally higher and urban services and facilities more accessible, poor town-dwellers may suffer more than rural households from certain aspects and vestiges of poverty. The urban poor typically housed in slums or squatter settlements, often have to contend with appalling overcrowding, bad sanitation, and contaminated water. The sites are often illegal and dangerous. Forcible eviction, floods, and pollution are constant threats. Some of these people are migrants from the countryside who are seeking better-paid work. The effect that migration to the towns has on rural poverty depends crucially on whether urban employment opportunities are better or worse than in rural areas. The general evidence suggests that urban areas do offer more opportunities for higher-paid work, and this implies that, on balance, urbanization helps to reduce poverty.

Analysis of Rural Poverty

People in poor countries are for the most part agrarian and pastoral folk. Poor countries are for the most part biomass-based subsistence economies, in that their rural folk eke out a living from products obtained directly from plants and animals. In 1988 rural people accounted for 65 % of the population of what the World Bank classifies as low income countries [5]. In sub-Saharan Africa, the per capita agricultural production has declined systematically over the past 30 years.

During the period from 1970 to 1984, food production per capita in sub-Saharan Africa fell at an annual average rate 1.3 % [11]. At that period, sub-Saharan Africa's world market share for three central agricultural exports, namely coffee, cotton, and cocoa declined by 13, 29 and 33 %, respectively [3]. Data available for 1982 and 1992 for the rural sector indicate that during this period, GDP per capita grew by only 0.3 % per annum. Agricultural income per head fell but average consumption per head grew at 1.3 % per annum (because of lack of saving). The percentage of population in poverty remained largely unchanged at about 47 %, but the depth of poverty increased—the shortfall of average income of the poor below the poverty line increased from 30 to 40 % in 1992. The poorest people tended to have less access to schools and health facilities.

Various country studies have showed varying degrees of poverty lines for rural areas. This is attributed to differences in the concepts and methodological approaches used to determine poverty levels. For instance, some studies have used subsistence farmer as the basis for establishing poverty line while others are based on income earnings falling below a certain threshold. The statistics for a country like Kenya indicate that rural areas are generally disadvantaged in terms of access to safe water [3]. Regardless of poverty, more than 50 % of Kenya's households do not have access to safe water although the proportion of the poor without access to safe water is higher. Regionally, provinces like Eastern and North Eastern are worse off, with less than a third of the poor households having access to safe water while within the districts some are so badly off that nearly no poor families have access to safe water. In terms of social profiles of rural poor in several poor countries, people in certain occupations such as subsistence farmers and pastoralists have a higher than average poverty rate. Poverty rates are also high among the unskilled workers in both private and public sectors.

In Sub-Sahara Africa, the rural poor cultivate, on average, more land and have more livestock than the non-poor; but the non-poor earn more than two and a half times the money from cash crop production and more than one and half times the money from livestock production. This cultivation and harvest pattern can partly be explained by differences in the fertility of land and affordability of farm inputs that improve production. As for livestock, cultural factors and lack of high-grade breeds for the poor could account for the low sales and income from livestock.

8.6 Concluding Remarks

The emergence of new conceptual analyses of poverty and the urban-rural poverty analytical dichotomy are positive developments with regard to precise policy formulation, and for developing well-targeted poverty reduction initiatives. At the same time, the emergence of new conceptual theories in the midst of persistence of the problem of poverty shows how challenging it has been and still is to find working combination of solutions. We can also infer that some of the past conceptual analytical tools had flaws and hence the consequent failures in the action plans crafted based on the analytical framework. As a matter of caveat, even some of the new and emerging analytical framework might eventually be found to exhibit similar flaws. Past history, reveals very destructive and expensive experiments based on what was at one period in time considered as the most sophisticated package of solutions crafted out of presumably sound economics and perfect conceptual models. Everyone now knows of the infamous World Bank's Structural Adjustments Programs (SAP) and the destruction left in the course and aftermath of the expensive economic experiments.

References

1. Dasgupta P (1993) An inquiry into well-being and destitution. Clarendon Press, Oxford
2. For constructing a least-cost diet, see Patrick J. Gormely, Are High-Protein Foods Economically Efficient? Food Policy 3 (November 1978): 280–88
3. Government of Kenya, Interim Poverty Reduction Strategy Paper 2000–2003, March 31, 2000
4. Ravallion M, Datt G, van de Walle D (1991) Quantifying absolute poverty in the developing world. Rev Income Wealth 37:345–361
5. See the World Bank (1989) World Development Report 1990 (New York: Oxford University Press, 1990) ix World Bank, World Development Report 1989. Oxford University Press, New York
6. Statement by Klaus Toepfer, Executive Director of UNEP, reported in Daily Nation, May 9, 2000 vii See the Daily Nation, May 9, 2000
7. United Nations-Habitat Annual Report (2006) UN-HABITAT, Nairobi, Kenya
8. World Bank (1990) World Development Report, Oxford University Press, New York
9. World Bank (1992) World Development Report, Oxford University Press, New York
10. World Bank (1993) World Development Report, Oxford University Press, New York
11. World Bank, World Development Report 1989, op. cit. p. 19 xi

Chapter 9
Poverty Measures and Dichotomy of Urban-Rural Poverty

9.1 Overview

Determination of the number of poor people in any society depends crucially on the poverty measures used. Different measures will give different figures. The most commonly used figure is the "poverty line", which itself is a difficult and controversial concept. Concerns with identifying people affected by poverty and the desire to measure it have at times obscured the fact that poverty is too complex to be reduced to a single dimension of human life. It has become common for countries to establish an income-based or consumption-based poverty line. The World Bank also uses income-measure method adjusted to purchasing power parity (PPP), albeit to also monitor income inequality within individual countries and regions. Since 1990, the World Bank has used one-dollar-a day measure until 2008, when a revised figure of 1.25 at 2005 PPP replaced the initial standard [4]. Although income focuses on an important dimension of poverty, it gives only a partial picture of the many ways human lives can be blighted. Someone can enjoy good health and live quite long but be illiterate and thus cut off from learning, from communication and from interactions with others. Another person may be literate and quite well educated but prone to premature death because of epidemiological characteristics or physical disposition. Yet in a third case, someone may be excluded from participating in the important decision making processes affecting her life. The deprivations described here cannot be fully captured by mere level of income. It is notable that income measure does not include any in-kind transfers of resources between individuals comprising households [2].

9.2 Why Poverty-Line Measure is Faulty

Also, people perceive deprivation in different ways—and each person and community defines the various deprivations and disadvantages that affect their lives in different ways. As such, careful consideration needs to be given even when analyzing urban

N. O. Agola and J. L. Awange, *Globalized Poverty and Environment*,
DOI: 10.1007/978-3-642-39733-2_9,

and rural poverty. There are obviously clear differences and similarities in the forms of deprivations faced by urban and rural poor. A good analysis of rural and urban poverty should reflect the diversity and complexity identifiable in the dichotomy. Such analysis hence leads to appropriate conceptualization of the deprivations, and consequentially, fitting remedial policies and action programs. A quick look at the use of poverty line as a poverty measure instrument of both urban and rural poverty fails to capture the underlying economic arrangements in the livelihoods of urban and rural poor. This is simply because the urban poor rely more on income earned from elsewhere as compared to rural poor whose economic livelihood relies considerably on non-market transactions that are hardly captured in the poverty line measure instrument. In addition, it is worth noting that even within the same country, the cost of living is much higher in urban areas than in rural areas. Also with the urban-rural poverty diversity, certain deprivations tend to be more pronounced as in the case insecurity in slum dwellings of the urban poor. Another front on which an attack can be waged against the poverty line measure instrument is its failure to capture the systemic nature and interaction of factors that feed into and sustains the ruthless poverty machine. Insecurity, for instance, feeds into economic relations of slum dwellers and shapes the very nature, cost and details of simple but vital transactions. Below is a true depiction of how pervasive rent seeking happens in one of the biggest urban slums in the world (and perhaps the same situation prevails in all slums around the major world urban centers in Latin America, parts of Asia and other African countries). The depiction is based on the personal observations and interviews of the authors in the course of the research for this book.

Kibera slum in Nairobi, Kenya, presents a case in point of the pervasive rent-seeking behavior in slum dwellings by persons who wriggle themselves into various positions of power and influence, with the result of considerably driving up the cost of most services. One of the instances we observed was the case of putting up a small house/shelter by a slum dweller. Because land rights and property ownership laws and security situation are weak in the slum, as soon as work to put up the shelter begins, some extortionists often turn up to stake their claim on the parcel of land on which construction has just begun, and forthwith demand small payment to let the construction proceed. But, as if that is not enough, another version of extortionists will again turn up offering protection by proxy, by way of close ties to the local administration official. It follows that this other version of extortionist goes on to demand his take for introduction to the administration official and on the other hand also demands the administration official's take. In what turns out like the pouring of an acoustic agent into a fresh wound, the slum dweller we interviewed narrated how he must also pay a guard at the gate to the construction site he visits daily to seek work on a daily basis. The experiences of our interviewee are but just the tip of the iceberg in what can be considered an extortionist-chain of rent-seekers in slum dwellings. Such experiences only serve to drive prices up for most if not all of the essential services and goods the urban poor need for their livelihood. However, such a scenario is quite a contrast to the experiences of the rural poor, thus underscoring the need for different conceptual analysis and remedial intervention. Life is relatively more orderly in rural areas where customary rules and family ties serve to reign in any social deviants.

Consequently, this chapter will deal with the new conceptual issues in the first part, and then proceed to discuss the dichotomy of urban-rural poverty and its implications for conceptual analysis, policy formulation and possible program actions.

9.3 Major Conceptual Issues in Poverty Measurement

There are four major conceptual issues in poverty measurement: the choice of poverty indicator, the determination of the poverty line, the unit of analysis, and the choice of equivalent scale. These four issues are frequently discussed in the literature. We shall review each one of these.

9.3.1 Choice of Poverty Index

Poverty may be defined in terms of a single broad indicator of economic resources. Such a definition may be restrictive. However, if the concern about poverty takes the form of concern about basic needs, such as food, housing, and clothing, the focus should be on individual items of consumption, and poverty would need to be measured in a multidimensional way, rather than in terms of a single indicator. The same consideration may apply if we are concerned about deprivation of commodities. Concern about poverty may relate to social or demographic indicators, such as life expectancy or literacy.

If the measure of disadvantage is limited to a single index of economic resources, then a natural choice may appear to be total consumption or expenditure plus home produced goods and services. A household is then said to be poor if its total consumption is below a specified amount. But most studies of poverty in advanced countries record poverty on the basis of total income rather than consumption.

In most poverty-stricken countries, however, the well being of individuals is measured by total consumption expenditures, not by total income. This is for a number of reasons. The most important of these is that consumption tends to fluctuate much less during the course of a month or a year than income. Moreover, experience has also shown that people tend to provide more accurate information about their consumption behaviour than about their income sources. Furthermore, if expenditure data can be used for welfare analysis, this has the compelling advantage that the poverty lines can be derived from the data itself and need not be adopted from other surveys.

A number of steps are required to arrive at usable consumption figures for households in countries with poverty incidence. First, information on household purchases of food items is converted into monetary aggregates based on households' reported quantities and prices paid. Second, the calorie equivalent of the observed food consumption per household is calculated. Third, non-food expenditures are evaluated and priced, paying particular attention to the valuation of water and the 'consumption' of durable goods such as refrigerators, houses, or cars. Finally, nominal expenditures for all households are adjusted to the variation in prices among different areas and regions within country.

9.3.2 Poverty Lines

A poverty datum line is used to draw a distinction between the poor and the non-poor in society. The World Development Report [6] used an upper line of US$2 (in purchasing power parity dollars) per capita for the poor as a cut-off point for absolute poverty, and a lower poverty line of US$1.25 for those in abject poverty. The poverty line is used to calculate the proportion of the national population falling below it. Two types of poverty are usually analyzed: absolute poverty and relative poverty.

9.3.3 Absolute Poverty

There are many types of poverty lines. Just to give a quick analysis of Welfare Monitoring Surveys (WMS), we take note that poverty lines are expressed in terms of equivalent adult consumption expenditure calculated using two popular money-metric concepts of poverty, namely, absolute and relative poverty. Three poverty lines are presented: food, overall, (food and non-food) and hard-core. A poverty line is one that is fixed over time in terms of the living standards indicator for the entire country. It does not take into consideration time or place. Proponents of absolute poverty concept argue that there are some basic elements of welfare that every human being has a right to and their attainment is not dependent on scarcity of local resources but is inspired by the universalistic valuation of human dignity.

Absolute poverty is below the income that secures the bare essentials of food, clothing, and shelter. Other essentials may be added. Thus, determining this level is a matter of judgment, so that it is difficult to make comparisons between countries. Moreover, what is considered poverty varies according to the living standards of the time and region. World Bank economists Martin Ravallion, Garav Datt, and Dominique van de Walle [5] show that national poverty lines increase with mean consumption, although poverty lines are below the mean in all cases. Accordingly, many people classified as poor by their government may be materially better off than many in the past decades or others today who are not considered poor.

Recognizing that the perception of poverty has evolved historically and varies tremendously across cultures, Ravallion, Datt, and van de Walle set an extreme poverty line and a poverty line. The lower line, the extreme poverty line, recognized as the absolute minimum by international standards, is based on a standard set in

India, the country with the most extensive literature on the subject, and close to the poverty line of perhaps the poorest country, Somalia.

The definition of poverty used by the World Bank is based on previous work by its economists, particularly that of Montek S. Ahluwalia, Nicholas G. Carter, and Hollis B. Chenery [3]. These economists, used Indian data (hence the use of I$) as their yardstick and assumed a population with a "normal" distribution by age and gender, define the extreme poverty line as the income needed to attain basic nutritional needs, that is, a daily supply of 2,250 calories per person. A figure of $275 purchasing power adjusted dollars or I$275 per capita in 1985. The 2,250 calories would be met by the following diet: 5 g of leafy vegetables, 110 g of other vegetables (potatoes, root vegetables, gourds, and so on), 90 grams of milk, 35 g of oil, 35 g of sugar, 10 g of flesh foods (fish and meats), 45 g of pulses (peas or other legumes), and 395 g of cereals (rice, corn, millet, or wheat). To illustrate, the 395 g of cereals might consist of about 2 cups of hot prepared rice, equivalent in weight to 54 % of the total diet.

The World Bank's upper poverty line, below which persons are designated as poor, was $370 purchasing power adjusted dollars or $370 per capita in 1985. This poverty line provides for consumption in excess of the bare physical minimum. However, it varies from country to country, reflecting the cost of participating in the everyday life in society.

Given information on income distribution, poverty is determined by finding the percentage of the population with income of less than I$370. The extreme poverty is found by finding the share of the population with an income of less than I$275. The assumption is that two persons with the same purchasing power adjusted income living in different countries will have the same measured poverty (not including non-income factors, such as access to public services and in-kind value received).

9.3.4 Accompaniments of Absolute Poverty

The more than 1 billion people living in absolute poverty suffer the following deprivations:

(1) Three- to four-fifths of their income is spent on food; the diet is monotonous, limited to cereals, rice, bread, cornmeal, yams, or cassava, a few vegetables, and in some regions, a little fish or meat. (2) About 60 % are undernourished and millions are severely malnourished. Energy and motivation are reduced; performance in school and at work is undermined; resistance to illness is low; and the physical and mental development of children is often impaired. (3) One of every 10 children born die within the first year; another dies before the age of 5; and only five reach the age of 45. (4) Beginning in 1975, the World Health Organization and UNCEF expanded immunization against the major diseases affecting the children around the globe. Immunization rates increased rapidly, and deaths from these diseases fell substantially from the 1980s to the 1990s. Still fewer than 50 % of the children in absolute poverty are vaccinated against measles, diptheria, and whooping cough,

which have been virtually eliminated in rich countries. (5) Two-thirds of the global poor lack access to clean and plentiful water and even a larger proportion lack an adequate system for disposing of their wastes. Lack of sanitation contributes to diarrhea diseases yearly. (6) Average life expectancy is also about 50 years, compared to 77 years in developed countries. (7) Only about one-third to two-fifths of the adults are literate. (8) Only about four of every 10 children complete more than 4 years of primary school. (9) The global poor are more likely to be concentrated in environmentally marginal and vulnerable areas, face higher rates of unemployment and underemployment, and have higher fertility rates than those who are not poor.

9.3.5 Relative Poverty

Definition of poverty in relative terms relates to the type of poverty analysis that endeavors to take into account the actual deprivation with respect to the average levels of satisfaction of needs in that society. Relative income/expenditure measures such as income share of the poorest 40 %, or the percentage of the population below a line defined in reference to the mean or median, e.g. at 2/3 of the mean income/expenditure are effective measures of inequality given that poverty need not fall as either aggregate income/expenditure or the income/expenditure of the poor or as both rises. By contrast, with an absolute poverty line, poverty must fall if the income/expenditures of all those who are poor increase. Given the arbitrariness of the relative poverty approach, one can easily over-or-under-estimate the population above or below the poverty line. Secondly, the method cannot give an indicative comparison among regions. For this reason, relative poverty measures are less relevant to developing countries in terms of policy interventions targeting the poor. The application of absolute poverty concept is therefore preferred in developing countries where the majority of the people live below the subsistence level, since in such situations one would be interested in knowing the size and distribution of those in absolute deprivation rather than relative deprivation. This is the generally expressed view by many governments and donors involved in poverty reduction programs rather than advocating more egalitarian strategies. Below are some other poverty metrics.

9.4 Measuring National Food Poverty Lines

Currently, many countries use the food poverty method to determine the poverty lines. These lines indicate the insufficiency of economic resources to meet basic minimum needs in food. There are three approaches to measuring food poverty.

9.4.1 Cost-of-Basic Needs Method

This approach sets the poverty line at the cost of a basic diet for the main age, gender and activity groups, plus a few essential non-food items. A survey then establishes the proportion of people living in households with consumption (or sometimes income) below this line. The basic diet may consist of the least expensive foods needed to meet basic nutritional requirements, the typical adult diet in the lowest consumption quintile or the investigator's notion of a minimal but decent diet. The choice of both the food and the non-food components included is necessarily arbitrary.

9.4.2 Food Energy Method

This method focuses on the consumption expenditure at which a person's typical food energy intake is just sufficient to meet a predetermined food energy requirement. Dietary energy intake, as the dependent variable, is regressed against household consumption per adult equivalent. The poverty line is then set at the level of total consumption per person at which the statistical expectation of dietary energy intake exactly meets average dietary energy requirements. The problem with this method is the caviar caveat: groups that choose a costly bundle of foods are rewarded with a higher poverty line than that for more frugal eaters.

9.4.3 Food Share Method

This method derives the cost of a consumption plan to acquire just sufficient nutrients. If the cost of basic nutrients is a third of total consumption, the poverty line is fixed at three times that cost.

All three approaches are sensitive to the price used to determine the cost of the bundle. And all three concentrate mainly on calories or dietary energy, because protein deficiency due to inadequate economic resources is perceived to be rare in most societies.

9.4.4 Poverty Lines for International Comparisons

A poverty line set at US$1 (1985 PPP) a day per person is used by the World Bank for international comparison. This was later changed to US1.25 a day (2005 PPP). This poverty line is based on consumption. A poverty line of US$2 (PPP) a day is suggested for Latin America and the Caribbean. For Eastern Europe and the ICS countries, a poverty line of US$4 (1990s) has been used. For comparison among industrial countries, a poverty line corresponding to the US poverty line of US$14.40 (1985 PPP) a day per person has often been used.

9.4.5 Equivalence Scales

In comparing household budgets, it is important to recognize that households differ in consumption and size. Thus, a simple comparison of aggregate household consumption can be misleading about the well being of individual members in a given household. Some analysts therefore use some form of normalization such as "consumption per adult equivalent". The key question is, "equivalent" in what sense? Earlier studies on poverty in most developing countries used equivalence scales that covered age groups of 0–4 (weighted by a factor of 0.24), 5–14 (weighted by a factor of 0.65), and ages 15 and above (weighted by a factor of 1.0). The general practice in developing equivalence scales is to undertake a comprehensive and intensive household consumption survey to monitor and measure actual quantities consumed by each household member of various commodities produced or purchased by the household. This practice recognizes the fact that both gender and cultural influences determine, to a large extent, individual entitlements at household level.

9.5 Global Urban Poverty

Urban poverty is one of the biggest challenges facing the world in the twenty-first century. The figures speak for themselves. According to UN projections, in 2025, the world's total urban population will rise to 4.54 billion, an increase of 1.69 billion people. Ninety four per cent of this increase will occur in developing countries. And the projections are that urbanization will increase at a great speed, as compared with the industrialized world.

In the industrialized world, it took more than a generation to urbanize. However, urbanization is happening with a great speed in Africa and Asia. Africa, which is currently the least urbanized has the highest rate of urban growth—4 % compared to 2 % for the rest of the world [1, 11]. In terms of sheer numbers, urban populations are greatest in Asia, where more than 1.3 billion people, half the world's total urban population, reside in urban areas. Klaus Toepfer, the Acting Executive Director of the United Nations Centre for Human Settlements (Habitat) has observed that: "Not only are we living in an urbanizing world, we are also experiencing an unprecedented urbanization of poverty" [7].

In many cities of the developing world, up to a half of the urban population live in slums or squatter settlements. Even though the twenty-first century has been dubbed the 'Urban Millennium', and there may be the assumption that urban dwellers have better living conditions and as such are better off as compared to rural dwellers, [9] report shows that in the actual sense and by most measures, the urban dwellers suffer from a set of conditions, called ‘urban penalty’, which makes them worse off than their rural counterparts. Despite the promise of increased opportunities and a better quality of life, city life for many urban residents is a negative and degrading experience. Slum-dwellers often do not have access to the most basic amenities such as water and sanitation and many are faced with constant threat of eviction. Cities, both in the developing world and in industrialized countries, are becoming breeding grounds for crime, congestion and pollution, while at the same time, the remain as the most critical socio-economic growth points, nationally and regionally. These social ills have been brought about by urban poverty caused by poor planning and poor governance, systematic socio-economic policies that benefits minority, mismanagement and corruption.

9.5.1 Poor Governance as a Cause of Urban Poverty in Asia and Africa

What is the solution? The answer is that there is need for better urban governance, inclusive economic policies targeting inclusive infrastructure to universally raise productivity, and pursuing environmentally sustainable city development strategies. For many years, governments have blamed urban ills, including poverty, on lack of financial resources to implement development programs. Recently, the blame has shifted to a global economy that favors the industrialized countries, and mounting debt payments to international lending institutions. But, as a recent United Nations report reveals [10], it is not so much the amount of resources that determine the level of poverty in a country or city, but how the resources are used. If anything, poor governance is the leading cause of poverty.

Embracing democracy is often not enough, and certainly not a panacea. In fact, representative democracy will have little or no impact on poverty, if it is not supported by mechanisms that enhance transparency and promote people’s participation. The bottom-line requirement for good governance is accountability in the use of public funds and the shifting of decision-making powers to those of most affected by such decisions—local governments, civil society organizations and the people themselves. This does not mean that local governments and civil society should take over the legitimate functions of government. But when the distance between government and the governed is too wide, abuse of power and resources is more likely. This breeds poverty.

In an effort to address these urban poverty issues, Habitat is now focusing on strengthening the role of local governments to facilitate transparency and accountability in urban governance structures and institutions. As part of its Global Campaign on Urban Governance, the UN Habitat, in cooperation with local government representatives from around the world, is currently engaged in formulating a World Charter for Local Self-Government which, if adopted, will increase local autonomy and promote decentralization of policies around the world. The ultimate aim is to get the World Charter adopted by the UN General Assembly. If the Charter eventually becomes a UN Convention, it will serve as a "Magna Carta?" of responsiveness to the needs of citizens.

Habitat is also working closely with the World Bank on an initiative that aims at improving the lives of 100 million slum dwellers by 2020, by building on successful community-based upgrading programs and by addressing the broader policy and institutional issues that often make such efforts unsustainable.

The call for decentralization and devolution of authority to the local level is based on the premise that when people are involved in decisions that affect their lives, they are more likely to support and sustain those decisions. They also have a greater stake in ensuring that resources are used more efficiently to meet actual needs. Besides, it is harder to mismanage funds or misuse authority in an environment where one's actions are more detectable. Participatory democracy and transparent local government structures have a great potential for improving the lives of millions of people around the world. Through a series of regional, sub-regional and thematic meetings, citizens scrutinize the past year's expenditure, agree upon current priorities, and allocate resources for new projects. Now that the urban millennium is upon us, the promotion of good urban governance will be at the forefront of all development efforts. How well or how badly cities perform will determine how well or badly nations perform. As countries develop, cities will account for a growing share of the national income. Already cities in both developing and industrialized countries account for between 55 and 85 % of Gross National Products [8].

9.5.2 *Example of Initiative to Deal with the Urban Poverty: Britain Launches an Urban Project*

A new initiative by the British Government seeks to improve lives for the poor in Kenya's urban centers, and reform most settlements currently marked by inferior housing, overpopulation and lack of the most basic amenities. The Nairobi Urban Poverty Partnership Project was launched in the city in the year 2000 with the British Government providing Sh325 million ($2.8 million) to start it off. The British government agency recently committed the same sum to a similar project in city of Mombasa that is currently getting underway. Through Britain's Department for International Development (DFID), which is responsible for administering the Sh2.8 billion annual development program in Kenya, the British Government plans to provide more funds to strengthen the Nairobi Informal Settlements Co-ordination Committee (NISCC)—the body to oversee the new project.

The British Government has laid out elaborate plans to get the project started in the city and has made it clear that it will increase its "close and constructive" work with the government and other stakeholders to come up with effective strategies to reverse these trends throughout the country. The project is a partnership between many different stakeholders and its main arm is the Nairobi Informal Settlements Co-ordination

Committee, which brings together Provincial Administration, the Nairobi City Council, non-governmental organizations, private sector and donor representatives. The plan is expected to be replicated in various urban centers once it proves its efficacy to address the growing challenge of urban poverty in most towns in the country.

The involvement of several organizations- public, private and non-governmental organizations underscores the gravity of the urban poverty problem. Similar initiatives are underway in most cities in Asia and Latin America, but this is only the beginning. It is still premature to measure the outcome and efficacy of the initiative, though we suspect that like in other poverty reduction measures that have been undertaken for decades, it might be also a case of negligible response a bit too late.

Other responses to urban poverty are such like that of the International Labour Organization proposal of strategy for increasing productive employment and Social Dimensions of Development Program. Despite this, poverty has increased over the years, particularly in the 1990s, as the economies of Least Developed Economies (LDC) continued to perform dismally. Both the real economic growth rate and real per capita income have continued to decline, leading to deterioration in access to basic needs such as food, clean water, clothing, shelter, health services and education. In majority of the LDCs, majority of the populations still live in rural areas, though urban poverty has increased significantly with recent estimates suggesting that 49 % of the urban populations are now below the poverty line as compared with 29 % in 1994. The majority of LDCs urban poor live in informal settlements, accounting for approximately 50 % of the urban population—millions people crowded on just 5 % of the total urban residential areas [11].

Most of these settlements are however characterized by poor housing conditions, high densities and a lack of the most basic urban services with residents exposed to severe health risks. Morbidity and mortality rates, especially among children, are also significantly higher than in other areas of the City. Other notable ills in these settlements are that these are temporary shelters and other housing structures that are at risk of fire and flood, the majority of residents are tenants, living in single rooms, the land tenure system is haphazard and prone to abuse and vulnerability, insecurity, drug abuse and crime.

It is notable that efforts aimed at tackling urban poverty around the world are based on partnerships of a wide range of organizations, while recognizing the importance of local participation and ownership in the intervention measures and action plans. This is a sharp contrast to the general poverty reduction measures in poor countries that for several decades were conceived and run by institutions in far-flung cities such Washington D.C.

9.6 Global Rural Poverty

At the turn of the twenty-first century, still the largest number of poor people in Asia, Africa and Latin America live in rural areas. It is notable that the incidence of poverty even in countries that have boasted the highest rates of sustained economic growth and development like China, still has majority of its poor in the rural areas. In addition, poverty as measured by low income tends to be at its worst in rural areas, even when we discount the often-substantial differences in the cost of living between cities and countryside. The problem of malnutrition, lack of education, low life expectancy, and substandard housing are also, as a rule, more severe in rural areas. The extent of poverty varies greatly among rural areas in Asia, Africa and Latin America the country.

Many of the rural poor are located in and confined to regions where arable land is scarce, agricultural productivity is low, and drought, floods, and environmental degradation are common. Such areas are often isolated in every sense. Opportunities for non-farm employment are few, and the demand for labour tends to be highly seasonal. Others among the rural poor live in regions that have a more promising endowment of natural resources but lack access to social services (education and health) and infrastructure (irrigation and other agricultural inputs, information and technical assistance, transport, and market centers).

Although urban incomes are generally higher and urban services and facilities more accessible, poor town-dwellers may suffer more than rural households from certain aspects and vestiges of poverty. The urban poor typically housed in slums or squatter settlements, often have to contend with appalling overcrowding, bad sanitation, and contaminated water. The sites are often illegal and dangerous. Forcible eviction, floods, and pollution are constant threats. Some of these people are migrants from the countryside who are seeking better-paid work. The effect that migration to the towns has on rural poverty depends crucially on whether urban employment opportunities are better or worse than in rural areas. The general evidence suggests that urban areas do offer more opportunities for higher-paid work, and this implies that, on balance, urbanization might help to reduce poverty, more if the HABITAT's inclusive socio-economic approaches are adopted.

9.6.1 Analysis of Rural Poverty

People in poor countries are for the most part agrarian and pastoral folk. Poor countries are for the most part biomass-based subsistence economies, in that their rural folk eke out a living from products obtained directly from plants and animals. In 1988 rural people accounted for 65 % of the population of what the World Bank classifies as low-income countries. As earlier depicted in the previous sections of this book, general high rate of urbanization in developing countries must also mean reduced labor available for rural farming, which remains highly labor intensive. So in a sense, rural zones in regions that still have a wide technology-gap (read technology input in economic activities), such as sub-Saharan African countries, the assumption of abundant low cost labor on the rural farms could as well be a fallacy. In reality, highly labor-intensive farming activities such as paddy rice cultivation do face labor scarcity in some parts of the African continent. It is notable that available and viable labor must be willing labor. While rice paddy cultivation happens under very highly labor-intensive and difficult wet conditions, it is always not easy to find willing labor even when family sizes may be large. As such, it is not tenable to simply do head count of total rural labor based on working age bracket. We therefore maintain that in certain instances of difficult working conditions, which also happen to be highly labor intensive, actual labor shortage conditions do obtain. We also note that in sub-Saharan Africa, the per capita agricultural production has declined systematically over the past 30 years. Perhaps this decline can be explained by urban migration of viable labor, and the lack of progressive labor-capital substitution.

During the period from 1970–1984, food production per capita in sub-Saharan Africa fell at an annual average rate 1.3 %. At that period, sub-Saharan Africa's world market share for three central agricultural exports, namely coffee, cotton, and cocoa declined by 13, 29 and 33 %, respectively. Data available for 1982 and 1992 for the rural sector indicate that during this period, GDP per capita grew by only 0.3 % per annum. Agricultural income per head fell but average consumption per head grew at 1.3 % per annum (because of lack of saving). The percentage of population in poverty remained largely unchanged at about 47 %, but the depth of poverty increased—the shortfall of average income of the poor below the poverty line increased from 30 to 40 % in 1992. The poorest people tended to have less access to schools and health facilities as well.

Various country studies have showed varying degrees of poverty lines for rural areas. This is attributed to differences in the concepts and methodological approaches used to determine poverty levels. For instance, some studies have used subsistence farmer as the basis for establishing poverty line while others are based on income earnings falling below a certain threshold. The statistics for a country like Kenya indicate that rural areas are generally disadvantaged in terms of access to safe water. Regardless of poverty, more than 50 % of Kenya's households do not have access to safe water although the proportion of the poor without access to safe water is higher. Regionally, provinces like Eastern and North Eastern are worse off, with less than a third of the poor households having access to safe water while within the districts some are so badly off that nearly all the poor families have no access to safe water. In terms of social profiles of rural poor in several poor countries, people in certain occupations such as subsistence farmers and pastoralists have a higher than average poverty rate. Poverty rates are also high among the unskilled workers in both private and public sectors.

In Sub-Sahara Africa, the rural poor cultivate, on average, more land and have more livestock than the non-poor; but the non-poor earn more than two and a half times the money from cash crop production and more than one and half times the money from livestock production. This cultivation and harvest pattern can partly be explained by differences in the fertility of land and affordability of farm inputs that improve production. As for livestock, cultural factors and lack of high-grade breeds for the poor could account for the low sales and income from livestock.

9.7 Concluding Remarks

The emergence of new conceptual analyses of poverty and the urban-rural poverty analytical dichotomy are positive developments with regard to precise policy formulation, and for developing well-targeted poverty reduction initiatives. At the same time, the emergence of new conceptual theories in the midst of persistence of the problem of poverty shows how challenging it has been and still is to find working combination of solutions. We can also infer that some of the past conceptual analytical tools had flaws and hence the consequent failures in the action plans crafted based on the extant analytical framework. As a matter of caveat, even some of the new and emerging analytical framework might eventually be found to exhibit similar flaws. Past history, reveals very destructive and expensive experiments based on what was at one period in time considered as the most sophisticated package of solutions crafted out of presumably sound economics and perfect conceptual models. Everyone now knows of the infamous World Bank's Structural Adjustments Programs (SAP) and the destruction left in the course and aftermath of the expensive economic experiments.

References

1. Ankerl G (1986) Urbanization overspeed in tropical Africa. INUPRESS, Geneva
2. Hederman RS (2006) Poor poverty yardsticks. The Heritage Foundation, Sept 8, 2006
3. Montek A, Carter NG, Chenery H (1976) Growth and poverty in developing countries. J Dev Econ 6:299–341
4. Ravallion M, Chen S (2009) Prem dollar a day. World Bank Econ Rev 23(2):163–184
5. The World Bank (2012) World Development Indicators. Washington, D.C. The World Bank (producer and distributor). http://data.worldbank.org/data-catalog/world-development-indicators
6. Toepfer K (1196) Foreword in cities and homes for all habitat agenda. In: United Nations conference on human settlements, Habitat II, Istanbul, Turkey
7. UN-Habitat 2012 Report (2012) State of the World's Cities 2012/13, UN-HABITAT
8. UN-Habitat 2006 Report (2006) State of the World's Cities 2006/7, UN-HABITAT
9. UN-Habitat (2012) Sustainable Urbanization: thematic think piece, UN-HABITAT
10. United Nations (2011) Population dynamics in the least developed countries: challenges and opportunities for development and poverty reduction. United Nations Population Fund
11. United Nations (2007) State of the World Population. United Nations Population Fund

Chapter 10
Poverty and Income Inequality: Global Perspective

10.1 Overview

The issue of poverty and poverty elimination has captured the attention of the international community in recent years. The present international resurgence of interest in poverty reduction in the world has probably been given impetus by the international pressure for poverty eradication. The World Summit on Social Development in Copenhagen in March 1995 was one of the notable beginnings of well organized international effort. The Summit called for an international commitment "to the goal of eradicating poverty in the world, through decisive national actions and international cooperation" [22].

The Declaration and Program of Action, adopted by the Summit, urged world leaders to commit themselves to the goals of eradicating poverty in the world, and to promote full employment as a basic priority of economic and social policies. It urged national governments to pay particular attention to the following:

(a) efforts and policies to address the causes of poverty and provide the basic needs for all. These include the elimination of hunger and malnutrition, the provision of food security, education, employment and livelihood, primary health care, safe drinking water and sanitation, adequate shelter and participation in social and cultural life, with special priority to the needs and rights of women, children, the vulnerable and the disadvantaged groups and persons; (b) creation of employment, reduction of unemployment and the promotion of appropriately and adequately remunerated employment as a strategic and policy focus, with full participation of employers, workers and their respective organizations, so as to give special attention to structural problems of enduring unemployment and under-employment of youth, women, and people with disabilities; (c) the promotion of basic social programs, in particular those affecting the poor and the vulnerable segments of society; (d) establishment of structures, policies and objectives that will ensure gender balance and equity in the decision-making processes at all levels; and, (e) broadening of women's political, economic, social and cultural opportunities, independence and empowerment, including the organizations of indigenous women at grassroots levels.

N. O. Agola and J. L. Awange, *Globalized Poverty and Environment*,
DOI: 10.1007/978-3-642-39733-2_10,

Furthermore, the international community was urged to develop and to implement techniques of debt conversion applicable to social development programs and projects.

Following from this Summit, the British Government issued a White Paper in November 1997 [5, 15], which set out the policies to help achieve sustainable development of the planet. The White Paper declares that "eliminating poverty is first and most importantly about the single greatest challenge, which the world faces". This is because "nearly one in four of the people of the world live in absolute poverty. We cannot build a just and sustainable world unless we tackle this problem". Then the Paper observes: "The problem is that the numbers of people living in absolute poverty are continuing to grow and the amount of human suffering is growing. This White Paper commits Britain to using its influence directly and in collaboration with others to mobilize a much stronger international commitment to poverty eradication".

Regarding poverty elimination, the White Paper argues that development must be sustainable. That means, the needs of today must be met and the needs of the future must also be met, but without sacrificing resources needed now and in the future. It points out that without sustainable development poverty will not be eliminated.

Another major development was the Millennium Summit 2000 [23], where 189 World leaders made the famous United Nations Millennium Declaration- containing eight-goal action plan, currently known as the Millennium Development Goals (MDGs). The eight goal-action plan almost impeccably exhaust and integrates the diverse and segmented development goals that had escaped inclusion in some of the previous analytical approaches and action plans aimed at providing solutions to poverty and the associated vices. The eight goal-action plan includes; Eradication of extreme poverty and hunger, achieving universal primary education, promoting gender equality and empowering women, reducing child mortality, improving maternal health, combating HIV/AIDS, malaria, and other diseases, ensuring environmental sustainability and lastly developing a global partnership for development.

It is also worth taking a critical note of the poverty reduction strategy initiatives of the World Bank and International Monetary Fund (IMF) in 1999. This is after the widespread protest against Structural Adjustment Policies (SAP), which left a hurricane-like economic destruction in majority of the Heavily Indebted Poor Countries (HIPC). The two institutions that have self-styled themselves as the most competent entities in macroeconomic and development policy introduced the Poverty Reduction Strategy Papers (PRSPs) as a replacement of SAP as condition for lending to the HIPC and other developing economies. The big question has been since the introduction of PRSPs—is this new conditionality the best practice in attaining poverty reduction and economic growth in the HIPC and other developing nations? Alternatively, one could ask as to whether PRSPs is a marketing gimmick by the two institutions given the embarrassing failure of the SAP in many countries? Taken just as given by the two institutions, PRSPs introduced the fundamental concepts of participatory approach, ownership by key stakeholders, sustainability of programs, and linking of program outcomes to poverty reduction. On close examination, the PRSPs can be quite controversial with regard to the efficacy of each of the fundamental concepts that were touted as marking a departure from the SAP era. Abugre [2], gives an incisive critical analysis of the PRSPs, in which he asserts that both participatory and ownership aspects do not hold water for PRSP cases in Kenya, Uganda, Ghana and other HIPC. While non-governmental organizations (NGOs) and Civil Society groups are widely consulted as stated in the PRSPs, the formulation of what actually matters, the macroeconomic plans, the two groups are quite voiceless. Abugre [2] cites the insistence by IMF and World Bank on a standard macroeconomic model that reflects an obsession with keeping factors such as inflation at its lowest. Also that the two institutions hold a veto power, read approval of the PRSPs for lending, shows that strict conditionality, a core aspect of SAP is still much around. If the participatory and ownership aspects were very much genuine, then the two institutions would have done away with their 'missions' to countries to help in the preparation of PRSPs. With regard to results, which should be measured against performance in poverty reduction in the recipient countries, it is still too early to gauge the outcomes of PRSPs, though it is nearly a decade since the inception of the programs. If the PRSPs were so much poverty reduction goal oriented, then it should have been feasible to have short-term, medium-term and long-term measurable poverty reduction benchmarks. The question remains as to whether we are in for another economic development experiment as it were in the case of SAP, but in a disguised easy to market version.

10.1.1 Issues for Consideration in Poverty Elimination

The British White Paper mentioned earlier argues that no progress can be made in eliminating poverty unless everyone's human rights including those of the poorest and most disadvantaged people are protected. Every man, woman and child has a right to be such things as security, freedom and dignity, regardless of their race, gender, religion or beliefs. Everyone has basic needs, which include the need for fresh water, food, basic education, and health care. These needs must be provided for so that poor people can take charge of their own future. Another important factor for achieving sustainable development and helping eliminate poverty is providing the right conditions for economic growth. Poor people must be given opportunities to work, generate income and share in the fruits of development.

10.1.2 Targets Set by the British Government for Eliminating Poverty

There is a remarkable similarity between the targets set by the British Government and those set by most developing country governments regarding the issues and dates for poverty elimination [4]. For instance, the year 2015 is common to both.

The following are the British targets:

(i) Economic Wellbeing-A reduction by one-half in the proportion of people living in extreme poverty by 2015.
(ii) Human Development Universal primary education in all countries by 2015 demonstrated progress towards gender equality and the empowerment of women by eliminating gender disparity in primary and secondary education by 2005.
(iii) A reduction by two-thirds in the mortality rates for infants and children under age five and a reduction by three-fourths in maternal mortality, all by 2015.
(iv) Access through primary health care system to reproduce health services for all individuals of appropriate ages as soon as possible and no later than the year 2015.
(v) Environmental Sustainability and Regeneration-The implementation of national strategies for sustainable development in all countries by 2005, so as to ensure that current trends in the loss of environmental resources are effectively reversed at both global and national levels by 2015.

The World Bank has lent support to this British policy on poverty. The President of the Bank has declared that the issue of poverty and the issue of equity are really everybody's problem [1].

10.2 Extent of Poverty in the World

On discussing the poverty problem in sub-Sahara Africa, the is often a tendency on the part of some observers to think that poverty is an exclusive African problem. In fact, when talking about Africa, most foreigners (and some Africans too) associate Africa with poverty—that is, Africa is synonymous with poverty, and vice-versa. The "Afro-pessimists" do not appear to think that poverty exists elsewhere in the world. This negative view of Africa is mistaken. Poverty prevails everywhere in the world; that is why the USA mounted the 'War on Poverty' in the early 1960s [11]. That war has yet to be won. That is also why the British established the Royal Commission on the Distribution of Income and Wealth, which issues its reports in 1975, 1976, and 1977. In most of the former Soviet Republics—now called the Newly Independent States or NIS—formed the Commonwealth of Independent States (CIS), poverty has been increasing dramatically since the collapse of the former Soviet Republics in 1989. In Asia, rampant poverty dominates much of the Asian continent. The South and Southeast Asia region has the largest number of poor people (716 million in 1996) in the world. This constitutes 73 % of the 1.2 billion people living on less than US$1 a day, in 1993. This situation has not changed in any radical way in the first decade of the twenty-first Century. On a regional basis, India has by far the largest number of poor people (29 % of its population between 2000–2006), according World Bank (Country Report 2006) [16]. Following Southeast Asia is Sub-Saharan Africa with 292 million poor people. Thus, the poor people in India alone are more than the total number of all poor people in Sub-Saharan Africa. Thus, without minimizing the problem of poverty in Africa, it is safe to say that poverty is not confined to Africa alone (See Fig 10.1).

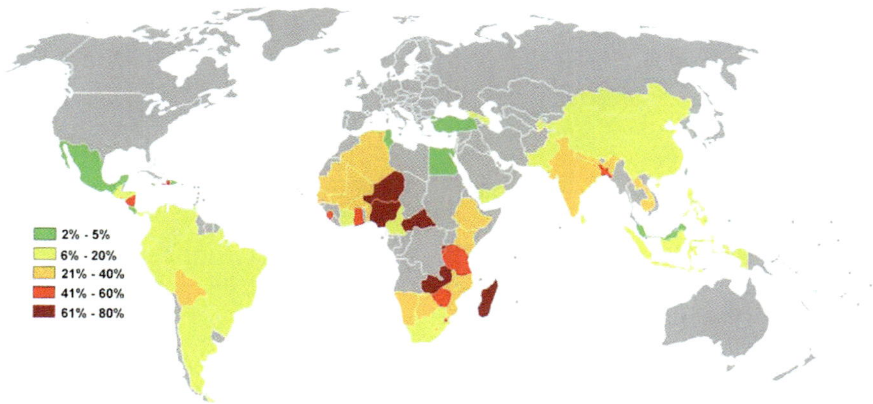

Fig. 10.1 Countries by Percentage of Population living under 1 dollar a day (Data Source: United Nations Human Development Report 2007–2008)

It may be quite easy and convenient to stereotype regions or countries with regard to poverty, but a thorough look at the fine details of income inequality within each country is necessary to get a complete picture of poverty problem. For instance, there are pockets of urban poor in almost all industrialized countries, yet these pockets of poverty may be conveniently passed over in terms of problem recognition—and as such can be wished away as tiny minority social misfits. There are also pockets of poor migrants in the major European economies—the question is, when documenting cases of poverty, the poor migrants also deserve to be part of the national poverty statistics, and problem recognition, that must be followed with due social and economic policy interventions. Again, given the expansive conception of poverty so far adopted in poverty research and thought underlying economic and social policy in most countries, one may not be surprised upon a discovery of a list of deprivations suffered by a large number of people even in the industrialized and rich nations. Altogether, a new set of deprivations in industrialized countries may have to be included in national poverty analysis. For instance, several adults in industrialized countries are unduly deprived of important time to be with family, while trying to hold two or three part-time jobs to barely make ends meet. On the other hand, children of middle income groups and even wealthy groups may be denied of any leisure time because they are forced to learn several skills at an early age, presumably to give them a head start in economic life. Viewed in a critical way, such Spartan-skill-drilling of children could be just as worse as child labor in the income poor countries. Almost 3 billion of the 6 billion people on earth live on less than US$2 a day, and over 1.2 billion people, nearly one quarter of the world's population, live on extreme poverty—under US$1 a day. Over the next two decades, this ratio could grow, as two billion extra people will be added to the planet, largely in poor countries. The world is increasingly divided with only 1.2 billion people living in the developed world and the rest in developing or transitional economies. Consequently, the "gap between rich and poor countries is getting bigger and within countries the rich are getting richer and the poor are getting poorer." By extension, poverty in industrialized countries also requires attention while noting that even though the industrialized countries have made great headway in having a large portion of their populations being materially well off, pockets of poverty still do exist, if not emerging. Japan can be singled out as a country with one of the most equitable income based society. This being the outcome of a combination of governmental, private sector and societal input and efforts towards the country's industrialization and economic development after the World War II. As such, the combination of governmental, private sector efforts of Japan leading to equitable income based society will be discussed in the section of this book dealing with possible new and innovative solutions to global poverty problem. Some caveat is necessary here—we need to point that Japan can only be a showcase of equitable income based society with regard to the post-war economy extending to the mid 1990s. In the late 1990s, the number of poor in what can be deemed 'pockets' of poverty or the loser group (makegumi) has started to emerge in Japan too. Simply discounting the poverty in industrialized countries as negligible and therefore not worth paying attention to would be a mistake in economic and human development history, more so at this juncture in human history when we embrace the importance of human rights of each individual. So the presumably 'small' number of poor or negligible incidence of poverty in industrialized countries should not be ignored in entirety.

The prevalence of poverty in the world as shown in Tables 10.1 and 10.2 has been a matter of concern for the international community, especially the international financial institutions. The World Bank has spearheaded the concern with poverty since the MacNamara days in the 1970s. The United Nations Development Program followed, with emphasis on the series of annual reports of Human Development Report prepared since 1990. The United Nations convened a major Social Summit Conference in Copenhagen in 1995 to address the problem of world poverty [22]. In September 1999, poverty eradication was a major issue discussed during the IMF/World Bank annual conference. Thus, poverty eradication has become an international issue, and the most—National Poverty Eradication Plans are but a part of this international poverty eradication agenda.

Table 10.1 2010 Poverty headcount ratio at $1.25 a day (PPP)—% of population.World Bank, Poverty and Equity Data Dashboard

East Asia and Pacific	12.5 %
Europe and Central Asia	0.7 %
Latin America and Caribbean	5.5 %
Middle East and North Africa	2.4 %
South Asia	31.0 %
Sub-Saharan Africa	48.5 %

Table 10.2 2010 Poverty headcount ratio at $2.0 a day (PPP)—% of population. *Source* World Bank, Poverty and Equity Data Dashboard

East Asia and Pacific	29.7%
Europe and Central Asia	2.4%
Latin America and Caribbean	10.4%
Middle East and North Africa	12.0%
South Asia	66.6%
Sub-Saharan Africa	69.9%

10.3 Setting the Global Poverty Agenda: Brief Historical Perspective

New efforts are afoot on the part of the international community to address the problem of poverty in the world. During their annual joint meeting in September 1999, the respective chief executives of the International Monetary Fund (IMF) and of the World Bank launched a new co-operation in anti-poverty efforts. They challenged the world community to launch a fresh assault on global poverty. Michel Camdessus, the IMF Director, said: "The extent of poverty still present at the end of a century of affluence is intolerable It is time to respond [13]." The IMF renamed its much reviled "enhanced structural adjustment facility" as the Poverty Reduction and Growth Facility. James Wolfensohn, President of the World Bank, added: "This is a moment in history when we can set a new course to a world of greater peace, equity and security. It is a time not just for review but for action [27]." He spoke of trying to "turn the Bank around" and pledged to redouble the Bank's efforts against poverty. Both the IMF and the World Bank called for industrialized nations to open their markets to exports from the world's poorest nations. They said that developing countries should receive full benefits of membership in the World Trade Organization (WTO).

The World Bank president pointed out, however, that trade and financial liberalization is not the solution to the increasing poverty worldwide. He noted that despite all efforts by governments, official institutions and Non-Governmental Organizations (NGOs), the global economy will never significantly reduce poverty in the world. Consequently, there was need for the world economies to build dynamic coalitions of governments, civil society, and the private sector to construct a global economy that benefits all people.

The US Treasury Secretary, Lawrence Summers, supported such moves, saying that economic aid must be seen to be used to reduce poverty, "rather than simply making it easier for the country to engage in excessive military spending [3]." Gordon Brown, Britain's Chancellor of the Exchequer, called for "a new global alliance against poverty" with a "virtuous circle" of debt relief, poverty alleviation and foreign aid [3].

The call against poverty came after the IMF and the World Bank sealed financing for a program aimed at slashing US$70 billion in nominal terms from the US$217 billion debt of the 40 world's poorest countries. Mr. Wolfensohn warned that debt relief would not be enough to prevent the number of poor people in the world rising to four billion by 2025. He observed that the debt relief packages "is the beginning of our challenge, not the end."

However, the chief executives noted that they could not offer immediate solutions to political problems that threaten their anti-poverty crusade. The rival claims of poverty and politics are an "eternal dilemma" for the international lending institutions. The problem is that some poor countries would rather spend the loan money on buying arms than spend it on poverty-alleviation programs. Said Camdessus: "In many countries, good policies have not survived the resurgence of armed conflict, which is often fed by massive arms sales to the world's poor countries, paid for by loans from arms exporters in the industrial world [13]."

These sentiments were strongly supported by the former US President Bill Clinton, who backed his support by action. On 29 September 1999, while addressing the joint annual IMF/World Bank meeting in Washington, DC, he offered to cancel 100 % of the US debt owed by the world's poorest countries, provided the money was used to relieve poverty. He told the meeting that he was directing his administration "to make it possible to forgive 100 % of the debt these countries owe to the United States when needed to help them finance basic human needs and when the money will be used to do so [10, 14]". He continued: "I don't believe we can possibly agree to the idea that these nations that are so terribly poor should always be that way. They should not have to choose between paying debt interest and investing in their children's health and education.... Any country committed to reforming its economy, to educating and vaccinating its children, should be able to make those commitments and keep them."

Some of the most indebted countries were slated to be eligible for United States debt relief. These were countries such as—Uganda, Tanzania, Ethiopia, Sudan, and Somalia The criteria must have been targeted at the bottom of the bottom economies as it was later explained by the Unites States Government that countries such as Kenya did not qualify for such debt relief because they were more advanced compared to those that were considered.

Tremendous progress has been made in reducing poverty in the developing world. Over the past four decades social indicators have improved in all regions. Over the past two decades, East Asia has achieved dramatic reductions in poverty: from 6 out of 10 living on less than US$1 a day in the mid-1970s to 2 out of 10 in the mid-1990s [28]. There have also been declines in recent years in the incidence of poverty in most of South Asia and in parts of the Middle East and North Africa and Latin America. Despite this progress, however, much remains to be done. In 1995 more than 9 million children in developing countries under the age of five died of avoidable causes. Except in East Asia, the number of poor people actually increased between the late 1980s and the early 1990s. In most countries there are major disparities in income and access to education and health care, sometimes associated with broader social exclusion.

In 2000, at the Millennium Summit of the United Nations, eight priority international development goals that have come To be famously known as Millennium Development Goals (MDGs) were set within a time frame of 2015 [23]. The eight goals were; eradicating Poverty and extreme hunger, universal primary education, gender equality and women empowerment, reduction of child mortality rates, improving maternal health, environmental sustainability, and global partnerships for development. The big question about MDGs is on post-2015 sustainability. Again, it is puzzling that agricultural sector development does not feature within the MGs, yet majority of the poor depend of agriculture.

10.4 The Population Factor in Poverty Analysis

The world population has grown rapidly in recent decades, with serious implications for increase in poverty. According to the United Nations Population Fund (UNFPA) [25], the world's population reached 6 billion in October 1999, and then increased to 7.1 billion in 2011 [26]. Despite increased awareness on the need for family planning, the global population has quadrupled since the beginning of the last century, growing faster than at any time in previous history. At the beginning of the twentieth century, the world's population was approximately 1.5 billion. In 1927, it reached 2 billion; in 1974, it was 4 billion; 1987, it reached 5 billion; and in October 1999, it hit 6 billion. It is projected to reach between 7.3 billion and 10.7 billion by the year 2050. People are living longer and healthier lives than ever before. Modern medicine has dramatically lowered the global death rate, particularly among children. In addition, better living conditions have increased the average life expectancy, and people are living longer to a ripe age of 60 years of age.

However, the sad reality is that a large percentage of the world's population still lives in poverty. According to the World Bank Development Indicators [18], the highest growth rates of population are taking place in Sub-Saharan Africa (where annual growth rates are 2.5 %). In the midst of greater wealth than the world has ever seen, 22.2 % of the global population still live without the elements of human dignity—clean water, enough food, secure housing, basic education and health care. In majority of the poverty-ravaged countries, the basic services are not available to everyone and unemployment rates are high. As we enter the twenty-first century, over a billion people are still deprived of basic needs. Of the 4.8 billion people in developing countries, nearly three-fifths lack basic sanitation. Almost a third have no access to clean water. A quarter does not have adequate housing and a fifth have no access to modern health services. A fifth of the children do not attend school.

There is also increasing pressure on the planet due to wasteful and unbalanced consumption patterns and growing numbers of people, raising demand for food and water. In addition, the effect of global warming remains to be seen. Possible geographical changes, such as sea level rises, increased storms and floods, could affect billions.

At the same time, HIV/Aids is taking a heavier toll than had been anticipated by demographic experts, especially in Sub-Saharan Africa where it rivals malaria as the leading cause of death. In many countries, it has erased decades of progress in reducing child mortality and increasing life expectancy. In 29 African countries (including Kenya), the average life expectancy is now seven years less than it would have been without Aids.

The cumulative effects of continuing poverty, gender discrimination, HIV/Aids, environmental change and shrinking resources for development have the potential to wipe out the benefits of lower birth rates.

Developing countries continue to record high birth rates, and the population is, therefore, not expected to decline. Nearly half of all people on earth are under 25 years of age. Over a billion people are between the ages 15 and 24. In 62 countries in

Africa, Asia and Latin America, over 40 % of the population is under age 15. Africa, which is the world's youngest and most rapidly growing region, has a median age of only 18. These age groups are the potential parents of the next generation.

More and more women in Asia, Africa and Latin America are gaining access to family planning, improved health care and education. As a result, they are having fewer children and families are increasingly becoming smaller than ever before. Since 1969, the birth rates in developing countries have dropped by half—from almost six children per woman to fewer than three in 1999 [7].

Meanwhile, population growth has slowed or stopped in Europe, North America and Japan. United States is the only industrial country where large population increases are still projected, largely as the result of immigration. Countries with aging populations, particularly in the more developed regions, will face challenges of providing support and medical care for the elderly. With fewer young people, they will look to active older people and immigrants to supply some needed services and contribute to the economy.

10.5 Global Poverty in the Midst of Growing Wealth

Even though the world has grown richer as time progresses, widening income inequality has been the hallmark of such growth. Unlike the previous economic growth marking industrialization of Western countries and those of East Asia, which lifted almost every citizen out of poverty into middle class category, recent and present economic growth among OECD countries seems to leaving many out. In essence increase in income inequality has been the outcome. OECD [8, 9] report on income distribution and poverty within its group of countries shows increasing inequality as depicted in Table 10.3, for the periods between 1980 and 2010. Likewise as mentioned in the earlier chapters of this book about failure of trickle-down economics in most low-income economies, we can point to a trend of consistently rising income inequalities.

Other than income inequality within countries, there is also glaring inequalities between regions of the world. Such inequality is even more paradoxical given the shifting global demography. The poorer parts of the world have become the majority, and should this remains so, then this constitutes real risk for global instability and conflicts. For instance, the urge to migrate to richer parts of the globe cannot go away so long as this situation remains. Trying to shut out multitudes of illegal immigrants by erecting expensive fences will not solve the problem. The case of Mexico and the United States of American exemplifies this situation.

Table 10.3 2010 European population at-risk-of-poverty or social exclusion, 2006-2011

	Percentage of the total population (%)						Number of persons (1,000)					
	2006	2007	2008	2009	2010	2011	2006	2007	2008	2009	2010	2011
EU-27(1)	25.2	24.4	23.6	23.1	23.6	24.2	122,688	119,281	115,694	113,773	116,309	119,568
Euro Area(1)	21.7	21.7	21.4	21.2	21.6	22.6	69,432	69,760	68,995	68,696	70,259	73,799
Belgium	21.5	21.6	20.8	20.2	20.8	21.0	2,247	2,261	2,194	2,145	2,235	2,271
Bulgaria(2)	61.3	60.7	44.8	46.2	49.2	49.1	4,734	4,663	3,421	3,511	3,719	3,693
Czech Republic	18.0	15.8	15.3	14.0	14.4	15.3	1,832	1,613	1,566	1,448	1,495	1,598
Denmark	16.7	16.8	16.3	17.6	18.3	18.9	896	905	887	962	1,007	1,039
Germany	20.2	20.6	20.1	20.0	19.7	19.9	16,444	16,760	16,345	16,217	15,962	16,074
Estonia	22.0	22.0	21.8	23.4	21.7	23.1	293	293	291	312	289	307
Ireland	23.3	23.1	23.7	25.7	29.9	:	991	1,005	1,050	1,150	1,335	:
Greece	29.3	28.3	28.1	27.6	27.7	31.0	3,154	3,064	3,046	3,007	3,031	3,403
Spain	23.3	23.1	22.9	23.4	25.5	27.0	10,155	10,257	10,340	10,652	11,675	12,371
France(2)	18.8	19.0	18.6	18.5	19.2	19.3	11,184	11,382	11,195	1,100	11,693	11,840
Italy	25.9	26.0	25.3	24.7	24.5	28.2	15,256	15,412	15,099	14,835	14,757	17,112
Cyprus(2)	25.4	25.2	23.3	23.5	23.5	23.7	193	195	181	188	192	199
Latvia(2)(3)	41.4	36.0	33.8	37.4	38.1	40.4	930	803	757	834	846	829
Lithuania	35.9	28.7	27.6	29.5	33.4	33.4	1,217	967	928	985	1,109	1,080
Luxembourg	16.5	15.9	15.5	17.8	17.1	16.8	74	73	72	85	83	84
Hungary	31.4	29.4	28.2	29.6	29.9	31.0	3,121	2,916	2,794	2,924	2,948	3,051
Malta	19.1	19.4	19.6	20.2	20.3	21.4	76	78	80	82	83	88
Netherlands	16.0	15.7	14.9	15.1	15.1	15.7	2,603	2,558	2,432	2,483	2,483	2,598

(continued)

Table 10.3 (continued)

	Percentage of the total population (%)						Number of persons (1,000)					
	2006	2007	2008	2009	2010	2011	2006	2007	2008	2009	2010	2011
Austria	17.8	16.7	18.6	17.0	16.6	16.9	1,454	1,376	1,532	1,406	1,373	1,407
Poland(2)	39.5	34.4	30.5	27.8	27.8	27.2	14,938	12,958	11,491	10,454	10,409	10,196
Portugal	25.0	25.0	26.0	24.9	25.3	24.4	2,640	2,653	2,757	2,648	2,693	2,601
Romania	:	45.9	44.2	43.1	41.4	40.3	:	9,904	9,418	9,112	8,890	8,630
Slovenia	17.1	17.1	18.5	17.1	18.3	19.3	343	335	361	339	366	386
Slovakia	26.7	21.3	20.6	19.6	20.6	20.6	1,439	1,150	1,111	1,061	1,118	1,112
Finland	17.1	17.4	17.4	16.9	16.9	17.9	886	907	910	886	890	949
Sweden	16.3	13.9	14.9	15.9	15.0	16.1	1,489	1,264	1,367	1,459	1,418	1,538
United Kingdom	23.7	22.6	23.2	22.0	23.1	22.7	14,193	13,527	14,069	13,389	14,209	14,044
Iceland	12.5	13.0	11.8	11.6	13.7	13.7	36	38	36	36	42	41
Norway	16.9	16.5	15.0	15.2	14.9	14.6	780	764	701	724	716	709
Switzerland	:	:	18.6	17.2	17.2	17.2	:	:	1,372	1,288	1,291	1,308
Croatia	:	:	:	:	31.3	32.7	:	:	:	:	1,321	1,382
Turkey	72.4	:	:	:	:	:	48,934					

(1) Eurostat Estimate, 2006 and 2011
(2) Break in series, 2008
(3) Break in series, 2011
Source Eurostat (http://epp.eurostat.ec.europa.eu/statistics_explained/index.php/Social_inclusion_statistics)

10.6 Comparing the Per Capita Incomes of Countries of the World

The richest countries in the world do have small populations , and therefore it is not surprising that, nations such as Monaco, Liechtenstein, Luxembourg, Norway, Qatar and Bermuda Take the top positions. Perhaps, it is relatively easier to plan and provide for smaller populations, more so when a country?s sources of wealth are natural resources or some specific economic activities. The Poorest Countries According to the World Bank, 1.5 billion people live below the world poverty line of US$365 a year. The world's poorest countries are in Africa. Somalia (US$139 in 2011) leads in the poorest country category. Congo, Democratic Republic follows with with a per capita income of US$231 in 2011 and Liberia and Malawi follow with US$281 and US$371 respectively. It is also notable that though the various African economies are at the lowest rank, Africa also hosts the top six countries among the global ten fastest growing economies.

In terms of regional inequalities, Western hemisphere and Asia-Pacific remains the wealthiest, while slow shifts are starting to happen regarding the position of low income economies. It can be expected that the fastest growing low-income economies will take less than ten years to double their economies, and perhaps in another decade be able to move to middle income economy category. This path is however not guaranteed, and there is also fear of some of these countries getting caught in the middle-income trap from which them may not altogether go past. Alternatively, in the next three decades, the line up of the largest and wealthiest economies might change so dramatically to the extent of the current top leaders being replace by the likes of Brazil and Turkey.

Income Poverty in Sub-Saharan Africa In Sub-Saharan Africa, income poverty runs deep, and it poses a serious threat to economic and social stability. Some observers have recognized this situation. For instance, following a visit to Nigeria in October 1999, Wolfensohn, the then President of the World Bank told the former President Obasanjo of Nigeria: "Two out of three people in Nigeria live in poverty. You cannot have stability, you cannot have social justice, you cannot have equity, you cannot have the sort of Nigeria that you want unless you deal with that issue [27]". Most recent World Bank data gives Sub-Sahara African countries poverty headcount ration at US$2 a day (PPP) at 69.9 % of the population [17]. People in this region, along with those in South Asia (66.7 % poverty head count ratio at $2 a day- PPP) , are among the poorest in the world. Even back then in 1992, about 45 % of Sub-Saharan Africa's population were income-poor according to national poverty lines [12]. In Gambia and Zambia nearly two-thirds of the people were income-poor; in Cameroon, Guinea-Bissau and Uganda, more than half; and in Cote d'Ivoire, Kenya and Nigeria, more than a third. The poverty gap in Sub-Saharan Africa, stood at 15 %, which was nearly twice that in East Asia and South East Asia and the Pacific (excluding China), at 8 %.

The growth of income in Sub-Saharan Africa has been dismal in recent decades. Between 1970 and 1992, per capita GDP (PPP$) grew by only $73, compared with growth of $420 in South Asia and $900 in East Asia, regions with incomes comparable to those of Sub-Saharan Africa in 1970.

Between 1981 and 1989, the region saw a cumulative decline of 21 % in real GNP per capita. This decline extended both to countries undertaking structural adjustment and to non-adjusting countries, exacerbating the conditions of the poor. Of the 35 countries in the region for which data are available, per capita GNP fell in 27. The most severe declines were in Gabon (58 %), Nigeria (nearly 50 %), Cote d'Ivoire (42 %), Mozambique and Niger (more than 30 %), Republic of the Congo (more than 25 %) and Zambia (more than 20 %).

Countries with adjustment programs were more successful in improving macroeconomic balances towards the end of the 1980s, after making greater income sacrifices early in the decade. Nonetheless, 11 adjusting countries had falling per capita incomes in 1985–90, as did the region as a whole.

Thus, income poverty was high and increasing in many countries of Sub-Saharan Africa in the 1980s. The situation remains alarming by the close of the decade (2000). On addressing the problem of poverty in Africa, Mr. Goodall Gondwe, Director of the International Fund's Africa Department, said that the IMF must modify its strategy to take more account of social development and poverty reduction, not just macroeconomic adjustment [7]. He argued that there was clearly a need for a more fully articulated social dimension that takes account of the interrelation between growth, social development and poverty reduction. He urged the IMF and the World Bank to address "very forcefully" the need to combat corruption and ensure that the money loaned by these bodies reached the people it was intended to help.

People in Sub-Saharan Africa (SSA) remain, along with South Asia, among the poorest in the world. However, in the past decade, the Sub-Sahara African Countries have made gains in poverty reduction. Thus, the proportion of those living on less than US $1.25 a day decreased from 56.5 % in 1990 to 47.5 % in 2008 [24] The basic causes of poverty are related to problems of access and endowment as follows:

- inadequate access to employment opportunities for the poor—caused by low rates of economic growth, and does not generate large increases in employment for the poor. This fact is worsened by the weak manufacturing sector in the region. It is manufacturing sector where living wages can be obtained, in addition to ongoing skill development of workers;
- inadequate assets such as land and capital for the poor, closely linked to the absence of land reform and minimal opportunities for small-scale credit;
- inadequate access to the means for fostering rural development in poor regions-caused by preference for high potential areas and strong urban bias in the design of development programmes;
- inadequate access to markets for the goods and services that the poor can sell—caused by the remote geographic location or other factors;
- inadequate access to education, health, sanitation and water services—caused by lack of social service delivery, resulting in the inability of the poor to live a healthy and active life and take full advantage of employment opportunities;

- destruction of natural resources endowments which has reduced the productivity of agriculture, forest and fisheries- also exacerbated often by the desperate survival strategies of the poor as well as inadequate and ineffective public policy on natural resource management;
- inadequate involvement of the poor in the design of development programmes—caused by donor/government discussions which do not involve representatives of poor communities;
- inadequate access to assistance by those who are the victims of transitory poverty such as drought, floods, pests and war—caused by lack of well-conceived strategies and resources.

In recent years, participatory poverty assessments in Africa have provided local communities own perspectives on what it means to be poor. In the case of Ghana, to take one example, participants were eager to distinguish between the poor who could feed themselves and their families, and the poorest, who could not due to disability or infirmity. Poverty is seen as being related to being isolated, lacking water supply, inadequate education and health services, and an inability to participate in the economic and social life of the community. Both wealth and poverty are frequently conceived of as potentially dynamic and transient conditions. Secure access to food, income, adequate housing and (to a lesser extent) education, emerged as major dimensions in local poverty criteria at all sites. It also became clear from the surveys that in Ghana, as elsewhere, men and women have substantially separate income streams and that the level of access to a secure livelihood varies within, as well as between, households.

Income Inequality the Industrialized Countries

There is a debate in the West about the rising income inequality within the individual countries of the OECD. While some people maintain that the rising inequality is inevitable, given the nature of the market forces as well as technology, others argue that inequality is not inevitable, given that inequality is driven by sociological factors which can be redressed by political decisions. The debate is not yet over.

Recent research has indicated that income inequality is rising in a large number of industrialized countries [8]. This phenomenon first attracted attention in the United States, where inequality began to rise in the 1970s. But the increase in the United Kingdom is even greater. The figures show data on the Gini coefficient—the most widely used measure of income inequality. This varies between 0 when everyone has the same income to 100 % when one person has all the income. In the case of the United Kingdom, the Gini coefficient rose by some 10 points from around 23 % in 1977 to 33 % in 1990. Income inequality has increased sharply in New Zealand, has risen in Germany and the Netherlands, but France shows little upward trend in the inequality of disposable income. This variety of experiences suggests that policy matters. Based on OECD study [7, 8, 18] of on rising inequality amongst its member countries as depicted in Fig. 10.2, only a few countries exhibit low level of inequality. Amongst these group are; Slovenia, Denmark, Norway, Czech Republic, Slovak Republic, Finland, Belgium, Sweden, and Austria. This impressive group of

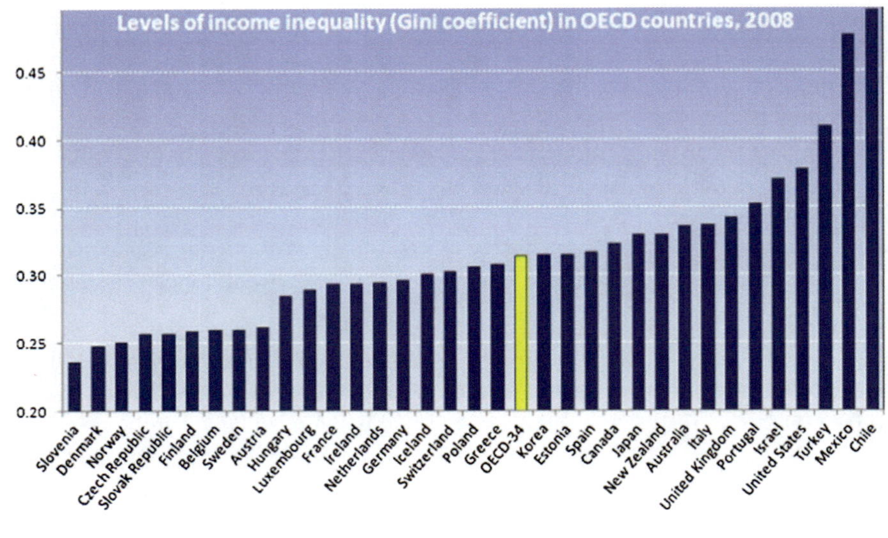

Fig. 10.2 Increase in Gini Coefficient of Income Inequality. *Source* OECD 2011, Divided we Stand. *Notes* The Gini coefficient ranges from 0 (*perfect equality*) to 1 (*perfect inequality*). Gaps between poorest and richest are the ratio of average income of the bottom 10 % to average income of the top 10 %. Income refers to disposable income adjusted for household size

countries is followed by another group of mid-level inequality, which comprises of; Hungary, Luxembourg, France, Ireland, Netherlands, Germany, Iceland, Switzerland, Poland, Greece, Korea, Estonia, Spain, Canada, Japan, New Zealand, Australia, and Italy. Countries with highest rising inequality are; United Kingdom, Portugal, Israel, United States, and Turkey. The worst in the group are Mexico and Chile.

There is a widely held belief that rising income inequality is inevitable; technological change and globalization of world trade are the most frequently cited culprits. This view is common on both sides of the Atlantic, particularly in key policy-making institutions. The 'Transatlantic Consensus' holds that increased inequality in the United States and high unemployment in Continental Europe are due to a shift in demand from unskilled workers to skilled workers, whether due to technology, trade or both. For some economists rising inequality is driven by the revolution in information technology. For others, trade—especially imports from low-wage developing countries—weakening the demand for unskilled labour in the developed world.

However, the Transatlantic Consensus is open to question. The rise in inequality is neither universal nor of the same extent across countries. Where there has been a rise, it has not happened at the same time. Moreover, in some countries the rise in inequality takes the form of steps—rather than a continuing trend. This indicates that the world is working in more complex ways than those described in simple technological and trade explanations.

The models themselves are too simple. For instance, separating workers into just two categories—skilled and unskilled—reflects the industrial economy of the nineteenth century, rather than that of the twenty-first century. Rather, there is a continuum of earnings reflecting a continuum of skills. Moreover, the models of the Transatlantic Consensus see wage-differentials as nothing more than the outcome of supply and demand—thereby ignoring the role of conventions and social norms.

More insightful economic theories draw upon the work of sociologists to argue that supply and demand only place limits on possible wage differentials, while social forces determine where wages actually lie between those limits. These models also indicate that support within society can at times shift away from pay norms.

Can Redistribution Offset Market Inequality

What people are paid in the market place is not the only factor determining the distribution of income. Market outcomes are significantly modified by income taxation and by social transfers financed out of the government budget.

Rising Inequality is Not Inevitable

The Transatlantic Consensus sees rising inequality as the product of exogenous, inevitable events with an emphasis on technology and trade—including trade with the developing world. There is nothing to be done about rising market inequality. It is pure supply and demand. But the rise in inequality in market incomes as the product of changing social norms resulting in a shift away from a redistributive pay norm to one where market forces dominate. Social conventions in the labour market have changed within countries, and these have spilled out into other spheres, most notably in national tax and benefit systems. But the fact that the driving force is social in origin—rather than trade or technology—means that there is more scope for political leadership. Social norms can be influenced by policy decisions, and thus rising inequality is not inevitable. The last part of this book will be about new proposals for stemming poverty, and the rising inequality in a model that proposes a win-win situation for both labor and owners of businesses.

10.7 Concluding Remarks

Though the world efforts at reducing poverty has been mixed in terms of mixed progress fro countries and regions, increasing global wealth, especially in new regions that were previously poor is an indicator that poverty can be reduced. This can only be tenable if the issue of inequality is addressed in economically sustainable way, perhaps by a combination of innovative market mechanism and public and international policies. The increasing inequality in the developed countries also presents a challenge to policy makers. Perhaps certain enduring assumptions on how to organize and grow economies, actually carry-overs from the twentieth century might have to be discarded.

References

1. A speech by the World Bank President James Wolfensohn to the Summit of the United Nations Conference on Trade and Development (UNCTAD) (2000) Reported in Daily, Nation, 17 Feb 2000
2. Abugre C (2000) Still Sapping the Poor: A Critique of IMF Poverty Reduction Strategies. World Development Movement, June 2000
3. Daily Nation (1999) 5 Oct 1999
4. Department for International Development (1997) Eliminating World Poverty: A Challenge for the 21st Century—A Summary, HMSO, London, p 5
5. Department for International Development (1997) Eliminating World Poverty: A Challenge for the 21st Century. HMSO, London
6. Eurostat (1997) Reported in International Herald Tribune, May 1997
7. Goodall Gondwe (1999) Reported in the East African Standard. Washington, D.C., 1 Oct 1999
8. OECD (2011) Divided We Stand Why Inequality Keeps Rising? An Overview of Growing Income Inequalities in OECD Countries: Main Findings, Dec 2011
9. OECD (2011) Society at a Glance 2011—OECD Social Indicators (www.oecd.org/els/social/indicators/SAG)
10. Part of President Bill Clinton's statement at the annual IMF/World Bank meeting, Washington, DC, 29 Sep 1999. Reported in the East African Standard, 1 Oct 1999
11. Public Papers of U.S. Presidents, Lyndon B. Johnson, 1963-1964 (Washington: G.P.O., 1965), 1, pp 375–380
12. See Peter Townsend (1979) Poverty in the United Kingdom: A Survey of Household Resources and Standards of Living. Penguin Books, New York
13. Statement at the Annual meeting of the IMF and World Bank in Washington, DC, in September 1999. Reported in the Daily Nation, 30 Sept 1999
14. The East African Standard (1999) 1 Oct 1999
15. The White Paper, op. cit
16. The World Bank (2006) Country Reports, World Bank Group, Washington D.C.
17. The World Bank (2010) Poverty Headcount Ratio for Sub-Sahara Africa
18. The World Bank (2012) World Development Indicators
19. This explanation was first given by President Moi during a public address, and then later by the Foreign Minister Godana. See the East African Standard, 1 Oct 1999
20. UNDP (1999) Human Development Report. Oxford University Press, New York, p 32
21. UNFPA, State of the World Population 1999, op. cit.
22. United Nations (1995) World Summit on Social Development in Copenhagen in March 1995
23. United Nations (2000) Millennium Summit of the United Nations, New York, 6–8 Sept 2000
24. United Nations Development Programme (2013) Poverty Reducion and the MDGs- Situation Analysis. UN, New York. Retrieved from: http://web.undp.org/africa/poverty.shtml. Accessed 1 July 2013
25. United Nations Population Fund (UNFPA) (1999) The State of the World Population. New York: UNFPA
26. United Nations Population Fund (UNFPA) (2012) The State of the World Population. New York: UNFPA
27. World Bank President James Wolfensohn in a speech to the Summit of the United Nations Conference on Trade and Development. Reported in Daily, Nation, 17 Feb 2000
28. World Development Indicators 1997

Chapter 11
Review of Global Poverty Eradication Solutions

11.1 Specific Observations

One of the most important messages to emerge from this study is that the pace of poverty reduction in the world has been unacceptably slow, and perhaps done in half-hearted and inept measures. Simply stated, the efforts of the past have been luckless, lackluster, too little, too slow, and inept. Pervasive global poverty makes the search for effective actions all the more imperative and urgent. Changes required for poverty eradication include the following considerations:

(a) Need for Unprecedented Commitment to Poverty Reduction
Greater commitment to poverty reduction as the central theme in all development programs is direly necessary. As things currently are, few governments have clear incentives pegged to poverty reduction for private sector entity activities, infrastructure development that is targeted at poverty reduction, per capita income improvements based on the concept of living wages, and most of all we still lack poverty reduction monitoring system that zeroes in on per dollar returns on some of the much publicized public expenditure on poverty reduction such as youth employment creation and rural development programs. It is notable that governments and donors have recognized the need to commit resources towards job-creation economic and social programs.
A good example of an expression of such commitment is the Kenyan government's Youth Empowerment Project, (KYEP) which aims to provide loans, training for capacity enhancement both at institutional and individual levels, and availing massive opportunities for labor-intensive jobs for the youths. This program can be a poster-child for commitment to poverty reduction given the nature of partnerships involved in it. It brings together the government of Kenya providing the institutional and administrative and legal framework, the Youth themselves as individuals and as community members, Kenyan private sector players imparting technical skills through internship-based training and the World Bank providing some of the requisite funding (Office of the Prime Minister of Kenya, Kazi Kwa Vijana, 2011: http://www.kkv.go.ke). We can glean a multi-pronged

N. O. Agola and J. L. Awange, *Globalized Poverty and Environment*,
DOI: 10.1007/978-3-642-39733-2_11, © Springer-Verlag Berlin Heidelberg 2014

attempt and commitment to work at poverty reduction. However, this program is still at a formative stage having been initiated in 2010. Lots of questions need to be posed such as on the capacity of the governmental institutions from ministerial level to the local district levels to administer the funds without incidence of fraud and corrupt dealings, the ability to competently process the socioeconomic needs of the youth, and the ability to select sustainable and transformative economic activities.

On the issue of project selection of the KYEP, it is notable that the emphasis is on labor-intensive projects, which can be quite effective in addressing the youth unemployment problem in the short-term. However, if projects selected do not have high economic and social impact and sustainability component, then most of the projects will remain viable in the short-term and soon wither out leaving the goal of job creation unsolved in the long term consideration. On the other hand, the over-emphasis on labor-intensive aspects negates the need for deploying economically transformative methods in economic activities. One tempting economic theory justification for the near-obsessive preference for labor-intensive method could be that the supply of low-cost labor (unemployed youth) remains elastic against an inelastic demand (limited number of short-term KYEP projects), but we need to remember that the fundamental goal of the project is employment creation.

Against this backdrop of supply and demand of labor, there can be rational justification for entirely opting for labor-intensive projects. However, employment creation can only be tenable when individuals gaining employment are kept in gainful employment over long-term periods, otherwise seasonal project-based employment will at best contribute to seasonal unemployment, and at worst the youths will rejoin the permanently unemployed class. The KYEP is most likely to create unemployment vicious circle, and not the targeted virtuous one of sustainable employment creation. In the short-term, the KYEP may be politically expedient for the political leaders associated with initiating the project as it signifies a credible attempt to address the problem of poverty and unemployment among the youthful population. In essence it is a short-run, stopgap solution based on tinkering with labor component of productivity. A comprehensive solution must seek to address the issue of labor-capital (K/L) ratio necessary for improving productivity to create sustainable expansion, growth and further employment creation. In Kenya, the reality is that the ratio of capital (K) to labor is low and so is the productivity of labor (L), which we take as the returns to L, and real per capita income.

An ideal situation (production of commodities X and Y as depicted in Fig. 11.1) should have been where (K) and (L) equally increase under constant returns to scale to enable the country to change its productivity for a wide range of goods and services using both capital and labor in various combinations. In such a scenario, the production-possibility frontier (PPF) will register positive change, actually a shift marked by higher productivity. We can derive vital lesson here that for actual growth in per capita incomes to happen in low-income countries, the K/L ratio must be the object of intervention. In essence, poverty is a product

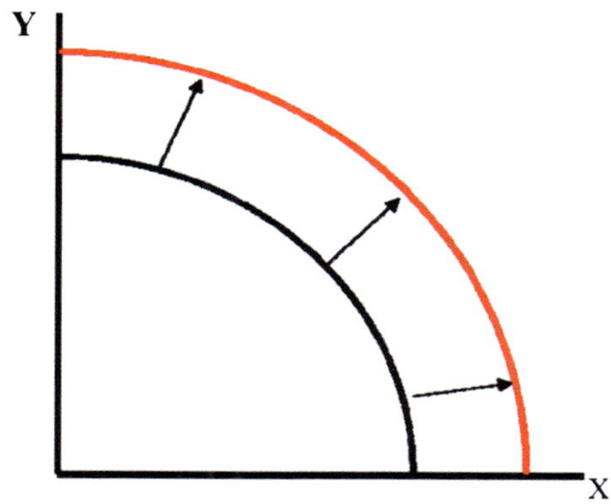

Fig. 11.1 Ideal balanced growth from (K/L)

of low productivity, and therefore any measures devoid of productivity gains can only be classified as short-run welfare maximizing mechanisms. Therefore amongst the multitude efforts aimed at solving poverty and unemployment, the question of productivity must be always raised. However, it is worth noting that productivity does not result from mere infusion of new technologies, but must also be supported by human capital development associated with capacity building. For instance, productivity gains can be realized in the service sector due to adoption of superior and more efficient organizational methods, which shorten time to accomplish tasks and improving quality of services.

As such, the KYEP program and those that could be initiated in other countries must address the need to improve productivity through deployment of transformative technologies accompanied by appropriate capacity building measures. Labor-intensive methods cannot be a panacea for high rate of unemployment in the long run, this being more so when low productivity characterizes economic activities in a country. Improved productivity is what contributes to expansion of economic activities and further creation of employment and growth in incomes ([4] on relations between poverty and low productivity). A further clarification is necessary on the issue of technology and productivity. While consideration the infusion of transformative technologies accompanied by capacity building efforts can contribute to increased productivity, it is worth noting that organizational and managerial adaptations and innovations go hand in hand with the technologies in use to contribute to improvement in productivity. It is also notable that a considerable portion of the entire unemployed youth population either has tertiary education or university level education. This modestly educated category of unemployed youth population is the one most likely to create entrepreneurial enterprises using transformative technologies. This group is capable of adopting and managing some of the standardized modern technologies for productive purposes. Let me emphasize that in any country, entrepreneurs have never been the majority, yet as few as they could be as a portion of an entire country's population, they are major change agents of economic activity. In retrospection, commitment to poverty reduction requires a concerted and innovative effort comprising of governments, poverty-victims, private sector players and international financial institutions. Choice of transformative economic activities matched with transformative technologies is what is required to create a virtuous circle and not a vicious circle of poverty and unemployment.

(b) Paradigm Shift—New Rules of Capitalism for Economic Development

A nascent paradigm shift on the part of private sector to fully understand that profitability and sustainability of business endeavors are very much connected to the incomes of individual members of society- simply put, private business depend on society for thriving profitable business, therefore poverty reduction should be a critical part of the main concerns of private business based on the idea of sustainability and not just short term profit gains. However, for this paradigm shift to be completed and become mainstream way of thinking about business and even conception of markets, current ways of thinking must give way. There is no doubt that there is a nascent global paradigm shift in the conception of markets for goods and services produced by multinational companies. Up till the end of the twentieth century, the main market conception paradigm has been that corporations go after thriving and expanding contemporary national and regional markets. Under such market conception, companies are simply focused on identifying countries with high growth rates accompanied by an expanding middle class with high purchasing power. In essence, well-formed and actively functioning markets, viewed simply as one of the strategic assets pursued by multinational corporations.

The common thinking among corporate executives and managers in charge of operations tended more towards an oversimplified refrain from engaging in business operations in small markets where majority of the population have low incomes. However, at the beginning of the twenty-first century, we have seen the beginning of a portent paradigm shift in the conception of markets for goods and services produced by multinational corporations. It is important to emphasize that this shift is at an infancy stage, and as such a wait-and-see approach to whether this shift will go full cycle and perhaps gain legitimacy and full currency among multinational corporation decision makers. The emphasis and efforts to promote the role of private sector in economic development in poor countries by the World Bank and other international development agencies has been a major precursor to the paradigm shift. The main eloquent proponent of this newly shaping paradigm was [8] in his seminal work, the Bottom of the Pyramid (BPO). Prahalad strongly argued that selling to the poor at the bottom of the pyramid (an expression carrying the meaning that the poor are still the majority on planet earth) made sense in terms of the potential profits, and therefore multinational businesses should not ignore low-income groups and markets. That the BPO proposition became quite a game-changer is not in doubt, at least judging by the wide attention followed by actual business undertakings by many multinational companies such as Unilever and SC Johnson , and lastly that the ideas formed part of the intellectual discourse at the Word Economic Forum sessions [6].

However, BPO model has not been without its critics, the most notable being [6], whom I have always concurred with on the ideas that the market size of the BPO is exaggerated, and that the idea is based on an over-romanticized conception of the poor as entrepreneurial and as avid well-informed value seekers. In addition, the cost of marketing to the poor could be quite prohibitive, specifically given the absence of infrastructure, which is a global common denominator in all physical locations occupied by the poor. If the BPO market exhibits weaknesses with regard to business impracticalities, then its intellectual persuasion, and thus wide adoption by businesses on a global scale is already compromised. Therefore the legitimacy of an ongoing global search for a new business model, both at the intellectual, discourse, and business practitioner levels is beyond doubt. The current ongoing shifts in global economic power, and the state of markets in industrialized countries vis-avis that of emerging markets reinforces the legitimacy of the need for new business model upon which companies from the industrialized and emerging economies can effectively participate in the slowly-but-surely formative markets, specifically that of the low-income countries. It is common knowledge that the markets for a wide range of consumer goods and services in industrialized countries are either saturated or in decline. On the contrary, the markets in emerging economies are on the rise, and with underserved swelling young population the emerging markets cannot be ignored any more by any serious corporation intent on being a player on the global stage. This scenario forms the stage on which companies both large multinationals and medium-sized enterprises must sit up and pay attention to the emerging markets. Such attention would only be rewarding if based on a well-conceived business models. This section is meant to provoke a discourse in search of pragmatic business models for doing business in emerging low-income markets.

It is worth pointing out the dire need to innovate to create new markets as an integral aspect of the current search for a business model for active and effective participation in emerging low-income markets. Perhaps, it is this aspect that makes more sense than even the final outcome in the form of a well-conceived and pragmatic business model. Simply put, it is the recognition of the need to innovate to create new markets instead of the troubled-old-approach characterized by a 'wait-and-see' approach. The dominant practice by companies big and small is to outdo each other in an obsessive bandwagon market entry behavior upon certain realization that markets in some countries and regions are in a growth and expansion phase. Usually an expanding urban middle class and country economic growth statistics are taken into considerations leading to foreign direct investments in low-income emerging economies. Markets and low-cost labor then immediately become the key strategic assets sought after by foreign corporations upon a statistically evident market emergence in the low-income emerging economies. Our contention is that the foreign corporate winners in the low-income emerging economies as further economic transition phases are experienced are not going to be those who wait for the markets to happen, but definitely those who innovate early enough to create new markets in these places. Essentially, this position suggests a new role for corporations—a radical change from being bystanders with folded arms waiting for markets to evolutionary emerge, then jump into head-to-head competition with other firms, to proactive innovative corporations that can create services and goods in low-income emerging markets where none existed before.

Counter-arguments against the proactive and innovative approach to creating markets could be that this approach would distract the efforts and resources of firms from the developed markets where they currently draw most of their revenues. This counter-argument does not hold when subjected to the realities of stagnating if not shrinking markets in most of the industrialized world. In addition, recent experiences of some multinational corporations show that innovating and developing products for low-income market customers can bring advantages needed to compete on price on offerings of similar products in the industrialized markets. A good case that amplifies this argument is Unilever's water-efficient detergent developed for low-income communities in India where water access is a daily problem. It is obvious that such products though originally intended for low-income populations lacking water can positively resonate with environment and water-cost conscious buyers in industrialized markets. Water footprint is a major issue for increasingly environment conscious consumers in industrialized economy markets where most multinational corporations originate. For low-income consumers in low-income countries, water-efficient products are a matter of necessity and common sense.

(c) Accepting and Re-clarifying Pivotal Role of Public-Private Sector Partnerships (PPPs)

Greater true willingness by governments and donors to coordinate their activities for poverty alleviation is badly needed. This calls for a new level of transparency on both sides with regard to poverty reduction monitoring, and hence efficacy of programs. This concern is related to the critical question of whether donor and government funding is actually ending up in actual poverty reduction efforts or goes towards payment of "good salaries" of economic development workers or towards bureaucracy.

(i) Greater true willingness by governments, multilateral financial institutions (MFI) and the donors to rely more on the judgment of intended beneficiaries-this is an acknowledgement of the limitations of top-down approach to poverty problem solving. This requires greater decentralization of decision-making, planning and implementation activities through increased involvement of poverty victim communities.

Some of the above issues are somehow addressed in the various National Poverty Eradication Plans. However, the Plans do not yet have financial backing required for their effective implementation. A quick run down through the statistics of available funding allocated to poverty eradication and other Millennium Development Goals (MDG) affirms this observation. According to the World Health Organization (WHO), additional funding is required considering a shortfall in requisite funds for 2011. WHO estimates that while the required funding for 49 developing countries is US 31 billion, the shortfall is at least USD 15 billion, but could actually be USD 19 billion ([9] see also [2, 3, 5], United States Agency for International Development (USAID), 2001). The true gravity of the funding shortfall can be understood when a breakdown of specific targeted funding goals are given. The targeted funding goals are in human development areas, which cannot wait—for example women and children's health, vaccination, provision of essential drugs, fighting HIV/AIDS, TB, malaria, water and sanitation. It is also important to point out that funding deficit in poverty eradication and meeting other MDG is not perhaps the only critical problem. Rather, it is pertinent to point out that a change in mindset at all levels (government, private sector and society), based on a shift from casual treatment of poverty problem to a full consciousness that tackling poverty problem benefits the entire society. For governments, tackling poverty means an achievement of the critical goal of creating a stable society based on meeting critical material, and socio-psychological needs of individuals. For the private sector, successful poverty measures means expanded and stable markets for goods and services. Even as individual citizens, we immensely benefit from enhanced socio-economic welfare of others. There is no better illustration of this assertion than the socio-economic dynamics and daily life experiences in the United States of America.

Nevertheless, it is significant that the current global-thought standard is slowly shifting and is moving toward a recognition that poverty is a serious issue in many countries as can be justified by national poverty eradication plans and intermittent charity activities being a proxy for the awareness and commitment. In the past, despite anti-poverty measures being adopted by international social and charity groups, many previous governments were not prepared to admit the magnitude of a poverty problem as discussed in this book. A concerted effort in varied forms of public private partnerships (PPPs) is what is required.

(d) Paying Attention to Floating Middle Class: Vital Lessons from the Televised Arab Revolution

The Arab revolution started in Tunisia and in a little while became contagious. Egyptians came second in order of televised revolution. These were major events with tremendous psychological effect not only on the socio-economic and political fabric of the Arab world, but also that of Sub-Sahara African nations. By extension, we can say that the events in the North African nations were very much global in terms of impact because even as far as China and perhaps in Latin American countries, the events have been viewed with fear and consternation by governmental leadership in these far flung parts of the world. The Chinese authority took censorship measures to prevent online access to unfolding revolutionary event information. Tunisia and Egypt became unsearchable terms online on any of the search engines used in China. The political fears aside, the events perhaps opened up a new area till up to now untouched by poverty studies—the floating middle class and its significance for poverty reduction measures.

It is pertinent to study major events with regard to what causes such events to understand the direct socio-economic and by extension political implications of such events to us. So far pundits and even lay commentators have rushed to sound warnings to Sub-Sahara African and other governments around the world of the possibility of an Arab revolution-like events happening within their borders. The legitimate fear and concerns about the possibility of such an event lies in the magnitude of socio-economic and political disruption. A revolution as a chaotic and violent form of change constitutes a major sovereign risk to foreign investors. Even worse is the wanton destruction and looting of both public and private property, which sets back perhaps by decades the countries' economies, institutions and infrastructure. Take for instance the burning of the central train station building in Tunis. A train station building is a vital public infrastructure that has nothing to do with despotic rulers whom any citizenry may hold long-standing grudges against.

We need to dissect and critically analyze the economics of the Arab revolution in order to fully grasp the cause and dynamics of the events in both Tunisia and Egypt. At the same time we need to draw from theories of chaos and complexity to understand the dynamics of the events. Perhaps it is common and general knowledge that twin sisters—poverty and authoritarianism largely contributed to the revolutions. This position is however not precise enough to the point of yielding knowledge on causality and dynamics of the events. Why is this so? Lets start by turning the spotlight on the vital statistical figures on income class compositions from the African Development Bank.

We can glean from Tables 11.1,11.2 and 11.3, the statistical data that African floating middle class constitutes the largest portion of middle class population in Africa (both for North Africa and Sub-Sahara Africa). Large floating middle class as a percentage of a country's population signifies the danger of a large part of the population falling into poverty. As such, monitoring the size of floating middle class and devising remedial economic measures to help the floating middle class move on to stable lower and upper middle class are critical to meaningful poverty reduction and creation of stable markets for goods and services. Contemporary concerns are focused on monitoring those who graduate from living below the poverty level, and celebrating the size of a country's middle class. Even private sector players and investors are duped by the blanket view to middle class size in country populations. Our strong contention following the revolutionary events in North Africa is that attention and analysis of the issues related to floating middle class need to be brought to the center of both economic policy and policy implementation addressing poverty reduction. Paying keen attention to the size of floating middle class brings the valuable insights on the progress of poverty reduction efforts and points to the likelihood of any reversals to poverty. In essence, we can see a large size of floating middle class as an early warning sign of limited success in poverty reduction programs and for the private sector we can read sovereign risk vulnerability, and high market risk even when all else seems normal.

Tunisia takes the first position in Africa as the country with the largest middle class population, actually standing at almost 90 % of its total population. Morocco follows with 85 % of its population falling within middle class income category. Egypt is third with approximately 80 % of its population in the middle class category. Sub-Saharan African countries' middle class composition

Table 11.1 Summary of middle class in africa by sub-classes

Year	Floating class[a] (in thousands)	Lower middle[b] (in thousands)	Upper middle[c] (in thousands)	Middle class without floating class		Middle class with floating class	
				(in thousands)	Percentage of population	In thousands	Percentage of population
1980	49,311	39,984	21,961	61,945	14.6	111,256	26.2
1990	70,771	51,362	29,309	80,672	14.4	151,442	27.0
2000	101,680	58,056	36,563	94,619	13.1	196,299	27.2
2010	190,585	79,785	42,910	122,695	13.4	313,280	34.3

[a] Floating class ($2–$4)
[b] Lower-middle class ($4–$10)
[c] Upper-middle class ($10–$20)
Source AfDB Statistics Department Estimates

Table 11.2 Summary of population, class size, by region (1990 and 2008 based on household survey means)

Region	Total population (million)	Population			Aggregate annual income/expenditure (2005 PPP $ billion)			
		Poor (<$2/person)	Poor ($2–$20) per person/day	High (>$20 per person/day)	Poor (<$2 per person)	Poor ($2–$20) per person/day	High (>$20 per person/day)	Total
Year 1990								
Africa[a]	562	69	27	4	77	225	64	336
Sub-Saharan Africa	274.8	75	24	1	70	109	44	224
Middle East and North Africa	162.3	18	80	2	16	247	39	303
Developing Asia	2,692.2	79	21	0	843	721	42	1,605
Developing Europe	352.3	12	84	4	23	638	141	802
Latin America and Caribbean	352.5	20	71	9	31	641	480	1,153
OECD	639.0	0	24	76	0	735	9,636	10,371
Year 2008								
Africa[a]	873	62	33	5	110	449	121	680
Sub-Saharan Africa	393.5	66	33	1	100	206	69	376
Middle East and North Africa	212.8	12	86	3	14	365	66	445
Developing Asia	3,383.7	43	56	1	696	3,285	350	4,331
Developing Europe	356.6	2	87	11	4	974	425	1,403
Latin America and Caribbean	454.2	10	77	13	22	1,008	924	1,953
OECD	685.4	0	16	84	0	542	12,617	13,159

Source African Development Bank Statistics Department Estimates for Africa

Table 11.3 Summary of middle class in African countries by sub-classes in 2008

Year 2008 countries	Floating class		Lower middle		Upper middle		Total	
	Percentage of population	Pop (million)	Percentage of population	Pop (million)	Percentage of population	Pop (million)	Percentage of population	Pop (million)
Algeria	49.2	16.9	18.0	6.2	9.3	3.2	76.6	26.3
Angola	24.9	4.5	7.1	1.3	6.1	1.1	38.1	6.9
Benin	6.9	0.6	5.9	0.5	4.8	0.4	17.7	1.5
Botswana	18.3	0.4	9.4	0.2	20.0	0.4	47.6	0.9
Burkina Faso	10.2	1.6	2.3	0.3	0.9	0.1	13.3	2.0
Burundi	2.4	0.2	1.9	0.2	1.0	0.1	5.3	0.4
Cameroon	23.1	4.4	8.6	1.6	7.5	1.4	39.2	7.5
Cape Verde	29.7	0.1	11.7	0.1	5.0	0.0	46.4	0.2
Central African Republic	6.9	0.3	3.8	0.2	3.8	0.2	14.6	0.6
Chad	10.3	1.1	5.1	0.6	2.5	0.3	17.9	2.0
Comoros	15.7	0.1	6.0	0.0	3.1	0.0	24.8	0.2
Congo D R	7.7	4.9	2.8	1.8	1.9	1.3	12.4	8.0
Congo, Rep	22.6	0.6	7.2	0.2	1.5	0.0	31.4	0.8
Cote dIvoire	18.2	3.8	11.8	2.4	7.1	1.5	37.1	7.7
Djibouti	25.4	0.2	8.8	0.1	5.9	0.1	40.1	0.3
Egypt	48.2	39.3	20.6	16.8	11.0	9.0	79.7	65.0
Ethiopia	13.2	10.7	5.3	4.2	3.0	2.4	21.5	17.4
Gabon	37.6	0.5	23.0	0.3	14.7	0.2	75.4	1.1
Gambia	22.0	0.4	12.3	0.2	3.7	0.1	37.9	0.6
Ghana	26.8	6.3	13.5	3.2	6.2	1.5	46.6	10.9
Guinea	6.3	0.6	2.8	0.3	1.5	0.1	10.6	1.0
Guinea Bissau	10.2	0.2	6.4	0.1	1.2	0.0	17.8	0.3

(continued)

Table 11.3 (continued)

Year 2008 countries	Floating class		Lower middle		Upper middle		Total	
	Percentage of population	Pop (million)	Percentage of population	Pop (million)	Percentage of population	Pop (million)	Percentage of population	Pop (million)
Kenya	28.1	0.4	15.2	0.2	1.6	0.0	44.9	0.7
Lesotho	26.3	0.7	7.2	0.1	3.9	0.1	37.4	0.8
Liberia	2.9	0.1	1.2	0.0	0.7	0.0	4.8	0.2
Madagascar	5.6	1.1	1.8	0.4	1.6	0.3	9.0	1.7
Malawi	4.0	0.6	2.6	0.4	2.0	0.3	8.5	1.3
Mali	17.0	2.2	4.9	0.6	3.2	0.4	25.1	3.2
Mauritania	37.9	1.2	2.5	0.1	2.5	0.1	42.8	1.4
Morocco	57.3	18.1	15.6	4.9	11.7	3.7	84.6	26.7
Mozambique	6.9	1.5	1.7	o.4	0.9	0.2	9.4	2.1
Namibia	38.2	0.8	3.8	0.1	5.3	0.1	47.4	1.0
Niger	8.7	1.3	3.3	0.5	2.0	0.3	14.0	2.1
Nigeria	12.9	19.5	6.2	9.3	3.8	5.7	22.8	34.5
Rwanda	5.1	0.5	1.5	0.2	1.1	0.1	7.7	0.8
Senegal	23.9	2.9	7.3	0.9	4.5	0.6	35.7	4.4
Sierra Leone	11.4	0.6	4.6	0.3	2.6	0.1	18.6	1.0
South Africa	23.4	11.6	14.0	7.0	5.7	2.8	43.2	21.4
Swaziland	11.2	0.1	6.5	0.1	2.9	0.0	20.6	0.2
Tanzania	9.2	3.9	1.5	0.6	1.3	0.6	12.1	5.2
Togo	11.6	0.7	7.3	0.5	1.6	0.1	20.4	1.3
Tunisia	43.9	4.5	29.0	2.9	16.6	1.7	89.5	9.1
Uganda	10.6	3.4	4.8	1.5	3.3	1.0	18.7	5.9
Zambia	8.8	1.1	4.3	0.5	1.3	0.2	14.4	1.8
Total	20.0	174.3	8.3	72.2	4.8	41.8	33.0	288.4

Notes Floating class = \$2–\$4; Lower-middle class = \$4–\$10 and Upper-middle class = \$10–\$20
Source AfDB Statistics department estimates for African Development Bank, ESTA Estimates

figures compare very poorly with the North African countries. With the exception of Gabon with 75 % of its population in the middle class category, the rest of Sub-Saharan countries have significantly less than half of their population in the middle class category. The worst Sub-Sahara Africa performers—Mozambique, Madagascar, Malawi, Rwanda, Burundi, and Liberia actually have single digit percentage figures. Kenya has about 45 % of its population falling within the middle class, while Ethiopia, Uganda, and Tanzania have approximately 22, 19, and 12 % respectively. These dismal figures communicate the stark reality that majority of the population in these countries live in poverty.

We assert here that the situation is so dire to the extent that the globally referenced-pyramid structure simply does not apply. We can accurately talk of a majority of the population in these countries being stuck at the base with a tiny minority immorally sitting at the top. If the North African nations of Tunisia and Egypt have a significant majority of their population falling within the middle class, then what caused their televised revolutions? The large number of floating middle class is the answer. The floating middle class is an extremely vulnerable group to economic shocks and is constantly in flux. Literally their life can be depicted as oscillating between time periods of picking crumbs from the national banquet table to moments of beggary at the gates of oasis of abundance.

Within the floating middle class we can find a majority of modestly educated, but seasonally or perennially unemployed young people. This group can be politically conscious and can at times have the ability to organize and galvanize others to their cause depending on the shared disgruntlements, and discontents of the others. Tunisia's floating middle class is slightly more than 50 % of the whole middle class category. This huge proportion is further buttressed by 32 % being confined to the lower middle class. It is notable that only 19 % of Tunisian belongs to upper middle class category. The same scenario applies to Egypt, and is shared by Sub-Sahara African countries. In reality, therefore, Africa has a very disappointingly small stable upper middle class population.

The probability of a televised revolution happening in Kenya or in any of the neighboring countries owes much to the floating middle class factor than the usual suspects—groups living below poverty level. Now here is the ultimate shocker on Kenya! 62 % of the Kenyan middle class fall within the floating middle class category, while 34 % are lower middle class, and only 4 % are upper middle class! Other African countries share the dubious distinction of having a disproportionately large floating middle class, averagely 60 % for the entire African Continent. Nevertheless, Kenya is an island regarding its extremely small upper middle class composition. It is therefore obvious that by keenly reading, the income class statistics of Kenya for instance, you can immediately and very logically conclude that it is only easier for many Kenyans to move from poverty to the floating middle class, but much harder to move to stable upper middle class category.

Violent forms of change like the Arab-revolution can be explained by of a host of socio-economic and political factors acting in concert. Then, when you factor in a high rate of urbanization accompanied by high rate of access to electronic communication tools, albeit in stark contrast to a dismal national democracy index, you obtain a perfect recipe for mass violent form of change. The Arab North African nations and their Middle-East brothers score lowest on the global democracy index. These countries scores are significantly lower than most of the Sub-Sahara African countries. All East African region countries have higher scores well a head of Tunisia, Egypt, United Arab Emirates, Oman, and even Qatar. However, in this instance we are only comparing worse off situations. Though not a perfect and precise measure of democracy, democracy index covers electoral process and pluralism, functioning of government, political participation, political culture and civil liberties, thus it is a convenient approximation tool for a relatively subjective value as democracy. Authoritarian regimes are conceived and premised on misconceptions, which then build into a trend. The ensuing trend and continuing misconceptions reinforce each other and can be sustainable over lengthy time span. A few decades of existence of authoritarian regimes belies their stability. Eventually the misconceptions and trends become unsustainable, and are thus self-defeating. Perhaps you can now start making some informed guesses about which countries in the larger African region and other parts of the world that might in the short-term or longer-term future explode to give us some televised revolution.

11.2 General Observations and Issues to Review

11.2.1 The Role of Non-governmental Organizations

As the role of state changes, non-governmental organizations (NGOs) are acting more and more as intermediaries between the government and the poor. The Government is increasingly aware of the important role of NGOs in poverty reduction and is beginning to trust them to undertake collaborative activities. NGOs are a major potential force for promoting poverty reduction in the world but they are very diverse in their objectives and competence. NGOs have, for example, pioneered participatory approaches in project design and implementation, and are strong advocates of approaches that see the poor as economic and social actors, rather than passive recipients of welfare. On the other hand, NGOs have also been involved in most relief programs dealing with the effects of sudden 'shocks' such as drought and civil conflicts.

The importance of broader institutional development to foster better conditions for poverty reduction within civil society is also clear. Mature and effective non-governmental institutions in fields such as the national and local news media, religious organizations, trade unions, policy research and advocacy groups can strengthen mechanisms of accountability for public policy and ensure that diverse groups in society have a voice in the formulation of poverty reduction measures. Hence, NGOs are likely to play an increasingly important role in poverty reduction at the national and community levels.

11.2.2 Macro Public Private Partnerships Versus Micro Public Private Partnerships

Whereas public private partnerships (PPPs, both as an economic development and financial mobilization model has increasingly gained attention, the model has been overly focused on large infrastructure projects undertaken by international financial institutions (IFI) and large private investors such as global private equity funds and investment banks. However, the PPP model can be extended to smaller projects with the active participation of small and medium sized local financial investors, and even individuals. However for PPP model to work in small and medium-sized projects, an enabling local regulatory infrastructure with transparent and accountable local system must be in place. The next chapter is dedicated to a discussion on innovative solutions to poverty and therefore will provide further discussion on the potential of extending PPP model to small and medium projects. In conclusion, we can point here the need to extend the deployment of the model to even small and medium sized projects as a means of competent tackling of poverty.

11.2.3 Rise of Social Enterpreneurship

The rise of social entrepreneurship is one of the most significant changes in how business activity is viewed as an antidote to socio-economic problems and poverty. That business can gainfully mesh social goals with profit goals while not being confined to activities in corporate social responsibility (CSR) is a remarkable development. However, it would be an embellishment of the truth to assert that this change in conception and perception of the role of business constitutes a paradigm shift. Rather it would be accurate to designate this change as work in progress due the realities that in as much as social entrepreneurship is a departure from past thinking on the role of business, not a prominent space has been rightly given to social entrepreneurship with regard to public policy of countries. Hardly is it possible to find countries with clear based business policies such as tax exemptions, preferential access to financing, and other forms of assistance to encourage businesses whose core goals target poverty reduction. At this moment funding and support to social enterprises is restricted to the generosity of certain international foundations with their televised project contests to secure funds and giving wide social recognition to social entrepreneurs.

11.2.4 Core Principles for Anti-poverty Policy and Action

The starting point for more effective actions is acceptance of six basic principles. The first is that poverty reduction is good economics and good politics. An acceptance of by all key players in the public, civic and private sector that solving poverty is an integral part of good and beneficial economics can be explained by the direct and indirect benefits of improved economic welfare of the majority in any society. Consequently, poverty eradication must be at the heart of any economic and social development strategy. The post World War II economic ascendance of Japan, which saw the emergence of a very unique situation where almost all citizens of the country graduated from poverty to middle class incomes. The historical case of Japan will be later revisited and analyzed as one of the unrecognized vital lessons in economic development. We can therefore draw some pertinent lessons from Japan's past economic development experience, albeit with full cognizance of the difference in the global socio-economic ordering of the past and the current twenty-first Century realities.

The second principle is that full, unambiguous and unequivocal government commitment to poverty eradication is essential for poverty reduction. The recent position of the International Monetary Fund (IMF), that developing countries must come up with a poverty eradication policy papers that is acceptable to both the boards of the IMF and the World Bank before they can qualify for consequential credit from the two lenders. Thirdly, economic growth is necessary, but not sufficient for the sustainable poverty reduction. Therefore, anti-poverty policy must embrace both the promotion of economic growth as well as other allied policies. The fourth principle is that the design and implementation of efforts to reduce poverty must be guided by the needs of the poor as identified by the poor.

The donors want the government to publish poverty eradication policy papers and ensure that all stakeholders participate in its formulation. The donors want low-income country governments to build a national consensus and evolve the support of all interest groups and stakeholders on the issue of poverty eradication. This kind of approach is needed in order to prevent proliferation of policies, which have no roots in the poor community targeted by policies. However this is easier said than done. It is notable that the World Bank Group (WBG) as a lender must be concerned about sovereign lending risk, but how about when devolution of public financial management in devolved administrative units are put in place in a country as one of the mechanisms targeting poverty reduction and general economic development? Would it not be cost effective for the WBG to deal only with a centralized public finance management (PFM) system under a national treasury than with several devolved PFM system, which might increase transaction costs and lending risk? Only time will tell how WBG can balance their crucial operational interest against a virtually important economic development responsibility mechanism such as devolution, which places local community in charge of their economic development issues. Fifthly, poverty reduction is a gender issue. In many countries, women were, by law or by institutionalized discrimination, receiving only a small fraction of the salaries given to men for the same work by the same employer. They were not accorded the same opportunities for access to credit facilities, as were men. Although such legal discriminations have been abolished in several countries, there is still scope for improvement. There has been no shortage of studies indicating that households headed by women are generally poorer than those headed by men. Furthermore, there is discrimination against women in the public service. At the present, there are few women in the cabinet of most poor countries and also in the boardrooms of both large and medium-sized companies.

Finally, poverty is an environmental issue. With the increase in population, and no commensurate resources to provide electricity, poor people increasingly resort to cutting down trees and thus increasing the rate of deforestation and desertification. At the same time, the poor degrade the environment through water pollution, overcrowding the urban ghettos, etc. All these negative consequences represent further deprivation of the crucial basic resources the poor need for their basic survival. There is a clear divide between the poor, upper and lower middle classes with regard to characteristics of environment in which each of these classes live. In almost every country, leafy well-planned settlements are a reserve for upper class and the super-rich. The poor are consistently confined to the worst environments. The stark contrast is one where the rich revel in the bounty of nature, while the poor know only the niggardliness of nature. The last chapters of this work will present a focused discussion on poverty and environment with clear examples to buttress our main arguments.

11.3 Action at the National Level

11.3.1 Empowerment of People

The starting point is to empower poor women and men—and to ensure their participation in decisions that affect their lives and enable them to build their strengths and assets. Poor people and poor communities rely primarily on their own energy, creativity and assets. Such assets are not just economic. They are also social, political, environmental and personal—both for women and for men. A people-centered strategy for eradicating poverty should start by building the assets of the poor—and empowering the poor to win their fight against poverty. What does such a strategy entail? First, it entails political commitment to securing and protecting the political, economic, social and civil rights of poor people. People must organize for collective action to influence the circumstances and decisions affecting their lives. To advance their interests, their voices must be heard in the corridors of power. Secondly, it entails policy reforms and actions to enable poor people to gain access to assets that make them less vulnerable. For instance, security of tenure for housing and land is as important as access to credit and other financial services. Thirdly, empowerment implies education and health care for all, along with reproductive health services, family planning and safe water and sanitation. This needs to be achieved soon—not postponed for another generation. Finally, it entails "social safety nets" to prevent people from falling into destitution and to rescue them from disaster.

11.3.2 Gender Equality and Poverty Eradication

Already women are on the front line of household and community efforts to escape poverty and cope with its impact. But too often they do not have a voice in the community or in national and international arenas. Gender equality needs to be part of each country's strategy for eradicating poverty, both as an end and as a means to eradicating other forms of human poverty. This means:

(a) Focusing clearly on ending discrimination against girls in all aspects of health, education and upbringing—starting with survival.
(b) Empowering women by ensuring equal rights and access to land, credit and job opportunities.
(c) Taking more action to end violence against women, the all-too-pervasive hidden side of human poverty.

A creative and innovative commitment to gender equality will strengthen every area of action to reduce poverty—because women can bring new energy, new insights and a new basis for organization. If development is not engendered, it is endangered. And if poverty reduction strategies fail to empower women, they will fail to empower society.

11.3.3 Gender Equality and Poverty Eradication

Economic growth can be a powerful means of reducing poverty, but its benefits are not automatic. That is why pro-poor growth policies are required. Several key elements of pro-poor growth include:

(a) Restoring full employment as a high priority of economic policy.
(b) Lessening inequality and moderating its extremes.
(c) Accelerating growth in poor regions of the country.
(d) Adopting clear measures of economic performance, which requires departing from the standard rolling out of figures on annual gross national product without analyzing to whom the benefits of such growth goes to. It is notable that more often than note the trickle-down-effect economics don't work well in economies with pervasive duality in economic structure.

11.3.4 Actions for the Rural and Urban Poor

More than three-quarters of world's poorest people live in rural areas, and are dependent on agricultural activities for their livelihoods. It is also important to point out that the number of people locked in urban poverty is also fast increasing. For these people, pro-poor growth means raising agricultural productivity in rural areas and urban incomes, bearing in mind the necessity to create jobs paying living wages as a critical means to solving urban poverty. The key priorities include:

(a) Creating an enabling environment for small-scale agriculture, micro-enterprises and the informal sector.
(b) Fostering technological progress in agricultural production.
(c) Reversing environmental decline in marginal areas.
(d) Speeding the demographic transition.
(e) Skill capacity building and deployment of technologies to improve productivity.

11.4 Action at the Global Level

11.4.1 Globalization and Opportunities

Proceeding at breakneck speed but without a map or compass, globalization has helped reduce poverty in some of the largest and strongest economies—China, India and some of the Asian tigers. But it has also produced losers among and within countries. As trade and foreign investment have expanded, the developing world has seen a widening gap between winners and losers. Meanwhile, many industrial countries have watched unemployment soar to levels not recorded since the 1930s, and income inequality reach levels not recorded since the last century. The greatest benefits of globalization have been garnered by a fortunate few. A rising tide of wealth is supposed to lift all countries, but some can be lifted more easily than others. Some countries are rising in response to new opportunities, but many countries are not benefiting from the globalization process and are sinking. The ratio of global trade to GDP has been rising over the past decade, but it has been falling for 44 developing countries, with more than a billion people. The least developing countries, with 10 % of the world's people, have only 0.3 % of world trade—half their share of two decades ago. The list goes on. More than half of all the developing countries have been bypassed by foreign direct investment, two thirds of which has gone to only eight developing countries. The developing countries are generally vulnerable to commodity price swings. Real commodity prices in the 1990s were 45 % lower than those in the 1980s—and 10 % lower than the lowest level during the Great Depression, reached in 1932. Recently, an upturn in commodity prices has been witnessed, albeit with a sudden and powerful interruption by the global financial crisis in 2008. The terms of trade for the least developed countries have declined a cumulative 50 % over the past 25 years. Average tariffs on industrial country imports from the least

Table 11.4 Share of world's private consumption

Worlds poorest 20 %	Consume 1.5 %
Worlds richest 20 %	Consume 76.6 %
Worlds middle 60 %	Consume 21.9 %

Source World bank development indicators (2008)

developed countries are 30 % higher than the global average. Developing countries lose about US$60 billion a year from agricultural subsidies and trade barriers to textile exports in industrial nations.

The bottom line for global poverty and incomes can be understood from the share of consumption by the various classes. The share of the poorest 20 % of the world's people in global consumption stood at a mere 1.5 %, while that of the word's richest 20 % stood at a sharply shocking figure of 76.6 % as given in Table 11.4 below. Perhaps the small and disproportionate figure of (60 %: 21.9 %) for the middle class can be explained by the dominance of middle class category by floating and lower middle class category. The consumption amongst the floating middle class and lower middle class can be quite volatile and fluctuates in tandem with the state of macroeconomic performance of the various countries and the global economy. As noted earlier, like the poor, they are also quite vulnerable to economic shocks. Here again we realize the need for a focus on creating a sizeable stable middle class as part of poverty reduction, for a stable middle class is a sure measure of poverty problem well tackled in a country. Post World war II Japanese economic experience can be looked at as case in point of mass creation of stable upper middle class as a sure way of tackling poverty.

11.4.2 Middle Class Population is Predominantly in Employed Labor

That majority of middle class individuals are in employed and gainfully paid labor is an aspect of global economic reality we should not lose cognizance of against the backdrop of popular attempts to promote microenterprises for every poor person in many of the developing economies. Some recent research show that most of these microenterprises are mere symptoms of unemployment and even when they seem to prove some income generation ability, they remain mere stopgap measures at dealing with poverty problem. As such, the promotion and creation of small and medium enterprises capable of growth and active participation in modern formal economy is what is required to deal with poverty. To create jobs we strongly argue that a sustainable shared-value partnership is necessary between all categories of stakeholders—governments, private sector, community groups, and non-governmental organizations.

11.4.3 Partnerships for Change

All agents in society—trade unions, the media, community groups, private companies, political parties, academic institutions, professional associations—need to come together in a partnership to address poverty in all its dimensions. And that partnership must be based on common interests and brokered compromises. Society must be open enough to tolerate a complex web of interests and coalitions and to ensure stability and progress towards human development.

The unprecedented progress in reducing poverty in the twentieth century sets the stage for eradicating absolute poverty in the early twenty-first century as a moral imperative, and as an attainable goal. Poverty should be relegated to history—along with slavery, colonialism and nuclear warfare, and that countries and regions have been able to effectively deal with poverty simply means that the goal can be achieved.

References

1. African Development Bank (2011) The middle of the pyramid: dynamics of the middle class in Africa, African Retrieved from: http://www.afdb.org/fileadmin/uploads/afdb/Documents/Publications/The
2. Bloom DE, Canning D (2000) The health and wealth of nations. Science 287:1207–1208
3. Bloom DE, Williamson JG (1998) Demographic transitions and economic miracles in emerging Asia. World Bank Econ Rev 12: 419–455 (These findings may reflect the adoption of growth-enhancing policy reforms)
4. Guillermo EP et al (2006) Poverty reduction and growth: virtuous and vicious circles. The World Bank, Washington
5. Horton S, Shekar M, McDonald C, Mahal A, Brooks JK (2010) Scaling up nutrition: what will it cost?. World Bank, Washington
6. Karnani A (2010) Failure of the libertarian approach to reducing poverty. Asian Business and Management 9(1):5–21
7. Konomoto S (2009) Rebuilding competitive strategy aimed at emerging markets. NRI Papers, No. 145 Aug 1
8. Prahalad CK (2005) Fortune at the bottom of the pyramid: eradicating poverty through profits. Wharton School Publishing, Philadelphia
9. WHO (2010) Constraints on scaling Up the health MDGs: costing and financial gap analysis. WHO, Geneva

Chapter 12
Innovative Solutions to Poverty

12.1 Why Innovation Matters for Poverty Solutions

Innovation is widely regarded as the best mechanism for gaining competitiveness in private sector activities, and there is evidence that the most profitable companies around the world are also the top innovators. Extending the innovation concept to the problem of poverty, which itself is inseparable from low productivity and lack of economic growth in diverse sectors of economies of low-income countries, we can count on innovation to improve productivity in these countries to deal a considerable blow to the endemic problem of poverty. Baumol [1] observed that essentially, all economic growth since the eighteenth century is attributable to innovation. It is therefore ironical and shocking that innovation has not gained a center-stage position, or to put it more succinctly innovation has not become the platform concept from which poverty need to be attacked.

While the practice of innovation in private sector has gradually mopped up into well-formed concepts, known requisite organizational and resource conditions, rules and incentive structure, no consciousness exists among public organizations about the need for innovation (See [12]). At the same time there is little initiative on the part of private sector actors to innovatively provide solutions in anti-poverty measures that also generate profits for businesses. As things are now, governments truly gravitate towards inertia, and perhaps are obsessed with enforcing dysfunctional and outdated rules, and this is usually more so when it takes less effort and commitment to keep those same bad rules. On the other hand majority of private sector actors are locked into the past century brand of business practices that have come to define capitalism as a production system mired in crisis. Under these practices, companies selfishly pursue short-term gain and are only interested in going after existing resources and thriving markets. The end result here is our current global problem of underserved large population groups, large pools of untrained labor with limited skills, and exploited natural resources with no genuine attempts at restoration and sustainability.

N. O. Agola and J. L. Awange, *Globalized Poverty and Environment*,
DOI: 10.1007/978-3-642-39733-2_12,

It is important to note that innovative solutions presented here assume national environments already characterized by adherence to sound macroeconomic economic principles. As such, maintenance of stable macroeconomic policies, working towards creating acceptably efficient markets (market mechanisms efficiently steering resource allocation), and paying attention to raising the levels of human capital are all deemed as requisites for the implementation of twenty-first century innovative solutions to poverty. We can then assert that these requisites form the national platform upon which the innovative solutions to poverty can be launched. There is need for some caution here because in as much as stable macro-economic policies are given summary treatment here, there is still need to interrogate the issue of what constitutes the right macro-economic policy for each country based on individual country circumstances. It is important to note that even what is regarded as past successful cases in economic development experience such that of the high-performing Asian economies (HPAEs) cannot offer standard lessons forming a template, which could generally be applied to the vast number of countries and regions still struggling with poverty issues (for information on HPAEs, see "The East Asian Miracle: Economic Growth and Public Policy", [13]). The need for such interrogative process takes us back to some later conclusive research findings reflecting on the World Bank study to evaluate the long-term realities of the HPAEs experience. In contrast to the glowing and positive depiction of economic growth and resulting equitable income distribution as represented in World Bank study, these later research findings expose the defective position that high and sustained economic growth inherently leads to equitable income distribution (see [10, 11]). Quibria notes that the HPAEs did not uniformly experience reduction in poverty and equitable incomes. As such we need to go further to look at the issues such as the source of economic growth.

The main question is about whether economic growth is generated by the internal processes of an economy? Or is it stemming from external mechanism such as trade and foreign investment? At the same time, the composition of the national output matters as well. Is the growth in output being fueled by the agricultural sector, light manufacturing or services? The answers to these questions are quite critical to realizing equitable income distribution from any sustained economic growth process. Finding answers to these questions can judiciously inform macroeconomic policies targeting equitable income distribution through sustainable market mechanisms. Therefore we assume the presence of an effective macroeconomic regime as a basic requisite for the innovative solutions suggested in this chapter.

12.2 21st Century Innovative Solutions to Poverty

12.2.0.1 Cost Efficient Collaboration Models and Value Innovation

We have a pervasively fractious socio-economic environment in which businesses, governments and civil society organization are all pulling in different directions. Companies have simply not figured out how to craft sustainable and profitable symbiotic relations with society and environment in away that rationally takes care of costs and confer benefits to critical stakeholders. Social movements and non-governmental organizations (NGOs) have also been too eager to vilify private sector actors. Governments on the other hand have not found a formula to harmoniously deliver value to all stakeholders. This scenario demands doing things differently to reverse the undesirable status quo, thus innovation is a must so that productivity, efficiency, and social welfare goals can be served from the same bowl with private sector, governments, NGOs and society sitting at the same round table. We illustrate the requisite conditions and key attributes of the anti-poverty innovation model around this problem by analyzing and making propositions based on the space within which we try to tackle poverty. It is notable that innovation does not suggest doing things in a radically different way only, but also making small incremental changes and even borrowing from some less publicized, but significantly successful cases in countries that otherwise have not been able to gain attention. First we need to have a clear understanding of what innovation means for poverty solutions. Figure 12.1, illustrates the four key areas of innovative intervention to solve poverty problem. We deliberately christen the innovative solutions the twenty-first century Innovative Solutions to Poverty because innovation itself strongly suggests doing things in a new way, both radically and incrementally, by recognizing deficiencies in extant poverty solution approaches and practices carried over from the immediate past century. At the center of the twenty-first century Innovative Solutions to Poverty is the need for a paradigm shift requiring an overhaul in mental model regarding how poverty is perceived, specifically on the question of who bears the big responsibility for providing solutions to poverty.

Fig. 12.1 Key elements in collaboration and innovative pricing model

In the twentieth century, this responsibility was squarely placed on the shoulder of governments with sporadic support from local and international NGOs. Private sector actors and local communities have had very negligible moral and purposely delegated responsibility for solving global poverty problem. It is therefore quite rare to find well-articulated and authoritative opinions on the idea that businesses draw their revenues from local communities within which they are located and operate. The idea of giving back has only been popularized at the level of sporadic gestures typified by corporate social responsibility. The new order we suggest here is marketed by going beyond such sporadic acts to establish long-term mutually beneficial relationship to generate shared economic value. Porter and Kramer [8] present a new model for creating shared value as a means to reinventing or saving capitalism from the current crisis marked by poverty, inequality and desperation afflicting more and more people in both developed and developing countries. They conceive shared value to be rooted in corporations refocus on social problems to generate competitiveness while at the same creating social value. They posit that social problems generate externalities which translates into private and internal costs to corporations, and therefore solving the problems can unleash value either by saving the transferred internal costs or raising productivity and value creation both at company level and within the value chain. Their suggestion of shared value here confirms the need to go beyond arm-length transaction based on minimalist approach to dealing with other entities both within and outside the value chain of a companys value environment. Indeed leading companies in innovation in specific product and service categories have recognized the value of collaborative assistance to entities within the lower part of their value chain, such as working with farmers in training, improvement in access to and lowering costs of key inputs such as agricultural machines and fertilizers.

12.2.0.2 Where is the Money Going to come from? Value Ecosystem Model as the Answer

In the twenty-first century Innovative Solutions to Poverty, we propose an inclusive-shared-value model based on what we call Value- Ecosystem, which requires a paradigm shift in how value is perceived, and how judiciously costs and benefits are willfully allocated. In the low-income countries, the most pervasive question is always where money is going to come from whenever any attempt is made to practically solve any socio-economic problem. Affordability is therefore one of the most pressing problems when considering implementation of almost all shades of solutions conceived of by public, private sector actors and even more so NGOs. Value Ecosystem Model proposed here is a novel approach, which starts with the idea that there is always latent value locked up in the various entities and sub-systems that constitute entire economic and social systems. Discovery and optimal utilization of such latent value is therefore key to solving affordability problem. Discovery of latent value starts with asking the question of what is of value, to whom, and is the value located whenever any socio-economic undertaking is considered. It is also important to point out that unobvious, but possible relationships between diverse economic activities and actors must constantly be the subject of unending search and eventual discovery. This in itself is synonymous with discovering potential new sources of cash flow by all types of entities who could make sense of latent value and hence be willing to pay for something that would otherwise go unutilized. We can illustrate this concept with actual cases closely following the concept and model of Value Ecosystem.

Example 1: Optimal Leveraging the Value Ecosystem Model—Street Lighting Project

Current approach to innovation is grounded on efforts to create new products and services, which when ready, then a price tag is put on such products or services. Essentially innovative thinking is never extended to the issue of price and payment system. Value ecosystem envisages discovery of value even at the payment stage by addressing the question of how innovatively payment can be subjected to the discipline and creativity of innovation. Value Ecosystem Model departs from the conception of a single source of payment the end user. Instead, the issue of payment is subjected to innovation by way of efforts to discover how payment can be shared amongst all entities for which the final product or service holds some latent value. This is more so true with public services and utilities. A few cases of unconscious attempts at optimal leveraging of value ecosystem do exist so we use them to illustrate how this new model proposed in this book can become mainstream in addressing the problem of affordability.

Usually cities in developing and emerging economies face financial constrain and thus have difficulty providing adequate modern urban infrastructure and services. Street lighting is one such example. Many streets are dark, either as a result of broken-down street lights which cannot be replaced or complete lack of financial muscle to provide street lighting. The conventional thinking and practice points to the need for local authorities to pay street lighting construction service directly in order to have well lit streets. Some cities in South Africa and Kenya have innovatively solved this perennial developing country urban problem by optimally leveraging Value Ecosystem Model. Here is how it works. The cost of constructing and installing the street lights is shared between all private sector players who may discover latent value in street lights and are therefore willing to pay for such value. Each company literally sponsors or in the exact language used ‘adopt’ the streetlights, which actually bear company commercial advertisement.

Basically companies are in dire need of visibility and exposure of their products and services. Therefore there is a scramble for available public and even private spaces with promise of being seen. These commercial entities know too well that a product or service unseen and unheard is simply unsold. So they are wiling to sponsor construction, installation and maintenance of streetlights so long as their billboards can hang and light up on the stems of the streetlights. Unlike the conventional method of paying for construction, installation and maintenance of street lights by urban authorities, the example described here not only solves the problem of lack of funds, but also the issue of sustainability of this key urban infrastructure. The revenue from advertisements on the billboards are shared between the urban authorities and advertising companies partnering in this program. Public members benefit by having well lit streets with accompanying benefit of reduce crime. In essence there value ecosystem, and there is sustainable sharing of costs and benefits.

Example 2: Optimal Leveraging the Value Ecosystem Model in Prevention Healthcare

There is no doubt that there is a nascent global paradigm shift in the conception of markets for goods and services produced by multinational companies. Up till the end of the twentieth century, the main market conception paradigm has been that corporations go after thriving and expanding national and regional markets. Under such market conception, companies focused on identifying countries with high growth rates accompanied by an expanding middle class with high purchasing power. In essence, well-formed and actively functioning markets were simply one of the strategic assets pursued by multinational corporations. The common thinking among corporate executives and managers in charge of operations tended more towards an oversimplified refrain from engaging in business operations in small markets where majority of the population have low incomes. However, at the beginning of the twenty-first century, we have seen the beginning of a portent paradigm shift in the conception of markets for goods and services produced by multinational corporations. It is important to emphasize that this shift is at an infancy stage, and as such a wait-and-see approach to whether this shift will go full cycle and perhaps gain legitimacy and full currency among multinational corporation decision makers.

The emphasis and efforts to promote the role of private sector in economic development in poor countries by the World Bank and other international development agencies has been a major precursor to the paradigm shift. The main eloquent proponent of this newly shaping paradigm was C.K. Prahalad [9] in his seminal work, the Bottom of the Pyramid (BPO). Prahalad strongly argued that selling to the poor at the bottom of the pyramid (an expression carrying the meaning that the poor are still the majority on planet earth) made sense in terms of the potential profits, and therefore multinational businesses should not ignore low-income groups and markets. That the BPO proposition became quite a game-changer is not in doubt, at least judging by the wide attention followed by actual business undertakings by many multinational companies such as Unilever and SC Johnson, and lastly that the ideas formed part of the intellectual discourse at sessions at the Word Economic Forum ([3] : 6).

However, BPO model has not been without its critics, the most notable being Karnani, whom I have always concurred with on the ideas that the market size of the BPO is exaggerated, and that the idea is based on an over-romanticized conception of the poor as entrepreneurial and as avid well-informed value seekers. In addition, the cost of marketing to the poor could be quite prohibitive, specifically given the absence of infrastructure, which is a global common denominator in all physical locations occupied by the poor. If the BPO market exhibits weaknesses with regard to business impracticalities, then its intellectual persuasion, and thus wide adoption by businesses on a global scale is already compromised. Therefore the legitimacy of an ongoing global search for a new business model, both at the intellectual, discourse, and business practitioner levels is beyond doubt.

The current ongoing shifts in global economic power, and the state of markets in industrialized countries vis-à-vis that of emerging markets reinforces the legitimacy of the need for new business model upon which companies from the industrialized countries can effectively participate in the slowly-but-surely emerging markets, specifically that of the low-income countries. It is common knowledge that the markets for a wide range of consumer goods and services in industrialized countries are either saturated or in decline. On the contrary, the markets in emerging economies are on the rise, and with underserved swelling young population the emerging markets cannot be ignored any more by any serious corporation intent on being a player on the global stage. This scenario forms the stage on which Japanese companies, both large multinationals and medium- sized enterprises must sit up and pay attention to the emerging markets. Such attention would only be rewarding if based on a well-conceived business model [4]. There is need for a discourse in search of pragmatic business model, and therefore the case study used here is to lead to discussions and further search of a pragmatic business model for low-income markets.

It is worth pointing out an integral aspect of the current search for a business model for active and effective participation in emerging low-income markets. Perhaps, it is this integral aspect that makes more sense than even the final outcome in the form of a well-conceived and pragmatic business model. Simply put, it is the recognition of the need to innovate and create new markets instead of the old approach characterized by a 'wait-and-see' approach culminating in going after markets as a mere part of the strategic assets sought after by corporations after their evident emergence in the emerging economies.

We may briefly turn our attention to the extant situation of Japanese corporations global participation in emerging low-income markets. To understand this situation, we need to ask the question of whether Japanese corporations fall behind Western and emerging country multinationals in participation in emerging low-income markets as part of their strategy to not only participate in markets as they are, but also in creation of new markets. Creation of new markets requires either product/service innovation or adaptation to suit both the needs of consumers, but also be aligned to the price level at which majority of the low-income consumers can be unlocked. Generally, Japanese companies pride themselves in their competitiveness based on high quality products commanding high prices, but this strategy flies in the face of the rationale of low-priced and mass-consumed products suitable for those at the bottom of the pyramid [2].

Validating Case

Sumitomo Chemicals started manufacturing innovative mosquito nets made from resin fibers infused with permethrin, the insecticide used in deterring mosquitoes for the first time in China in the late 1990s. This was a new manufacturing operation requiring building manufacturing capability in terms of both processes and manufacturing equipment. Both the manufacturing capability and volume of exports rose in tandem, however the decision to produce directly in the African market came when the World Health Organization (WHO) strongly recommended Sumitomo Chemicals to manufacture the nets in an African country. The Olyset nets are both technologically and functionally superior because unlike the rival products on the market relying on dousing nets in insecticide chemical, Olyset net with the resin fibres infused with insecticide have the simple advantage of longevity of the chemical staying on, never wearing off, and thus are quite effective in keeping mosquitoes at bay. Effectively this net became the most effective in malaria prevention in the market. It is therefore does not come as a surprise that Olyset net gained not only the approval of WHO, but also the organizations support for the nets to be widely distributed in the most malaria prone regions of the globe. Given that the mosquito nets were targeted at people in low-income countries, low price was a critical part of the manufacturing locational consideration. It is against this backdrop that Sumitomo Chemicals began production of the nets in Tanzania in 2003, based on a royalty-free technology licensing to a local company, A–Z Textile Mills. The production was revamped in 2008 with a 50:50 joint venture factory. According to the companys reported figures, the output stood at over 29 million nets in 2009, among the highest in the entire global production of the Olyset mosquito nets.

12.2.0.3 Public-Private Sector and NGO Collaborative Model

Producing products directly in the low-income, but slowly emerging markets comes with several challenges as compared to exporting and selling to those markets. Export activity as a market entry strategy is less risky in terms of requisite resources, and it is possible to adjust the level of volume of exports to the market demand situation for products. In contrast, direct local manufacturing in the low- income economies is replete with deficit in requisite strategic asset challenges. For instance, lack of skilled managerial labor, lack of reliable infrastructure, high cost of marketing to dispersed rural population, and lack of competence in dealing with the local cultures are factors that can overwhelm any multinational corporation. Sumitomo Chemicals would have had to deal with these mentioned challenges had they chosen to be directly involved in manufacturing the Olyset nets in Tanzania. It is important to note that most multinational companies, and even medium-sized companies choose to refrain from direct foreign investment in the low-income economies based on the oversimplified understanding that the markets are way too small.

It is therefore noteworthy to analyze Sumitomo Chemicals mode of market entry and the ensuing market operations in Tanzania as this may provide some understanding of those markets, and perhaps point towards a new business model for engagement by Japanese companies in some of the low-income economies. Figure 12.1, provides a depiction of the five interconnected and mutually reinforcing market entry and operational factors on which Sumitomo Chemical's operations hinged on. First, it is important to acknowledge that business as usual in industrialized markets where mode of appropriation of technology assets is through licensing contracts may not be suitable as a joint venture structure with low-income market firms, though this is not in all cases. Hence licensing of technology may not be the appropriate point at which to capture value. In the case of Sumitomo Chemical, they chose royalty-free licensing with the understanding that they could capture upstream value through innovative pricing of its mosquito nets.

Perhaps it is important to point out that royalty-free technology licensing can only work when a company has clearly identified the point at which it can capture value. Nothing is free in practicality in the market. The second element in the market entry model is the choice of a local company to manufacture the nets, and interface with the consumers through distribution, and marketing communication. Manufacturing activity itself require an understanding of local cultural issues impacting management and organizational practices on which corporate competitiveness is based. The local company A–Z was better placed in terms of competency to deal with such issues in the shortest time possible. It is also notable that local distribution of nets requires making deep access into rural regions and also creating working relationships with a myriad of stakeholders, itself a daunting task that only a local company can undertake in liaison with governmental agencies.

The last part of the business model is the funding method requiring working with a wide network of international development agencies and foundations whose work is related to malaria prevention and eradication around the globe. Securing funding is an imperative given the weak local consumer purchasing power to the level where they could buy the nets at market price. The network of international development agencies and foundations therefore provide funds, which subsidizes the lower prices payable by the locals to acquire the nets. However, two weaknesses lurk in this funding model. The first weakness is the lack of sustainability in the flow of funds from the network of international development agencies and charitable foundations. Usually, the international development agencies and charitable foundations commit themselves to contributing specific amounts within a given time frame, and there is no assurance that they could continue to do so in the future. This uncertain situation is a contrast to market mechanism where demand and supply controls production. The second weakness lies in the possibility of market distortion and hence interference with suitable conditions for market competition in the production and supply of mosquito nets. Suppose other mosquito net producers exist in the market, they simply cannot compete unless they are subsidized by similar arrangements. As such, the business model might require some tweaking to incorporate market competition mechanism, which in itself can also contribute to the issue of business sustainability. Market mechanism in this case may involve payments being made by other businesses keen on deriving potential value related to the mosquito nest business. Advertising opportunities is an example of such derivable value.

One very important suggestive lesson that can be drawn from the Sumitomo Chemical business model for low-income market entry is simply that such entry can be made on a collaborative platform. On such a platform would be a number of value-generating entities, each performing a role congruent to her organizational and technological core competencies. Mutually beneficial goals among the entities also provide the glue that can keep the concerted business and social efforts beneficially functional. We can conclude that there is a good case to make for collaborative market entry in the low-income markets. The nature of such collaboration must then depend on the type of corporate activity, and it should be the role of corporate decision makers

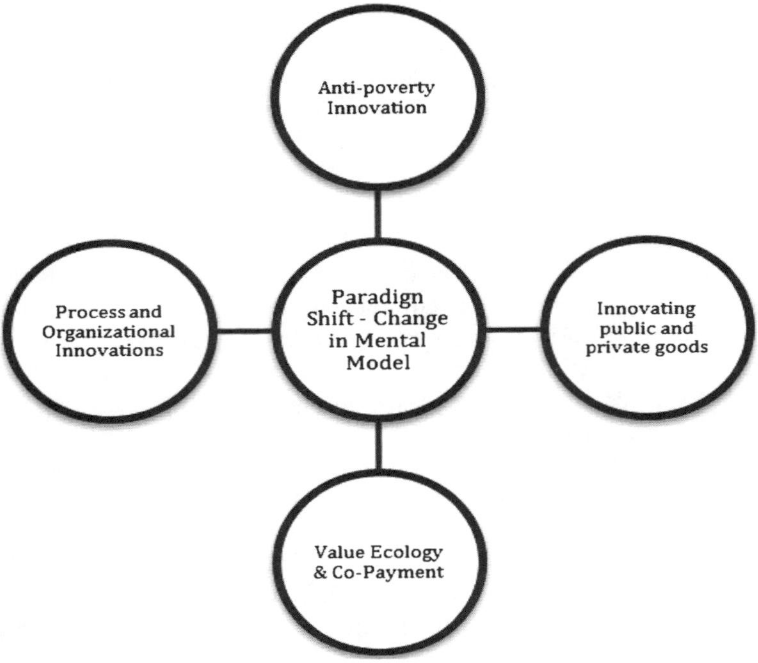

Fig. 12.2 Anti-poverty innovation space

to innovatively determine the intricate details of the collaborative arrangements based on value creation by each participating entity.

In this case, Value Ecosystem Model allows for an expansive conception of value in a new way not conventionally practiced by markets. Such expansive conception is based on poverty alleviation innovation space as the starting and point of departure from conventional practices as illustrated in Fig. 12.2. Regarding specific types of explicit or latent value, we can conceive of various value typologies—utility value conferred to direct and indirect users, reputational value, and market exposure-value as when companies perform advertising campaigns. In essence, value ecosystem thinking recognizes the reality that when some activities are performed, some of the values mentioned here go untapped by those to whom either explicit or latent value would make market and utility sense. Housing developments in slum areas for instance could be subsidized, though in a small way by commercial advertisements on walls. This is just a simple example of latent value, which could be captured or altogether go unexploited. It is also clear that value system is a network concept, thus it works best when actors actively work at discovering where value may lie and to whom it would make sense, therefore conferring real exchange value when none would initially have been inconceivable. This in itself requires donning an innovative thinking cap and thus defying industry conventional practices.

For instance, in pricing goods, companies predominantly think only of capturing their value from single end users. In reality, there could be some latent value tacked in the item on sale, which could be captured by some unintended user. What of if sellers of products that end up generating garbage, but which could be useful for instance as fuel stock for garbage powered plants, thought of factoring such utility value in pricing their product at the point where it is purchased and also at the point of disposal. This would represent a new concept of innovative pricing based on sharing payment at different points at which value becomes relevant and actually optimally extracted. This however can only work based on the awareness of the concept of Value Ecosystem, where participants are conscious of both explicit and latent value, and constantly search for different points at which value would be relevant and ready for extraction at an exchange vale or price. This innovative line of thinking might help to reduce cost of delivering some of the essential services such as housing, consumer goods and even services that ordinarily majority of the poor cannot afford if direct one-point pricing is the norm. To practice the concept of Value Ecosystem, both corporate players and even non- governmental agencies must break away from their conventional practices (paradigm shift) leading them to overpassing latent value and failing to see the disparate points at value would make sense for initially unintended users.

A paradigm shift denotes a fundamental change in way of thinking towards poverty problem among all the stakeholders—public sector actors, private sector actors, NGOs and communities. The change in way of thinking towards poverty problem rests on the pivotal common understanding that we are all in the same boat, which then defines our inclusive-shared-value. Significant poverty reduction and to a large extent eradication is possible, and we only need to deepen our understanding of the development experience of some the countries that have effectively played catch-up industrialization strategy. The specific case of Japan offers quite some vital lessons in economic development. This was mainly with regard to solving the two pervasive problems of poverty eradication and negative duality in economic structure. So the next section will discuss the Japanese economic development and its lessons for solving poverty problem in most of the developing world.

12.2.0.4 Attacking Economic Structure Duality and Raising Productivity

Duality in economic structure is an enduring characteristic of low-income economies, and within such economic structure, poverty lurks both in rural and urban settings. Duality in itself is defined by a glaring disparity in productivity as a result of factor combinations used. The most common forms of duality are a relatively productive manufacturing sector pitted against low-productivity agricultural sector, large companies using advanced technologies versus small enterprises with crude technologies, impoverished rural sector against urban sector with modern facilities, and underdeveloped primary sector versus progressive secondary sector. Whatever form of duality, the type of technology, capital labor ratio, level of economic development, annual growth rates and value added, per capita incomes are the premises upon which we can understand the problem of such economic structure. We can express duality as follows:

$$D = \frac{Q_1}{Q_2}$$

D is the duality index, Q_1 being product per worker in say agricultural rural sector, and Q_2 being the product per worker in the urban manufacturing sector. We can go a head to derive the values of Q_1 and Q_2 as follows:

$$Q_1 = \frac{V_1\ P_1}{L_1};\ Q_2 = \frac{V_2\ P_2}{L_2}$$

In this case are the real product per worker in agricultural rural and manufacturing urban sectors, P_1 and P_2 are the prices for products from the two sectors, and L_1 and L_2 indicates labor size in the two sectors. We take the relatively more productive manufacturing sector as having a higher capital-labor ratio, K_2^* compared to the agricultural sector K_1^*. The values of K_2^* and K_1^* can be expressed as:

$$K_1^* = \frac{K_1}{L_1};\ K_2^* = \frac{K_2}{L_2}$$

V_1 and V_2 are aggregate functions of capital and labor, so can be expressed as:

$$V_1 = f(K_1, L_1);\ V_2 = f(K_2, L_2)$$

Real product per worker can be expressed as:

$$\frac{V_1}{L_1} = \phi(K_1^*);\ \frac{V_2}{L_2} = \phi(K_2^*)$$

So long as the capital-labor ratio remains high for the manufacturing sector relative to the agricultural sector, the productivity and incomes will remain far higher than those for the agricultural sector, and so long as no technological progress accompanied by significant rise in factor prices, duality condition will continue to prevail. So how can we attack the problem of duality in economic structure based on the explanations? We must look at the causes of duality and its dynamics, which from the above cause-effect expressions, we can summarize as five-action points to obtain relative changes between any of two sectors defining duality in economic structure as given in Fig. 12.3.

Dynamics and Intervention Points for Solving Duality in Economic Structure

Japanese economic development in the 1950s and 1960s: Vital Lessons for Poverty Eradication and Reduction in Duality in Economic Structure

Fast and sustained economic growth is a fundamental requisite for poverty eradication and tackling of the problem of duality in economic structure. This is validated by Japanese economic development history, and recently that of China where near to half a billion people have been lifted out of poverty, while many rural locations

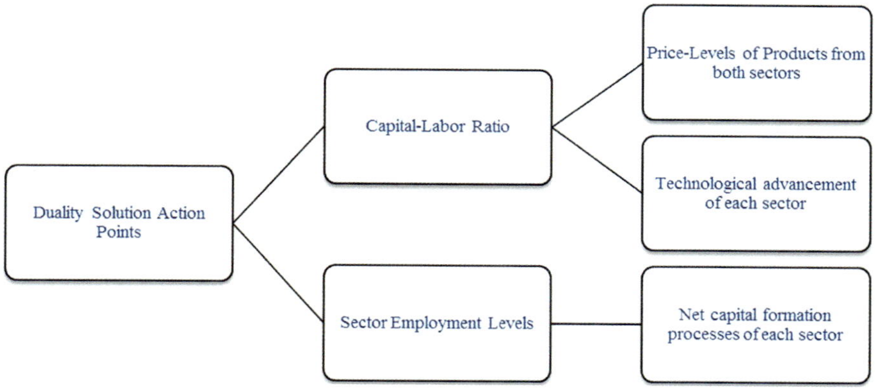

Fig. 12.3 Duality intervention points

have been opened up for modern industrial and infrastructure development. However, specific macroeconomic policies and private sectors practices are equally important in meaningful implementation of anti-poverty measures and reduction of problem of duality in economic structure. The role of private sector in economic and social development goals is indisputable, and therefore by extension the macro-level managerial practices and policies of corporations deserve scrutiny in economic development discourse. This is with regard to how macro-level private sector managerial practices may contribute to solving poverty, reducing income inequality, and even rural-urban income disparity. Valuable lessons abound in the Japanese post-war period marked by fast growth and distinctive corporate managerial practices rooted in the socio-economic context and dynamics of that time. The post-war period was marked by high unemployment rates, while at the same time shortage of skilled workers prevailed.

While the Korean War became a game changer triggering and ushering in post-war Japanese economic growth, companies had to either contend with poaching skilled workers from rival firms or had to develop skills of freshly hired workers in-house. Poaching skilled workers was simply unsustainable given the shortage of skills. However, in-house training of freshly hired workers posed its own challenge of capacity building cost, and sometimes even the lack of trainers in cases where companies had to engage in totally new production activities. Against this backdrop, it is notable that Japanese companies at a macro-level adopted in-house training and long-term nurturing of skills of their workers.

The concept of on-the-job training, continuous learning and upgrading of industrial and production skills was born and became firmly rooted in the Japanese managerial practices. This practice was pivotal in lifting a chunk of the countrys population out of unemployment, but even more significantly, it guaranteed improvement in income levels as the workers improved their skills and productivity. Obviously with productivity increase, wage increases become sustainable. The remuneration system of the Japanese companies during this era also deserves mentioning here. The companies adopted a useful and pragmatic concept of living wages, which meant that workers were paid based on the living cost needs and wage increments and adjustments conformed to real life needs of the workers. The wages peaked when workers financial needs also peaked i.e. when they needed to pay college tuitions for their children. An elaborate and carefully thought social benefits system accompanied the human resource training and remuneration system. A distinctive element of the social benefits system was the provision of housing and support of transportation needs of the workers most of whom were recruited from far-flung rural locations. In addition, majority of the companies also created cafeteria facilities for their workers.

The Japanese economic development experience also teaches us something about economic development process. Whereas the late Nobel laureate Arthur Lewis posited that surplus labor in the subsistence agricultural sector can smoothly transit to the manufacturing sector, and that through capital accumulation and reinvestments, an economy will eventually modernize, empirical evidence from many developing countries struggling with economic development process suggest otherwise [5]. Such smooth transition has failed to happen, and industrialization has remained at bay. However when we look at the Japanese economic development experience, we can draw a vital lesson in how labor transition actually happened. This was not due to an automatic market mechanism where expansion in the manufacturing sector and rising wages necessitated the absorption of the surplus labor from agricultural sector. Rather we do see that in Japanese economic development history, the pragmatic cooperation between labor and capital owners was one of the critical positive contributors to solving the perennial problem of labor resistance to adoption of technologies that negatively impact employment opportunities. Through such cooperation, it became possible for Japan to adopt new technologies in productive activities, while owners of capital took on the cost of training and retraining of workers. This represents an economic model beyond the confines of market mechanism. It is a model that conforms to the twenty-first century, the novel thinking of shared-value, both in terms of creation of value and allocation of the proceeds of value created. We need to move beyond obsession with markets mechanisms, because market mechanisms left to themselves dont work well and fast enough in ensuring a smooth labor transitioning and expected rise in productivity and technology progress. This viewpoint is to sound a clarion call for organizational innovations, both at micro-level, institutional and macro- levels, which is further expounded on in the next section.

Targeting the problem of duality as suggested above presents a simplified, but yet a strategically powerful model for attacking poverty for a number of compelling reasons. First, tackling the problem of duality in economic structure is a pro-growth model in the sense that it is premised on optimal allocation of resources to both urban and rural parts of the economy. Therefore, it facilitates the freeing of productivity-raising potential of both rural and urban sectors, while at the same time bringing positive social impact on rural urban migration. We note here that rural urban migration spawned by pervasively unequal development is a destabilizing phenomenon on both planning for provision of public urban facilities and while for rural sector, such migration is a drain on agricultural labor, and other potentially productive activities in the rural sector. Secondly other problems such as food insecurity can only be solved by tackling the problem of duality in economic structure once we are able to determine that there inter-causal relations. With an equally productive rural sector, it is not hard to see possibilities of a flexible labor market in which even some urban dwellers actually willingly move to the rural sector as agricultural labor. Lastly, we can argue that tackling the problem of duality facilitates competition in an economy, and competitive pricing of agricultural products in urban sectors become a reality.

To complete the picture of how Japan was able to solve the duality problem in economic structure, we have to add that provision of infrastructure such as roads, railways, water systems and telecommunications played a very important role in the leveling of rural- urban incomes through improved productivity. Again as stated in the introductory part of this chapter, sound macro-economic policies well implemented by competent national institutions is key to tackling the problem of duality in economic structure. It may sound like asking too much of any developing country given the near-mutual exclusivity between a developing country and competent institutions, but we must note the progressive human potential for learning and change. The situation is not permanently static. The need for change in policy and action approaches to problem of duality in economic structure is an urgent one—noting that urgency brings both positive and negative consequences. Positively, the urgent need for change can lead to putting in place measures leading to creation of sound institutions. On the other hand, urgent need can lead to desperate and social-economically expensive measures in a situation of complete lack of technical and even moral preparation by poor countries.

12.2.0.5 Social Innovation: Sustainable Resource Mobilization and Equity

Considering the Mondragon Model

In considering poverty solutions, much attention is given to funding issues and a host all sorts of policy interventions that needs to be adopted, yet social organizational innovations at both micro and macro levels get only scanty and accidental attention. This can only mean two things. First one is that how we organize ourselves at communal and national levels in various countries is perhaps perfect and therefore optimally serves us in attaining both economic and social goals. Secondly, it shows that we take it for granted that organizational forms we conventionally adopt or rather use without any questioning on efficacy and costs, are free instruments that do not involve any direct and indirect costs. Yet we know from extant global socio-economic problems as typified by poverty that our organizational systems meant to sustainability secure basic needs such as food, clothing and shelter simply do not work optimally. If cost effective use of resources is important, and at the same time the output from the use of resources is important, then we must consider the efficacy of conventional organizational systems used to achieve desired economic and social output. Innovation can happen at the level of positive transformation of tools we use (technology), but it can also happen at the level of how we organize ourselves while minimizing cost of doing so, but at the same facilitating maximization of both economic and social returns. So by social organizational innovation, we mean how people can be organized at a lesser cost in relation to higher economic and social value generated. Economic and social value generated through organizational innovation can range from lower cost-financial mobilization models involving members of society, methods of tapping into social systems for health improvement and awareness, employment creation efforts, to communal security provision, etc. Precisely, there is need to move beyond the view that economic development can only be achieved through the deployment of technologies, while ignoring the critical productivity role of organizational forms.

There are immense benefits of focusing search for economic development solutions on social processes of innovation given some of the results attributable to social innovation. From micro-credit as typified by Grameen Bank to modes of providing educational services, we can see the impact of social innovation. Invariably, social innovations can be extended to private sector business models, and public sector in governance, resource distribution and public policy. The widely successful case of Mondragon, a Spanish industrial cooperative in the Basque region is worth mentioning here [7]. These industrial cooperatives created to solve poverty problem in the Basque region in the first half of the twentieth century had very humble beginnings, but have over the decades flourished into a business empire comprising of high-technology companies, banking services, educational and retail network with the lowest failure rate that is unrivaled by any known model in the history of capitalism.

Even more remarkable is the resource mobilization abilities of this system and equity consideration on the wealth created within the system. They are simply owned, managed and controlled by the members, which is a solution to a problem conventional corporate governance system has failed to conclusively solve. Aligning interests of corporate owners and agents (managers/controllers) is a perennial problem that will never go away in conventional corporate ownership and control structures. Yet despite the success, lowest failure rate, and perfect alignment of ownership and control, and positive income distribution qualities of the Mondragon, very scanty attention has been paid to the model in economic development and management studies. There is real value in paying attention to the Mondragon model with a possible widespread adoption, albeit with modification in other parts of the globe to solve the problem of poverty. The Mondragon success if a perfect example of how social innovation works. Such social innovation is expressed in how people organize themselves to mobilize human capital, financial and even technological resources to attain sustainable social and economic outcomes.

12.2.0.6 Bringing Social Entrepreneurship to the Center of Public Policy

The power of private enterprises to make transformative contributions to society by way of efficiently producing and distributing goods and services is indisputable. It is notable that the whole concept of businesses as legal entities partly has its roots in being charged with fulfilling specific public policy goals such as conducting trade to secure the very much nationally needed raw materials and goods from distant oversea locations. Even when it became mainstream practice that businesses mainly the private interests of its owners, that of returning profits, it has not been a lost fact and an issue that will never wither away, that doing some good to society is something valuable. It is also evident that the minimalist school thought with regard to corporate social responsibility (CSR) has not gained the upper hand in public acceptance of role of corporations. Rather, more attention has been directed at corporate social responsibility as evidenced by the commitment of large companies to show their corporate social responsibility activities to gain positive public perception. It is also notable that with developments in information technology and accompanying growth in consumer awareness, ethical considerations by consumers will play a critical role in decisions they make in market transactions. However, in as much CSR has moved mainstream in the practice of market transactions today, standard best practices are hard to arrive at. Difficulty exists in distinguishing genuine CSR from mere public relations exercises.

Conversely tensions have reigned high between the pursuit of dividend maximization goal for shareholders on the one hand, and commitments to CSR on the other hand. Invariably, it can also be said that CSR, both as a concept and in practice has not gained full trust of society anywhere. This is simply because CSR perception at societal level is very much dependent on acts of diverse corporate players. As such we get a super mixed bag of CSR activities ranging from the genuinely socially beneficial acts to mere public relations exercise. However before the turn of the twenty-first century, we have witnessed the advent of corporations founded on the resolve to put social returns first before profits. A new breed of entrepreneurs, rightly dubbed 'social entrepreneurs' are currently changing how private businesses can solve societal problems while also being able to make profits. Social enterprises are a mark of progress representing moving beyond the threshold of CSR to a new paradigm in which solving socio- economic problems of society takes precedence, and is compatible with profit making motive. In addition, social enterprises bring in an important funding sustainability model, which has always eluded Non-governmental Organizations (NGOs) in their operations around the globe. We can also rightly allude to the distinguishing quality of social enterprises in that they address the weaknesses in governmental system as expressed in unmet socio-economic needs, and also the shortcomings of conventional corporations. We may even assert that social entrepreneurship represents capitalism in its most dynamic response to doubts about the system as raised by the 2008 financial crisis and widespread economic failures engendered.

Social Enterprises Filling the Void Created by Market and Government Failure

The emergence and role of social enterprises provokes the question of whether all businesses should be taxed and over all be treated equally in public policy. This question is premised on the idea that businesses making critical, distinguished, and clearly measurable contributions need to be treated differently with regard to taxation and any dispensation of subsidies. The logic here is simplified, but without incurring the risk of distortion inherent in oversimplification. Governments have the basic function of taking care of public goods and services, and this is justified by tax collection and accompanying expenditure patterns. Even when governments extend subsidies to private firms, the guiding rationale is to serve critical public interest goals (that is if you discount meddlesome interest groups). Consequently, there is strong logic in supporting deliberate and positively discriminative treatment of social enterprises in public policy because these enterprises already serve society in areas where tax money would have been ploughed, but in most cases without desired outcomes. Having made this argument, we need to examine the reality of public policy landscape in which social enterprises operate. This is more so in developing countries where there is dire need for socio-economic intervention to solve the problem of poverty. Whereas there is a definite rise in cases of social entrepreneurships in developing countries, local entrepreneurs face the serious challenge of finding funds to invest and grow their social-return focused businesses.

It is also a standard occurrence that most governments have never learnt to distinguish social entrepreneurs from the profit-first type of entrepreneurs. As such most governments do not have public policy framework within which social enterprises can be nurtured and even encouraged through diverse set of incentives to satisfy unmet socio-economic needs. This reality needs to be reversed because as private enterprises continue to expand into playing the role of provision of both public goods and services, roles previously fulfilled by the public sector, albeit inadequately, it is notable that conventional enterprises too have generally failed to measure up to the need to conscientiously harmonize social and private profit motives. There are only few notable exceptions of innovative companies globally and at local levels that have mastered the balancing of profit motive and conscientious social responsibility.

The global public backlash against capitalism is actually a testimony to the frustrations and failures of current market systems to sustainably generate and allocate wealth to the various stakeholders. With regard to wealth generation, the current system perhaps is not as efficient as purported to be. Given that accounting systems are not comprehensive in coverage of all costs actually incurred in the due course of corporate activities, there many costs and externalities generated by firms. Externalization of costs is still rampant within the current system. Lets take for instance the revered practice of outsourcing, which involves producing goods at far-flung locations based on comparatively lower factor of production costs. Hardly has conventional accounting system been able to factor in costs of Carbon emission that happens when goods are transported or factoring in cost of environmental destruction likely to happen at production locations where environmental legal regimes are more often weaker than in the markets of industrialized countries. So we can assert that the current corporate system is fixated on short-term results and on passing on costs, which by default are not easily captured by conventional accounting system. Externalized costs are borne by the public in cases such as when the public must deal with downstream socio-economic problems engendered by environmental destruction or widespread unemployment caused by relocations of corporate activities.

Mainstream media is filled with lamentations about failure of capitalism, more so following the 2008 financial crisis and the economic destruction stemming from the crisis. It is however important to note that capitalism is driven by uncertainty, which itself compels the participants to be adaptive, if not innovative. Such adaptive nature of capitalism is what ensures that we can always have better forms socio-economic systems and institutions, that though have limitations and flaws, still remain to beat any other known and tried systems of organizing economic, social and political life.

There has been historically traceable changes in the main forms of capitalism, starting with state-led capitalism, liberalization and limited state intervention, extreme push for market mechanisms even where market systems have not taken roots, and now a rethinking of the issue the need for a balance between the push for unfettered markets and judicious state intervention.

Obviously, as with any change processes, tensions exist as to what should constitute the right balance. Social enterprises' express goals are to simultaneously pursue profits and social returns and therefore represent the right model out of the present crisis. We can therefore view social enterprises as part of the evolutionary adaptation of capitalism, and like in the history of other economic paradigm shifts, entrepreneurs have again stayed a head of state mechanisms in place either in the form of taxation or regulation. Social enterprises perhaps will transform the idea of doing good, or ethical consciousness as a requisite for conducting business in most communities, at least to gain access to customer loyalty and low-cost marketing. This might be even more so given the liberating impact of information technology on customers and the ease with which they can exchange market information.

So social enterprises will increasingly play a leading role in the twenty-first century in solving problem of global poverty given the inherent organizational fit harmonizing the pursuit of both social good and profits. The mainstay of social enterprises, we can assert is the ability to innovatively come up with economic models, which serve social good while also conferring private gains to entrepreneurs. As such, innovation which until now has been very much limited to private corporate activities will have been ushered into the mainstream as the default model for most poverty alleviation efforts.

For the paradigm shift described above to take its full course, governments around the world and other socio-economic institutions would have to fully come round to accept social entrepreneurship confer upon it its right place, that of being the most effective organizational form to solve the problem of poverty and other socio-economic challenges. It is indisputable that governments through policy instruments have powerful resource allocation influence, though this has yet to be displayed in a comprehensively desirable form including resource allocation and efficiency concerns. All said and done, only a few governments have fully recognized the socio-economic pivotal potential role of social enterprises, and have already taken measures and put policies in place to provide an enabling environment for nurturing that potential. Actually any such steps worth writing home about are only limited to some of the industrialized counties, which of course dont view poverty as a scourge requiring utmost urgent public policy intervention. These infant stage steps are specifically limited to industrialized countries like the United Kingdom, the United States of America and some European Union countries.

Majority if not all of the countries most afflicted by poverty are yet to fully understand, embrace and commit to coherent and pragmatic public policy positions aimed at nurturing the potential of social entrepreneurship. Yet most of these countries still groaning under the weight of debilitating poverty and accompanying social ills have also already witnessed the exhibits of the potential power of social entrepreneurship. For instance, India has had experience with a number of successful social entrepre-

Table 12.1 Social entrepreneurship cases in India

Sector	Cases
Health	• Dr. Govindappa Venkataswamy and Thulasiraj D. Ravilla established Aravind Eye Hospital in 1976. Impact-To date, it has treated more than 2.3 million outpatients, about two- thirds of them free • Narayana Hrudayalaya Institute of Medical Sciences and its network of hospitals run by Devi Shetty. Impact- has performed about three-dozen surgeries a day, 60 % are carried out at nominal cost or free of charge
Finance/Micro-finance	• Bhartiya Samruddhi Investments and Consulting Services (BASIX) started by Vijay Mahajan is the first microfinance project to lend to the poor—Impact—sustainability—the organization is still active and achieving its original goals
Education	• Barefoot College, started by Bunker Roy in 1972, has made innumerable school dropouts in villages into "barefoot" doctors, engineers, architects, teachers, designers and communicators. Impact—sustainability because the organization is still active and achieving its goal
Multi-Sector	• Technology Informatics Design Endeavour (TIDE) operated by S. Rajagopalan and Svati Bhogle supports the development of financially rewarding and environmentally- friendly methods invented by leading research institutions into thriving enterprises • Self-Employed Women's Association (SEWA) founded by Ela Bhatt in 1972 and provides financial, health, insurance, legal, childcare, vocational and educational services to poor self-employed women, who are its members

neurship cases (see Table 12.1, showing some examples of social entrepreneurship icons in India), but the government and its bureaucracy has yet to fully commit to a pragmatic public policy position targeting the nurturing of social entrepreneurship. We have to take complete note of the word pragmatic, because it strongly suggests attainment of desired result/s. In the case of Indian government, for decades there has been in place a policy targeting priority sector lending, a vital financial resource allocation mechanism effected by the Reserve Bank of India (RBI), and historically the RBI has had the conventional practice of having all public sector banks committing 40 % of their capital to lending to rural-development-programs focused banks. RBI even added a requirement that such lending be at below market rates.

It is however notable that generally there is no capital or sources of finances targeted to benefit social entrepreneurs in the form of startup seed capital. Other dangers lurk within the Indian government public policy. Lower borrowing rates can be quite an effective tool for channeling financial resources into priority economic sectors, such as microfinance, and other social enterprise ventures. However, there is need for caution against falling into an obsessive trap of thinking that microfinance is the antidote for poverty. There is the danger of over- borrowing due to easy credit situation, which would be simply overwhelming to the poor population with inadequate knowledge of finance, their rights in such transactions, and therefore are extremely vulnerable in every way conceivable. The second risk might be the resulting increased likelihood of a race to hijack micro-financial services to the poor by traditional financial intermediaries whose goals may be purely profit making. Even initial public offerings by microfinance institutions may not be an indicator of more financial resources to be channeled to serve deserving poor. Questions needs to be posed whether there is certain danger of encroachment by traditional private sector actors to take over social enterprises whose mainstay are patient capital and social returns.

Whatever the fears and future prospects might be over social enterprises issuing initial public offerings upon their wide market success, it is important for governments to chart clear public policy positions aimed at encouraging social enterprise startups through policy instruments, taxation and perhaps even regulatory systems. Other than governments, International Financial Institutions (IFIs) also need to focus more of their resources towards supporting social enterprises not just by having one-time competitions for funding, but providing all-year round windows for applications for access to financial and technical advisory services for startup social enterprises. It is also notable that most of the competitive events aimed at providing funding either in the form of grants, loans or equity financing to social enterprises tend to happen when the enterprises have already shown great and proven success. This may be viewed as valuable support extended to social enterprises, but upon a more critical assessment, might be viewed as mere willingness to be part of the celebration when the party is ready, but all the same tacitly represents unwillingness to participate in the difficult preparation stages. So it is imperative to point the need for not only national governments, but also IFIs to formulate highly supportive pragmatic support positions towards social enterprises. Such pragmatic positions need to be informed by the developments and changes in both structure and form of social enterprises.

12.3 Conclusion

Innovative solutions to global poverty are urgently needed given the frustrations and decades of limited success of some of the mainstream methods often advocated both by governments and international development agencies. Though practice of innovation has been confined to private sector actors, innovation as an approach to providing solutions to problems has a very relevant role in alleviating poverty. Social enterprises are the epitome of innovation being applied to global poverty alleviation efforts, but innovation approach needs to be extended to other areas including public policy in as much as that would seem difficult. Extending innovation approach to public policy obviously faces some difficulty of public policy inherently being grounded on majority and representative consensus, while innovation traditionally require radical departure from extant practices and thought systems. All in all, innovation approach brings the advantage of being able to look at solutions as work in progress, and therefore it becomes possible to avoid the big hammer syndrome as was typical with the infamous World Bank structural adjustment programs and shock therapies.

References

1. Baumol W (2002) The free-market innovation machine: analyzing the growth miracle of capitalism. Princeton University Press, Princeton
2. Japan Bank for International Cooperation, Symposium: Rethinking Japanese Companies Business Strategy for Middle Income Segment in Emerging Market Countries. Retrieved from: http://www.jbic.go.jp/en/about/topics/2009/0224-01/index.html
3. Karnani A (2010) Failure of the libertarian approach to reducing poverty. Asian Bus Manag 9(1):5–21
4. Konomoto S (2009) Rebuilding competitive strategy aimed at emerging markets. NRI Papers, No. 145 August 1
5. Lewis WA (1954) Economic development with unlimited supplies of labor. Manchester Sch Econ Soc Stud 22:139–91. Retrieved from: http://www.nri.co.jp/english/opinion/papers/2009/pdf/np2009145.pdf
6. Melamed M, Hartwig R, Ursula Grant U (2011) Jobs, growth and poverty: what do we know, what dont we know, what should we know?. Overseas Development Institute, London
7. Monks RAG, Minow N (2008) Corporate Governance. Wiley, Chichester
8. Porter ME, Kramer MR (2011) Created shared value. Harvard Bus Rev 89(1/2):62–77
9. Prahalad CK (2005) Fortune at the bottom of the pyramid: eradicating poverty through profits. Wharton School Publishing, Philadelphia
10. Quibria MG (2002) Growth and poverty: lessons from the East Asian miracle Revisited, Asian Development Bank Institute Research Paper 33
11. Stiglitz JE, Yusuf S (2001) Rethinking the East Asian miracle. World Bank and Oxford University Press, Oxford
12. Tidd J, Bessant J (2009) Managing innovation: integrating technological, market and organizational change. Wiley, Chichester
13. World Bank (1993) The East Asian Miracle: Economic Growth and Public Policy

Part III
Environment Link

Chapter 13
Water in the Poverty-Environment Nexus

Accounting for water for environmental requirements shows that abstraction of water for domestic, food and industrial uses already have a major impact on ecosystems in many parts of the world, even those not considered "water scarce" [75]. In the absence of coordinated planning and international cooperation at an unprecedented scale, therefore, the next half century will be plagued by a host of severe water-related problems, threatening the well being of many terrestrial ecosystems and drastically impairing human health, particularly in the ***poorest regions*** of the world [49].

13.1 Water in Relationship to Poverty and Environment

Water plays a significant role both in poverty alleviation and environmental related issues. With regard to poverty, food policy contributes toward the overall goal of eradicating extreme poverty and hunger, and is dependent on water as one of the drivers for realizing such a goal (see, e.g., [42]). It provides a key component of agricultural requirement for food production as well as for domestic use. If people are able to provide food, then absolute poverty measure of a society could arguably be low. Indeed, that water and food security issues are closely linked has been pointed out, e.g., by Hanjra and Qureshi [42]. Rijsberman [75] is even more direct by stating that water will be the major constraint for agriculture in the coming decades, more so in the continents that experience high percentage of poverty such as Asia and Africa.

The problem, however, is that more focus is placed on the water crisis, which is likely to be fuelled by increased population leading to about 1.2 billion lacking water. To the contrary, less documentation exist for the large part of the population that live in rural areas below the poverty line. Duraiappah [31], identified through literature review, identified

- water shortage, and
- water pollution

DOI: 10.1007/978-3-642-39733-2_13,

as the two major issues within the water sector that plays an important role in the poverty environmental degradation nexus. Understanding the link between poverty and environment in relation to water, therefore, calls for a more closer look at the whole issue of water scarcity in relation to the poor rather than a holistic approach where such scarcity is viewed in terms of the total population. This in essence calls for poverty eradication measures to incorporate issues that will address water scarcity and insecurity. This is partly because if the poor are unable to access clean and safe water, food productivity will be hampered leading to hunger and malnutrition hence less productivity. Also, lack of access to safe drinking water by the poor will only aggravate the risk of water borne diseases such as cholera. When the low income group are faced by such water-borne diseases, their productivity deteriorates and they risk losing their jobs and hence sources of income. The expected outcome of loss in income is that the low income earners will experience economic and social hardship, which over time results in poverty, thus exemplifying how environmental degradation causes poverty (e.g., [31]).

13.2 Diminishing Fresh Water Resources

13.2.1 Status

Fresh water is one of the basic necessities without which human beings cannot survive since it is key to the sustainability of all kinds of lifeforms. Water has multiple uses namely; nutritional, domestic, recreational, navigational, waste disposal and ecological as it is a habitat for living and non-living organisms (biodiversity) etc. And, because it is indispensable to different sectors including manufacturing, agriculture, fisheries, wildlife survival, tourism and hydroelectric power generation, it is a vital factor of economic production. For many countries, most freshwater endowments encompass surface waters, groundwater, wetlands and glaciers. Surface water bodies include lakes, rivers, swamps, springs, dams and water pans dispersed within different basins. In general, people living in the vast arid and semi-arid parts of the world rely heavily on groundwater resources. Furthermore, groundwater is also an important supplementary source of water for many urban households in most developing countries.

At a global scale, although much of the Earth is covered by water, most of it is unsuitable for human consumption, since 96 % of it is found in the saline oceans. According to the U.N., only 2.5 % of the roughly 1.4 billion cubic kilometers of water on Earth is freshwater, and approximately 68.9 % of the freshwater is trapped in glacial ice or permanent snow in mountainous regions—the Arctic and Antarctica. Roughly 30.8 % is groundwater, much of which is inaccessible to humans, and the remainder 0.3 % comprise surface waters in lakes and rivers [95]. Of these 0.3 % available for human and animal consumption, much is inaccessible due to unreachable underground locations and depths [44]. Jury and Vaux Jr. [49] caution

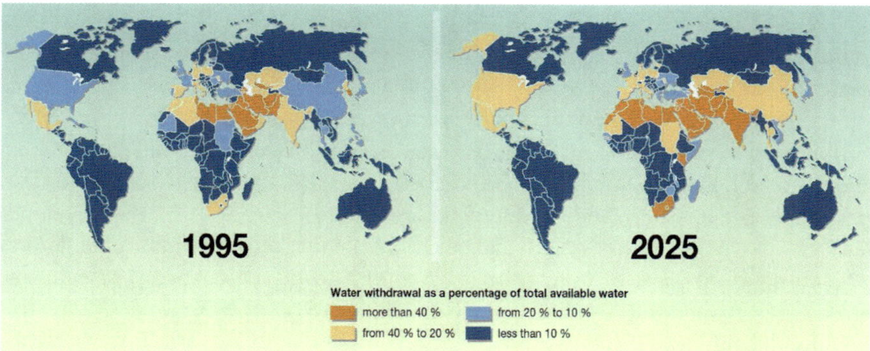

Fig. 13.1 Increased global water stress. *Source* [72]

that focusing on the global freshwater storage resource alone is misleading because much of the water is inaccessible, and suggest that humanity's freshwater resource consisting of rainfall used to grow crops, accessible groundwater, and surface water be considered.

13.2.2 Water Scarcity

Although no common definition of water scarcity exists, Rijsberman [75] defines a water insecure person as an individual who does not have access to safe and affordable water to satisfy his or her needs for drinking, washing, or livelihood. A water scarce area is then said to be an area where a large number of people are water insecure for a significant period of time (e.g., Rijsberman [75]). Rijsberman [75] points out that an area qualifies as a water-scarce area depending on, e.g., (i) the definition of people's needs and whether the definition takes into account the needs of the environment, the water for nature, (ii) the fraction of the resource that is made available (or could be made available) to satisfy these needs, and (iii) the temporal and spatial scales used to define scarcity.

Even though Rijsberman [75] argues that at a global scale, and from a supply and demand perspective, it is still debatable whether water scarcity is fact or fiction, it is incontestable that fresh water is increasingly becoming a scarce resource and shortages could drive conflict as well as negatively hit food and energy production [49] (see Fig. 13.1). To underscore the seriousness of water shortage, the World Bank in its 1992 World Development Report pointed out that 22 countries faced severe water shortage while furthe 18 were in danger of facing shortages if fluctuation in rainfall persisted [31]. Recent studies, e.g., Omondi et al. [65, 66] point to fluctuation in rainfall in parts of Greater Horn of Africa where most low income earners depend on rain fed agriculture for food production, thus suggesting that the water shortage problem is not fading away any time soon. A more gloomy picture is the estimate that by 2050, about two billion people will be short of water, a potential cause of conflict [47]. That a large population of the world will face water scarcity is supported by several studies (e.g., [2, 3, 97–99]), with the most likely to be affected dwelling in Africa, Asia and the Middle East (see, e.g., [104]).

By reviewing several publications on water in relation to poverty and environmental degradation nexus, Duraiappah [31] presents activities that lead to water shortage. ***First***, there is the issue of the commercial interest of the rich and wealthy driven primarily by power, greed and wealth that benefits from market and institutional failures (e.g., absence or misuse of water property rights). ***Second***, Duraiappah [31] attribute over usage of water supply by the small holders (the poor) to water subsidies that provided incentives. Of the two groups, the poor are more likely to be affected by water shortage as compared to the rich, a situation that could contribute to environmental degradation on the one hand, which would lead to poverty on the other hand, i.e., endogenous poverty causing environmental degradation (see, e.g., [31]).

This picture leads Jury and Vaux Jr. [49] to warn that "without immediate action and global cooperation, water supply and water pollution crisis of unimaginable dimensions will confront humanity, limiting food production, drinking water access, and the survival of innumerable species on the planet". They list the following four factors to support their hypothesis that the world is headed toward a future where billions of people will be forced to live in places where their food and water requirements will not be met [49]:

1. Unlike estimates of the global supply of scarce minerals or underground fuels, which are surrounded by uncertainty, planetary supplies of water are relatively well characterized. No large deposits of groundwater await human detection in readily accessible locations, so that any new resources discovered will be very expensive to develop.
2. Many vital human activities have become dependent on utilizing groundwater supplies that are being exhausted or contaminated.
3. Much of the population growth projected for the next century will occur in areas of greatest water shortage e.g., Africa and Asia, and there is no plan for accommodating the increases.
4. Global economic forces are luring water and land from food production into more lucrative activities such as biofuel (see, e.g., Payne 2010 and Neff et al. 2011), while at the same time encouraging pollution that impairs drinking water quality for a large and ever-growing segment of the population.

So vital are water resources that it is difficult to discuss any monitoring of the environment without it. Evidently, the management of water resources conflicts, focusing on negotiation, mediation and decision-making processes, in order to prevent, manage and resolve water conflicts is emerging as a contemporary and topical research issue.

Physical water scarcity is evident in densely populated arid areas in many parts of Central and West Asia, and North Africa with projected availabilities of less than 1000 m^3/capita/year [75]. This has a wide range of negative impacts and ramifications. For instance, it results in higher incidences of waterborne, water-related or sanitation-related diseases such as malaria, diarrhoea and skin infections. In addition, as has been pointed out before, increasing cases of water conflicts, especially between pastoralist and farming communities along lower and upper river basins, as witnessed in water stressed countries like Kenya [24] (with a renewable fresh water per capita endowment estimated at about 548 m^3/capita/year [102]), are also likely to heighten food insecurity. On the gender scale, since women are the primary collectors, users and managers of water for domestic use in most developing countries, water scarcity disproportionately affects them because it is they who have to trek long distances, often all day, in search of water. Against the above background, is the growing realization today that availability of sufficient, accessible and quality potable water is not only a matter of great socio-economic and political importance, but also one of fundamental human rights, see e.g., [20, 39, 82, 106] etc.

13.3 Need for Geospatial Monitoring of Water Resource

The importance of water as a resource, therefore, calls for sound environmental conservation measures that enhance its protection and management. It is in relation to this that the World Bank, as an emerging priority of its lending framework, decided to broaden the development focus in its 1993 "Water resource management policy paper" to include the *protection* and *management* of water resources in an environmentally sustainable, socially acceptable, and economically efficient manner [102]. The protection and management of water resources calls for an elaborate and well established management and monitoring program.

Information about water resources and the environment is inherently geographic. Maps, whether on paper or in digital Geographical Information System (GIS) formats, continue to be the medium for the expression of engineering plans and designs. This is because we are basically concerned about the spatial distribution and character of the land and its waters. Johnson [48] argues that weather patterns, rainfall and other precipitation, and resultant water runoff are primary driving forces for land development, water supplies, and environmental impacts and pollution. Our water resources systems comprise dams and reservoirs, irrigated lands and canals, water supply collection and distribution systems, sewers and storm water systems, and floodplains. These systems are designed in response to a complex mix of topography and drainage patterns, population and land use, sources of water, and related environmental factors [48].

In general, the planning and engineering design processes used in the development and management of water resources involve different levels of data abstraction. Data are collected and used to characterize the environment at some level of detail, or scale. In seeking to make decisions about plans and designs, data must be collected to describe the resource, and procedures or models must be developed to predict the resultant changes. These data and models help us understand the real world, and this understanding guides our decision making [48].

According to Taylor and Alley [92], essential components of a water level monitoring program include; *selection of observation wells*, *determination of the frequency of water level measurements*, *implementation of quality assurance*, and *establishment of effective practices for data reporting*. In selecting the observation wells, the authors state that the decisions made about the number and locations of observation wells are crucial to any water-level data collection program [92]. In regard to locations, Global Navigation Satellite Systems (GNSS) satellites could contribute in generating a fast and accurate survey of well location-based data. These data could then be integrated with other information such as water level in a GIS system to enhance the accessibility of water level data, where the GIS plays the role of depicting the locations of the observed wells relative to pertinent geographic, geologic, or hydrologic features, e.g., [92].

Taylor and Alley [92] present areas where the monitored ground water levels could be used. Some of these include: determination of the hydraulic properties of aquifers (aquifer tests); mapping of the altitude of the water table or potentiometric surface; monitoring of the changes in groundwater recharge and storage; monitoring of the effects of climatic variability; monitoring of the regional effects of groundwater development; statistical analysis of the water level trends; monitoring of the changes in groundwater flow directions; monitoring of the groundwater and surface water interaction; and numerical (computer) modeling of groundwater flow or contaminant transport.

Information on the spatial and temporal behavior of terrestrial water storage, therefore, is crucial for the management of local, regional and global water resources. This information will [74]:

- Enhance sustainable utilization of water resources by, e.g., farmers, urban consumers, miners, etc.
- Guide water resource managers and policy makers in the formulation of policies governing its sustainable use, conservation and management. In particular, state water managers are more informed in regulating the utilization of water, e.g., for industrial and irrigation purposes.
- Benefit local environmental monitoring, management policies and practices that ensures a balance between sustainable utilization and environmental conservation and protection. Changes in water availability impacts upon the environment in several ways, e.g., any significant imbalance in its level affects the ecological system by influencing salinity, land subsidence and the vulnerability of wetlands ecosystem among others.
- Benefit various government agencies at various levels (national, provincial, and local) by providing data that enhances and compliments their work. Such agencies include departments of *water*, *agriculture*, *weather forecasting* and *climate* studies, and so forth.

The conservation and management of water is of paramount importance in areas with arid or semi-arid climates, which include many parts of Australia, especially in times of severe drought, as experienced in Murray Darling Basin [74]. In 2006 for example, Australia faced its worst drought in a century as seen from daily reports that were emerging in both the local and international media. A more grim picture of the future of the water situation for Australia was to follow from the IPCC [45] report, which stated that Australia's water crisis will worsen in the coming years due to drought! There clearly exists an urgent need to have efficient monitoring technique(s) that will enhance the analysis of water scarcity at river basin or more localized scales. Indeed, this argument is supported by Rijsberman [75] who states that the global analysis of water scarcity is of very limited use in assessing whether individual or communities are water secure. To this effect, Rijsberman [75, and the references therein] states:

> The river basin is more and more adopted as the appropriate scale to understand the key processes with increasing water scarcity as human use goes up to the point where basins "close".

One such technique that has emerged supreme in monitoring changes in stored water at river basin scales, is the use of Gravity Recovery And Climate Experiment (GRACE) satellites (Sect. 13.4.2; see details in [12, 13]), which is demonstrated in the examples to follow.

Timely and precise information on the changes in stored water at smaller (localized) scales of economical values, e.g., urban consumption, agriculture, industries, and mining to within 10–14 days (so far achievable by GRACE satellite) will enhance sustainable conservation and management of this precious *dwindling* resource.

The availability of techniques that delivers information on the changes in stored water at a more local scale, is the first step towards realizing an efficient water society. Water resource managers are able to make decisions based on timely and accurate knowledge; thereby saving considerable resources that are often spent as a penalty of inefficient decisions based on a lack of information. In the south-western wheat belt of Australia, for example, accurate knowledge of changes in stored water will be beneficial to the sustainable utilization of water, while at the same time realizing the economic contribution of wheat farming to the overall Gross Domestic Product (GDP). A blind focus on the GDP's growth without paying attention to the state of salient contributors such as water stored in aquifers is detrimental, since a fall in the amount of the available water in such areas would definitely mean reduced yields.

Since the entire system of stored water is coupled within the hydrological cycle (Fig. 13.2), hydrologists will be in a position to better understand their local hydrological cycle, thanks to information at localized levels. Hydrologists will also be able to use such information to refine and calibrate local-scale models, e.g., rainfall runoff models [32], for further improvement in their hydrological cycles. This will also contribute to our understanding of the impacts of climate change on regional and global hydrological cycles.

Environmental studies have a chance of greatly benefiting from information about changes in stored water. It is widely acknowledged that stored water (surface and groundwater) plays a key role in sustaining natural biodiversity and the functioning of the environment as a whole. Knowledge of the changes in water level is therefore essential for the very survival of the entire ecosystem, which could be adversely

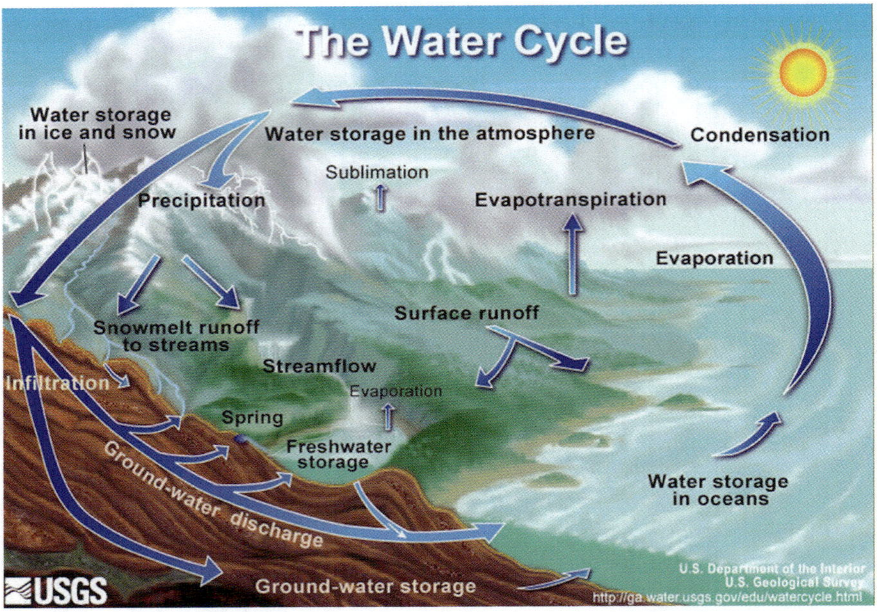

Fig. 13.2 Components of hydrological cycle that lead to temporal variations in the gravity field. *Source* US Geological Survey (USGS)

affected by extreme change in stored water. In wetlands, for example, some vegetation and ecosystems have been known to respond to water level fluctuations [25].

Accurate monitoring of changes in stored water at smaller wetland scales will thus help in the preservation and conservation of such wetland ecosystems. Changes in water level also brings with it environmental phenomena such as salinity, compacting of aquifers due to the removal of water causing land subsidence, and changes in the properties of the top 5 cm of soil. Information on changes in stored water thus contributes enormously to the environmental conservation and protection. Remote sensing and GIS can be used to monitor the water quality. This is possible by employing multispectral, multitemporal image data and analyzing parameters such as the distribution of suspended sediment, turbidity and chlorophyll. These indicators can be determined throughs regression analysis.

13.4 Geospatial Monitoring of Stored Water at Basin Scales

In Awange [12] and Awange and Kiema [13], the *geoid* is introduced as a fundamental physical surface to which all observations are referred to if they depend on *gravity*, and whose shape is influenced by inhomogeneous mass distribution within the interior of the Earth [55, p. 29]. In the discussion that follows, the concept of *gravity field variations* is related to hydrological processes. Measurements of the time-varying gravity field by LEO (low earth orbiting) satellites, e.g., GRACE discussed in Sect. 13.4.2 are the key to the contribution of space monitoring of changes in water levels at basin scales. Such techniques now enable the monitoring of groundwater recharge, which is the most important element in groundwater resource management and could also be applicable to monitoring salinity management measures at the catchment level (see Sect. 15.5.2). For example, in 2009, GRACE satellites showed that north-west of India's aquifers had fallen at a rate of $0.3048\,\mathrm{m\,yr^{-1}}$ (a loss of about $109\,\mathrm{km}^3$ per year) between 2002 and 2008 (see, e.g., Fig. 13.3).[1]

13.4.1 Gravity Field Changes and the Hydrological Processes

The hydrological cycle (Fig. 13.2) refers to the pathway of water in nature, as it moves in its different phases through the atmosphere, down over and through land, to the ocean and back to the atmosphere [23]. The associated variations in gravity field are therefore caused, e.g., by

- the redistribution of water in the oceans, including, e.g., El Niño and Southern Oscillation (ENSO) events,
- movement of water vapour and other components in the atmosphere,
- seasonal rainfall; snow and subsequent drying and melting,
- groundwater extraction, or
- drying and filling of lakes, rivers, and reservoirs.

13.4.2 Sensing Changes in Stored Water Using Temporal Gravity Field

The potential of using the relationship between temporal gravity changes and hydrology (Fig. 13.2) was first recognized by Montgomery [58] who estimated specific yield through a correlation between gravity and water-level changes [56]. In 1977, Lambert and Beaumont [53] used a gravity meter to correlate groundwater fluctuations and temporal changes in the Earth's gravity field. Goodkind [40] recorded observations from seven super conducting gravimetric stations to examine non-tidal variations in gravity and noted that at one of the stations (Geysers geothermal station), much of the variation could be correlated with rainfall and seismic activity. Such measurements had not been possible before the advent of super conducting gravimeters, thus providing evidence of the existence of temporal variation in gravity.

In 1995, while estimating the atmospheric effects on gravity observations around Kyoto, Mukai et al. [59] noted that changes in gravity around the station could

[1] The Economist, September 12th 2009, pp. 27–29: Briefing India's water crisis.

have been caused by changes in underground water. In the same year, Pool and Eychaner [69] assessed the utility of temporal gravity-field surveys to directly measure *aquifer-storage* changes and reported gravity changes of around 100–134 μGal, equivalent to 2.4–3.2 m of water column, considering infinitely extended sheet approximation. Their results from the analysis of changes in stored water in the aquifer indicated an increase in the gravity field of 158 ± 6 μGal when the water table rose by about 17.7 m, providing further evidence of the possibility of using temporal gravity-field surveys to monitor changes in stored aquifer water. In fact, according to Bower and Courtier [22] who analyzed the effect of precipitation on gravity and well-levels at a Canadian absolute gravity site, 90 % of the gravity variation was found to be due to the effects of precipitation, evapotranspiration and snow-melt.

The last decade has also recorded increased use of temporal gravity field studies in monitoring changes in stored water, see e.g., [33]. It saw the beginning of satellite missions dedicated to monitoring temporal variations in the gravity field. Smith et al. [85] investigated the ability of ground-based gravity meters to monitor changes in soil moisture storage.

Moving from local tests to regional, a different application of gravity surveys was investigated by Damiata and Lee [29], who simulated the gravitational response to aquifer hydraulic testing. The synthetic system was composed of an unconfined shallow aquifer and the purpose of the investigation was to assess the potential of the gravity measurements for detecting groundwater extraction. Draw-down due to pumping causes a decrease in mass and consequently in gravity measured at the surface. The results showed that the gravitational response to aquifer testing could be used to monitor the spatial development of the draw-down cone. For the configuration considered in the investigation, the signal was of the order of tens of μGals and could be detected up to several hundred meters away from the pumping well.

Water storage changes, such as changes in soil moisture, snow and ice cover, surface and groundwater, including deep aquifers, can be monitored either by in-situ observations or indirectly through changes in gravity [91]. While in-situ observations provide valuable localized information, they suffer from limited spatial coverage for regional to continent-wide studies [77]. Any change in water storage also manifests itself in a change in the gravity field. This property can be used to infer water-storage changes from time-variable gravity observations as demonstrated by Rodell and Famiglietti [76] for 20 globally distributed drainage basins of sizes varying from 130,000 to 5,782,000 km^2 to assess the detectability of hydrological signals with respect to temporal and spatial variations. Space-borne techniques can provide time-variable gravity observations on a regional and global scale, thus allowing for large-scale water storage monitoring and the ability to close the 'gaps' between locally limited in-situ observations [32].

Since the launch of the GRACE satellite mission in 2002 (see [12, 13]), a new powerful geospatial tool for studying temporal gravity field changes has become available, and numerous articles assessing the potential of GRACE recovering hydrological signals have been published, see e.g., Awange et al. [10, and the references therein]. Tapley [91] provided early results of the application of the GRACE products for detecting hydrological signals in the Amazon-Orinoco basin. Following these results, numerous other authors have subsequently applied GRACE to detect hydrological signals in various situations and locations, see references in [10].

For instance, Ramillien et al. [70, 71] and Andersen et al. [4] investigated the potential of inferring inter-annual gravity field changes caused by continental water storage changes from GRACE observations between 2002 and 2003, and compared these changes to the output from four global hydrological models. It was possible to correlate large scale hydrological events with the estimated change in the gravity field for certain areas of the world to an accuracy of 0.4 Gal, corresponding to 9 mm of water, see also [4, 62, 103].

Syed et al. [90] examined total basin discharge for the Amazon-Orinoco and Mississippi river basins from GRACE, while Rodell et al. [78] estimated groundwater storage changes in the Mississippi basin. Crowley et al. [28] estimated hydrological signals in the Congo basin, while Schmidt et al. [83] and Swenson et al. [87, 89] used GRACE to observe changes in continental water storage. Winsemius et al. [101] compared hydrological model outputs for the Zambezi River Basin with estimates derived from GRACE. Monthly storage depths produced by the hydrological model displayed larger amplitudes and were partly out of phase compared to the estimates based on GRACE data. Likely reasons included leakage produced by the spatial filtering used in the GRACE data, and the difficulty to identify the time of satellite overpass as opposed to simply averaging over the whole period. Awange et al. [9] used GRACE to study the fall of Lake Victoria's water level in Africa. This last example will be elaborated upon in more detail in Sect. 13.5.4.

As already stated, GRACE satellites detect changes in the Earth's gravity field by measuring changes in the distance between the two satellites at a 0.1 Hz sampling frequency. The variation in the distance between the two twin satellites caused by gravitational variations above, upon, and within the Earth all have an effect on the satellites. This variation in gravity could be due to *rapid* or *slow* changes caused, for example by the redistribution of water in the oceans, the movement of water vapor and other components in the atmosphere, the tidal effect of the Sun and Moon, and the displacement of the material by earthquakes and glacial isostatic adjustment. The data therefore must be processed to isolate these effects so as to retain only those which correspond to the process of interest, in this case, terrestrial water storage changes, see e.g., [18]. In the following examples, the application of GRACE satellites to monitor stored water resources are illustrated.

13.5 Monitoring the Dwindling Water Stock

13.5.1 Groundwater Depletion in Northern India

Northern India and its surrounding is home to approximately 600 million people and probably the most heavily irrigated in the world [93]. India's water crisis has been reported in various forums. For instance, the New York Times of 29th September 2006 reported that the water crisis, which is decade in making, has seen an increase in recent years fuelled by exploding population growth, high sprawling cities, and a vast and thirsty farm belt.[2] If this water situation report did not worry the ruling elites in India, then the news reported on water situation by the Economist on 12th September 2009 should have sounded more alarm. In it, the Economist wrote:

> India's water future was worrying even without climate change. Despite daunting seasonal and regional variations, it should have ample water for agricultural, industrial and household use. But most of it falls, in a remarkably short time, in the wrong places. India's vast task is therefore to trap and store enough water; to channel it to where it is most needed; and, above all, to use it there as efficiently as possible. And on all three counts, India fares badly. Without huge improvements, according to a decade-old official estimate, by 2050, when its population will be a shade under 1.7 billion, India will run short of water.[3]

Failure to properly manage rain water led to another more serious problem, the depletion of groundwater. The unsustainable use of groundwater by both the rich and the poor is partly due to poor water policy from the government thus echoing the sentiments of Duraiappah [31] in Sect. 13.2.2. To pump the water, electricity is required. The government has facilitated the availability of free or cut-price electricity that has seen groundwater extraction escalate at a rate that led to nearly a third of India's groundwater blocks to be defined as critical, semi-critical or over exploited.[4] In addition to free or cheap electricity, farmers do not pay for the groundwater they extract. Other causes of water loss are attributed, e.g., to leaking pipes and taps, over-exploitation of canal waters, the success of the 1960 green revolution, and inefficient and corrupt water-board officials.[5]

Such depletion of groundwater resources only mean that the poor of India, most of who rely on it for irrigation and domestic use are the ones mostly affected. No wonder the Economist continues:

> This is storing up trouble. As bore-holes run dry, as those over the hardrock aquifers of southern-central India do on a monthly basis, many poor people may be deprived of safe drinking water. Currently, 220 m Indians lack this. Not all India's groundwater is potable anyway; in places, it is getting seriously polluted. And India's groundwater reserves will be especially missed when climate change makes surface-water sources even more sporadic. Their depletion will accentuate this, with springs, which could have provided a trickle of run-off during the extended dry seasons, increasingly failing.[6]

As one of the solutions, The Economist points out to sharing of information on rainfall, groundwater levels, and cropping by farmers so as to regulate the use of water and cites an endeavour that brings together about 25,000 farmers in Andhra Pradesh

[2] The New York Times 29 September 2006: In Teeming India, Water Crisis Means Dry Pipes and Foul Sludge.

[3] The Economist, 12th September 2009; India's Water Crisis.

[4] The Economist, 12th September 2009; India's Water Crisis.

[5] The Economist, 12th September 2009; India's Water Crisis.

[6] The Economist, 12th September 2009; India's Water Crisis.

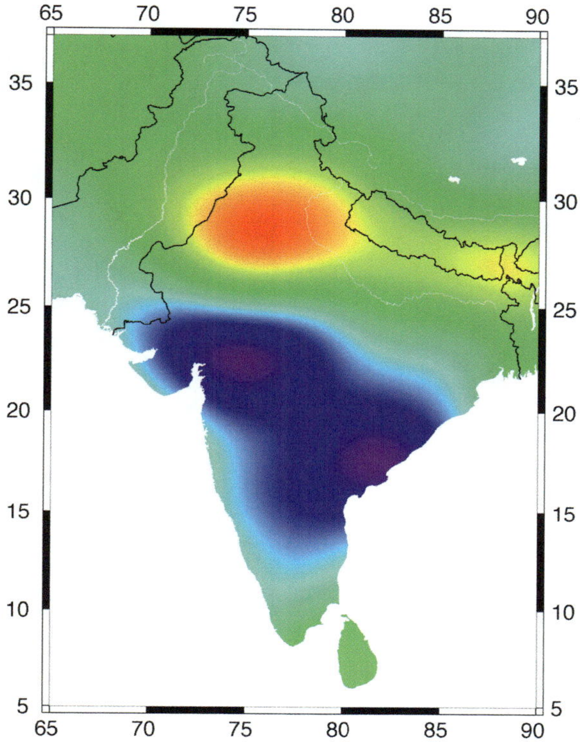

Fig. 13.3 Groundwater changes in India (2002–2008) as seen from GRACE satellites' data. Groundwater recharge is indicated by *blue* while depletion is indicated in *red*. *Source* NASA (I. Velicogna/UC Irvine)

in India.[7] Geospatial GRACE satellite technique discussed in Sect. 13.4.2 is useful in revealing changes in groundwater at larger spatial coverage that can influence policies on water for a highly populated country like India that relies heavily on groundwater for its agricultural production. In 2008, using the GRACE satellites, Rodell et al. [79] established that groundwater depletion from August 2002 to October 2008 was equivalent to a net loss of 109 km^3 (see Fig. 13.3), a fact attributed to unsustainable use in irrigation and other anthropogenic activities. The findings of Rodell et al. [79] were supported by those of Tiwari et al. [93], which revealed a steady large scale mass loss that is also attributed to groundwater extraction.

13.5.2 Understanding the Decline of the Ramser-Lake Naivasha, Kenya

13.5.2.1 The Lake Naivasha Basin

Lake Naivasha (00° 40′ S-00° 53′ S, 36° 15′ E-36° 30′ E) is the second largest fresh water lake in Kenya with a maximum depth of 8 m. It is situated in the Central

[7] The Economist, 12th September 2009; India's Water Crisis.

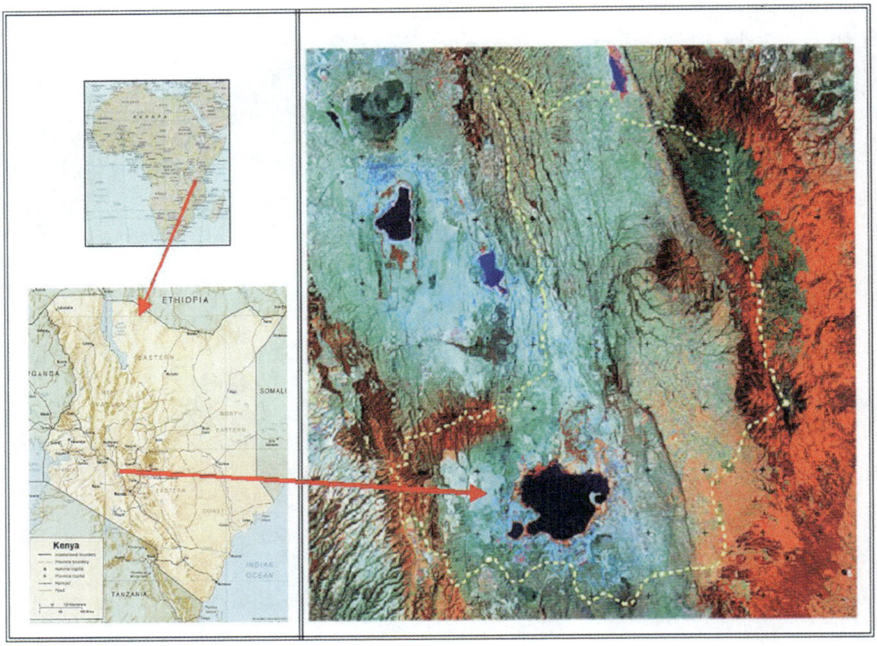

Fig. 13.4 Location map of the Lake Naivasha Basin. *Source* Becht et al. [16]

African Rift Valley at an altitude of 1890 m above sea level and is approximately 80 km northwest of the Kenyan capital, Nairobi. Its basin (Fig. 13.4) lies within the semi-arid belt of Kenya with mean annual rainfall varying from about 600 mm at the Naivasha township to some 1,700 mm along the slopes of the Nyandarua mountains, with open water evaporation estimated to be approximately 1,720 mm/year [16]. Mount Kenya and the Nyandarua Range capture moisture from the monsoon winds, thereby casting a significant rain shadow over the Lake Naivasha basin [16]. Unlike Lake Victoria, which has its highest rainfall during the March-April-May (MAM) wet season, e.g., [8, 9], Lake Naivasha basin experiences its highest rainfall period during April-May-June (AMJ). There is also a short rainy season from October to November. The lake's levels therefore follow this seasonal pattern of rainfall cycle, with changes of several meters possible over a few months. Imposed upon this seasonal behaviour are longer-term trends, for example, there has been a change in the lake's water level of 12 m over the past 100 years [16].

The lake is fed by three main river systems: the Gilgil, the Malewa and the Karati, the last of which only flows during the wet season (see Fig. 13.5). There is also a groundwater inlet into the lake from the north, and an outlet to the south, which when combined with the river systems and the biochemical and geochemical sedimentation processes that remove ions such as sulphates and carbonates from the water, results in the freshness of the lake [34, 43]. Becht et al. [16] state that whereas a small portion of the groundwater evaporates and escapes in the form of fumaroles in the geothermal

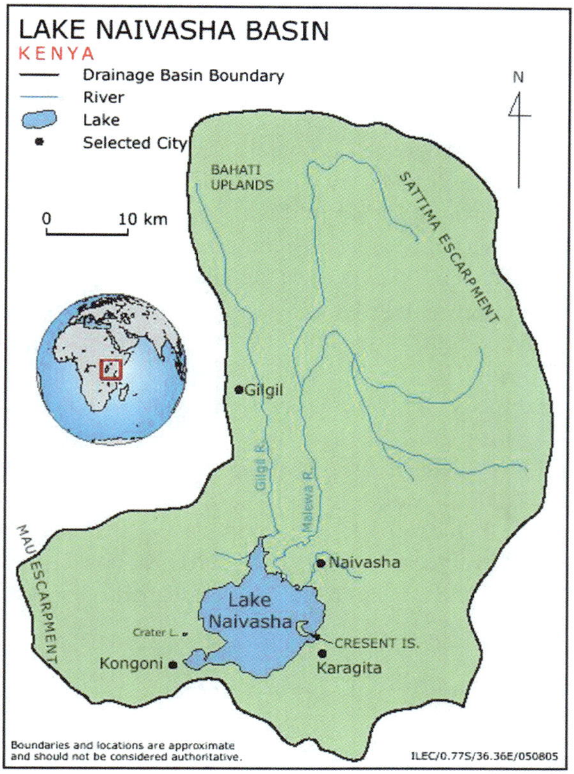

Fig. 13.5 The Lake Naivasha drainage system. Three rivers flow into the Lake; Gilgil, Malewa, and Karati. The Lake's outlet is through an underground system to the south. *Source* Becht et al. [16]

areas, the remaining water flows into Lakes Magadi and Elmentaita, taking thousands of years to reach them. The basin's water balance has been calculated from a model based upon long-term meteorological observations of rainfall, evaporation and river inflows [14]. This model reproduced the observed level from 1932 to 1982 with an accuracy of 95 % of the observed monthly level, differing by 0.52 m or less [46]. This pattern was, however, noticed to deviate after 1982 and by 1997, the difference had reached 3–4 m [16]. In fact, the onset of this reduced ability to model the lake's level coincides with the increase in horticultural and floricultural activities.

In general, three contemporary global water issues can be identified as occurring in this region, namely *water scarcity/availability*, *water quality*, and *water security*. Several previous works have focused on the problem of water quality and competition for water resources within the area. Water quality studies have endeavored to analyze the physical, chemical and biological characteristics of the water, which represents a measure of the condition of water relative to the requirements of one or more biotic species and/or to any human need or purpose. Most studies have concluded that the main causative factors for the deterioration of the water quality of Lake Naivasha are the large quantities of sediment inflow from the catchments of the Malewa and Gilgil Rivers, polluting inflows from Naivasha town and the intensive floriculture enterprises adjacent to the lake, see e.g., [15]. These pollutants include high levels of phosphates, nitrates, pesticide residues and other agro-chemicals.

The use of GNSS in monitoring water pollution discussed in Sect. 14.2 could also be applied to Lake Naivasha to map the sources of point pollutants. Although water security issues are a reality in the Lake Naivasha basin, few studies have been done to better understand the underlying issues. Carolina [24] asserts that the area of Lake Naivasha basin is of high economic and political importance to Kenya, which presents a wide variety of economic activities around the water resources with many different stakeholders often competing for the water resources.

The flower industry in Kenya has experienced a phenomenal growth, maintaining an average growth rate of 20 % per year over the last decade. It is an industry that is the second largest export earner for Kenya, employing between 50,000–60,000 people directly and 500,000 others indirectly through affiliated services [51]. Although flowers are now grown in many areas with temperate climate and an altitude above 1,500 m in Kenya, the region around Lake Naivasha still remains the nation's main floriculture farming center. The foremost categories of cut flowers exported from Kenya include: roses, carnations, statice, alstromeria, lilies and hyperricum. Indeed, Kenya is arguably the largest exporter for flowers in the world, supplying over 35 % of cut flowers to the world's largest market—the European Union [51].

13.5.2.2 Impacts of Flower Farming

Lake Naivasha (Kenya) is the only freshwater lake in the Great Rift Valley of East Africa in an otherwise soda/saline lake series [35]. In fact, it is the freshness of the water of Lake Naivasha that is the basis for its diverse ecology [43]. However, recent years have seen a rapid decline in its extent to the point where questions are being raised in the local media as to whether the lake is actually dying.

So unique is Lake Naivasha in the chain of East African Rift Valley lakes that in 1995 it was declared a Ramsar site due to its importance as a wetland. Lake Naivasha and its basin supports a rich ecosystem, with hundreds of bird species, papyrus fringes filled with hippos, riparian grass lands where waterbuck, giraffe, zebra and antelopes graze, dense patches of riparian acacia forest with buffaloes, bushbuck and other species, swampy areas where waterfowl breed and feed and, at the same time, magnificent views of the nearby volcanoes [16]. The lake also supports local fishery and tourism, and is used for recreation, water sports, subsistence farming and hunting. The surrounding lands are dominated by the cultivation of flowers, vegetables, fruit and cereals, as well as power generation [14].

In fact, the *floriculture* industry in this area provides large quantities of flowers that are exported to Europe and other countries of the world. The growth in the flower industry has been favored by the permeable and fertile soils, low rainfall, reliable supply of good quality water, favorable climatic conditions, availability of cheap labor, and easy access to Nairobi airport [16]. Since much of the water used in the flower farms comes from irrigation, the only source, therefore, is the lake and its basin. The lake and its basin also supply drinking water to Nakuru (of which

Naivasha is part of, with a population of about 1.6 million as per the 2009 census) located 160 km north west of Nairobi.

The study of fluctuations in Lake Naivasha's water levels has been carried out, e.g., by [63, 73]. Richardson and Richardson [73], for instance, state that the lake was nearly twice as extensive in the 1920s as it was in 1960–1961. Nicholson [63] noted trends of lower levels during the first half of the nineteenth century, very high levels during the last decades of the nineteenth century, with a rapid decrease occurring during the twentieth century. He further points out that the lake returned to a relatively large extent during the 1960s, but this ended in the 1970s, a fact also stated by Richardson and Richardson [73] who point out that the wetter years beginning in 1961 saw a sharp rise in the levels of Lake Naivasha, as well as of Lakes Elmenteita and Nakuru. The decrease in the lake's water level between the 1920s and 1960s is attributed by Richardson and Richardson [73] to a slight trend of decreasing rainfall during this period, averaging 5 mm/year over the basin, between 1920 and 1949 (see also [81]), as well as an increase in human consumption from river influent and boreholes.

In the 1980s, the fall of the lake's level continued, with the local Olkaria geothermal power station and subsurface drainage thought to be the main culprits [30]. But then, there was little notice taken of the influence of the flower farms, since the first farms had just started in the early 1980s, see e.g., [16]. However, during the 1990s, over 100 km^2 of land around lake Naivasha was converted to floriculture for the European-cut flower trade, e.g., [46]. With this growth came an influx of workers leading to a greater extraction of water from the lake, local aquifers, and the inflowing rivers for the agriculture, floriculture and domestic use by the rapidly increasing population [46].

At this point, the impact of such development on the lake's resources begun to be felt, with its size shrinking due to this direct extraction from the lake and also indirectly from closely connected aquifers. In the work of Abiya [1], the exploitation of the resources of Lake Naivasha is said to pose serious threats to the fragile lake ecosystem and its biodiversity. Abiya [1] considered the dynamics of the changing lake ecosystem, the imminent threats to this system, and the community-based approach towards the sustainable utilization of the lake. Their study showed that the sustainable use of the lake was not going to be fully realized without a sound management plan, and recommended the enactment of consolidated environmental legislation in Kenya, which will enable the strengthening of environmental conservation and the protection of natural resources [1]. This in turn has led to other proposals for the sustainable use of the lake and basin (e.g., [34, 43]).

In the last decade, the lake's level has continued to drop with floriculture being blamed for excessive water extraction from the lake and aquifers, and the small holder farms in the upper catchment being blamed for nutrient loadings, see e.g., [46, 57], leading to outcry in both the local and international media that this Ramsar site could be dying as a result of the very resource that it supports, see e.g., [57]. For example, Mekonnen and Hoekstra [57] observed that the total virtual water exported in relation to the cut flower industry from the Lake Naivasha basin was 16 Mm^3/year during the period 1996–2005. Other factors that have also been proposed as influencing Lake

Naivasha's water changes include irregular rainfall patterns [43], and trade winds [96]. All of these discussions therefore point to the need for the *reliable mapping* of the lake and its basin in order to properly understand the dynamics of this area. This need for accurately monitoring the lake was captured by Becht and Harper [14], who state that there is an urgent need to accurately measure all abstractions and provide consistent, reliable, hydrological and meteorological data from the catchment, so that a 'safe' yield may be agreed upon by all stakeholders and a sustainable use of the lake waters achieved.

However, lack of reliable basin mapping techniques hampered the proper mapping of changes in the lake basin, while also not allowing accurate predictions of the likely future situation, despite modeling methods being used to calculate its water balance, e.g., [14]. The situation is compounded by the fact that Lake Naivasha has no surface outlet that could assist in hydrological monitoring, and the fact that changes in its water level occur rapidly, over the order of several meters over just a few months, shifting the shoreline by several meters [16].

The emergence of satellite based methods offers the possibility of providing a broader and more integrated analysis of the lake and its basin. Using products from the GRACE gravity mission, changes in the stored water in the region extending from the Lake Naivasha basin to Lake Victoria was assessed by Awange [7] to determine whether the changes are climatic or human induced. Changes in precipitation can be examined by the analysis of products from Tropical Rainfall Measuring Mission (TRMM), allowing the determination of how much of the fluctuations in Lake Naivasha are related to changes in precipitation behavior.

The fluctuations in the water level of Lake Naivasha can be determined using both ground-based tide gauge observations and satellite altimetry data (TOPEX/Poseidon and Jason-1; see e.g., [12, 13]). These results may in turn be related to the use of satellite imagery (e.g., Landsat) and change detection techniques to map the shoreline changes of Lake Naivasha, analyzing the trend of changes over the period of interest, and correlating shoreline changes with proposed causes. In general, the combination of different data sets, both space-borne and ground based, provide a valuable geospatial contribution to understanding the hydrological behaviour of the East African Great Lakes region, e.g., [17].

Example 13.1 (Satellite-based monitoring [7]).

Several different types of space-borne observations were used in Awange et al. [7]: (1) GRACE gravity-field products (2) precipitation records based on TRMM products (3) satellite remote sensing (Landsat) images (4) satellite altimetry data and (5) flower production data. In addition, data from an in-situ tide-gauge station were used. The results of Awange et al. [7] confirm that Lake Naivasha has been steadily declining with the situation being exacerbated from around the year 2000, e.g., Fig. 13.8, with water levels declining at a rate of 10.2 cm/year and a shrinkage in area o 1.04 km^2/year (see e.g., Fig. 13.6 where GNSS was used to *geo-reference* the satellite imagery). This

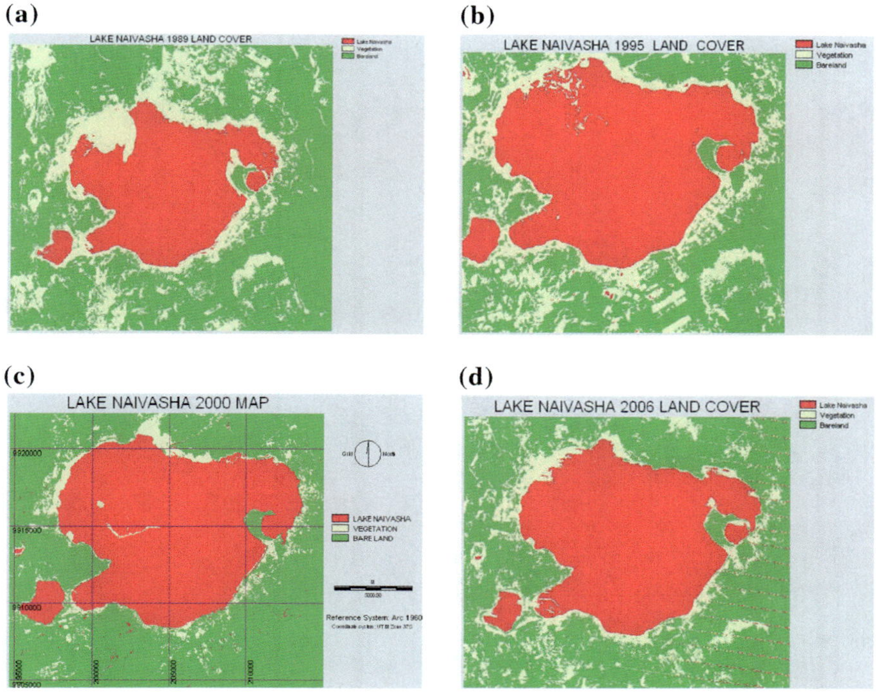

Fig. 13.6 Surface-type classification results for the considered Landsat images. **a** 1989, **b** 1995, **c** 2000 and **d** 2006. GNSS satellites were used to georeference the satellite images. *Source* Awange et al. [7]

rapid decline can be traced largely to anthropogenic activities, an argument supported by a coefficient of correlation value of 0.68 between the quantity of flower production and the lake's level for the period 2002–2010 at 95 % confidence level (Fig. 13.7), a period during which such production doubled, see e.g., Fig. 13.9. These results, supported by the use of GNSS show that there is therefore a need for those different communities and interest groups that depend upon Lake Naivasha to better formulate their management plans, a need which can exploit results such as those presented in [7].

13.5.3 Water Challenges Facing the Nile Basin

The Nile Basin (Fig. 13.10) is one of the Earth's most impressive examples of the influence of topography and climate on the flow conditions of a water system. The Nile has two major tributaries, the White Nile and the Blue Nile, the latter being the source of most of the river's water. The White Nile rises in the Great Lakes region of Eastern Africa, and flows northwards through Uganda and the South Sudan. The Blue Nile starts at Lake Tana in the Ethiopian highlands, flowing into Sudan from

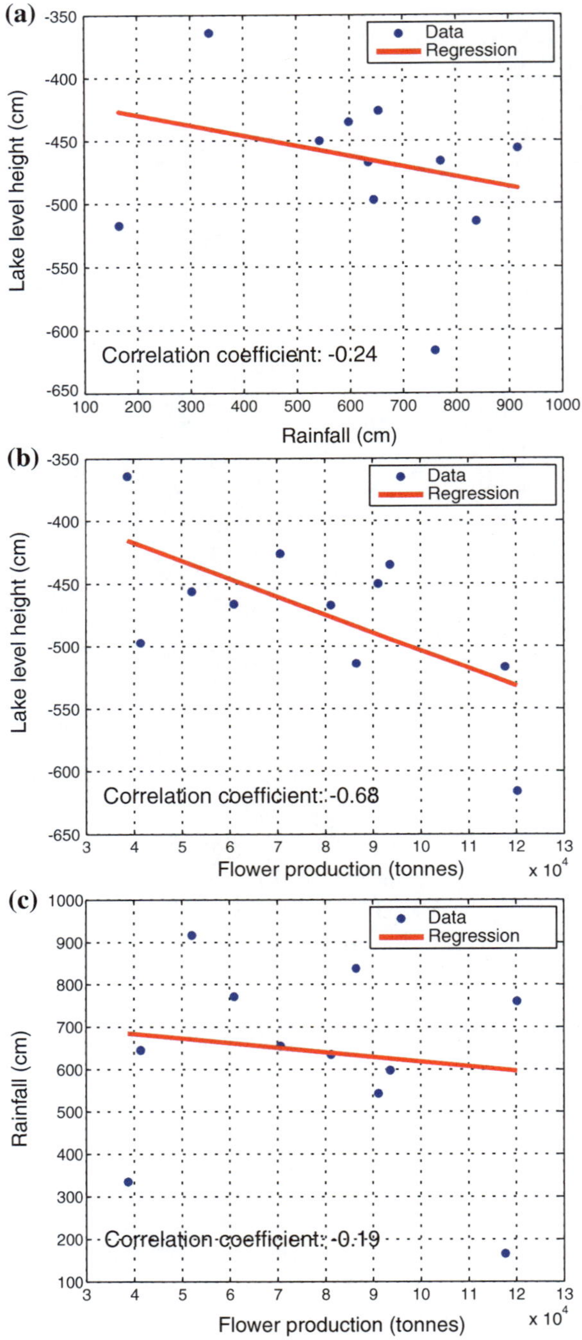

Fig. 13.7 Comparing annual average lake levels with **a** rainfall observed at the Naivasha station and **b** flower exports. **c** Comparing annual average rainfall of the Naivasha station and flower exports. The solid lines are fitted linear trends, along with the correlation coefficients

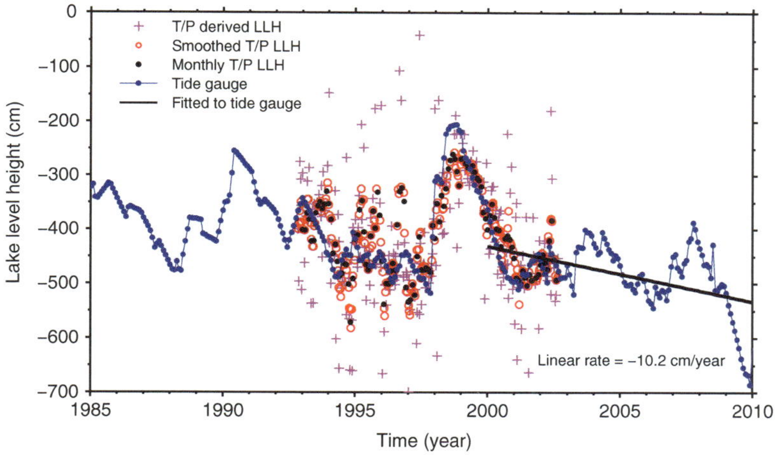

Fig. 13.8 Time Series of lake level height (LLH) changes for Lake Naivasha as provided by satellite altimetry (T/P), and a tide gauge. GNSS satellites support the satellite altimetry as discussed in [12]. *Source* Awange et al. [7]

Fig. 13.9 Annual flower exports from Kenya. *Source* Awange et al. [7]

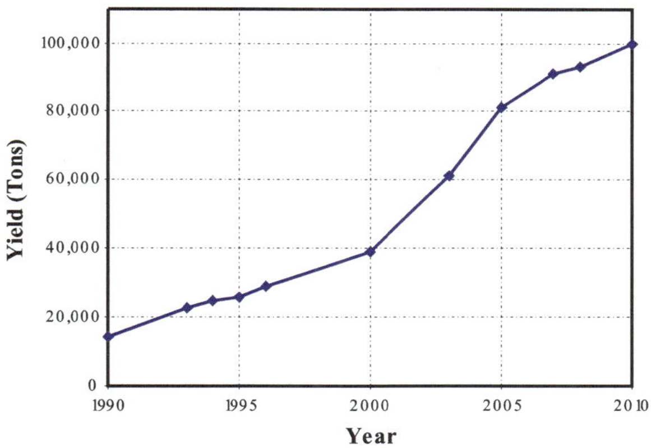

the southeast and meets the White Nile at Khartoum in Sudan. From there, the Nile passes through Egypt and ends its journey by flowing into the Mediterranean Sea.

A basin as large as the Nile, which crosses such a wide latitude range (from ∼5 °S to *ca.* 31 °N) cannot be expected to experience homogeneous climatic and rainfall patterns over its extent. In addition, variations in the geology and soils of the basin strongly influence groundwater availability. Significant rainwater deficits and the variable duration of the rainy seasons over yearly to decadal time scales results in hydrological deficits that are not necessarily reflected in a direct response of the base flow, e.g., [67].

The East African lake region includes the countries of Burundi, Rwanda, Uganda, Kenya and Tanzania, and is the home to Lake Victoria, the world's second largest freshwater lake, and the source of White Nile [8]. From Lake Victoria, the waters

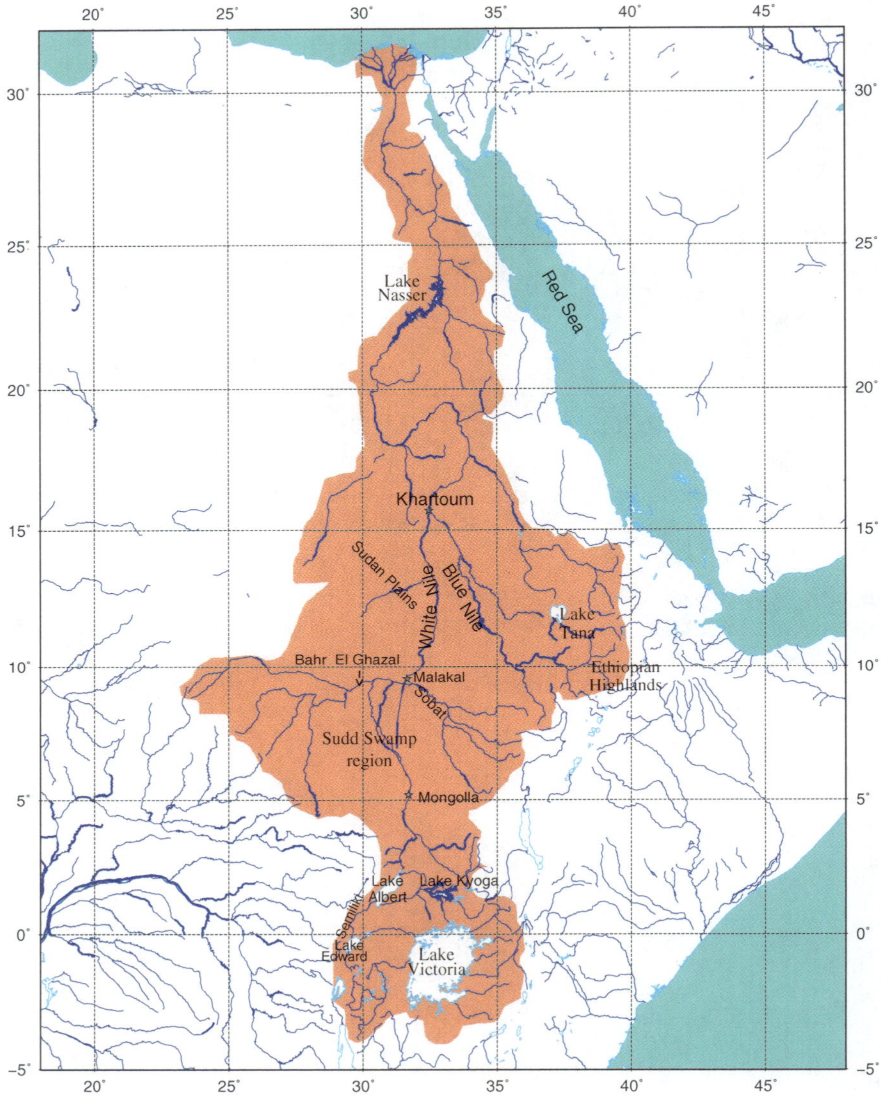

Fig. 13.10 The Nile Basin (brown shaded region) with the major features and place names as discussed in the text

are discharged to Lake Kyoga, which also receives water from its surrounding 75,000 km^2 catchment before flowing on to Lake Albert. In addition to the waters received from Lake Kyoga, Lake Albert is supplied by its upstream Semiliki basin and the Lake Edward sub-basin (Fig. 13.10). Together, Lakes Edward, Albert, and George form the western edge of the Nile Basin, comprising an area of 48,000 km^2, of which 7,800 km^2 is open water [105]. In the Sudd swamp region, the supply to the Nile benefits from two other basins, the Bahr-El-Ghazal (500,000 km^2) to the west and the Sobat (150,000 km^2) to the east, before exiting at Malakal.

The Ethiopian highlands are comprised of twelve significant sub-basins aggregated into four primary basins; Lake Tana (20,000 km^2, the main headwaters of Blue Nile, although it contributes less than 10 % of the total Blue Nile flow), the upper Blue (150,000 km^2), the lower Blue (60,000 km^2), and the Dinder-Raghad (70,000 km^2). All together, they cover almost 300,000 km^2 and contribute approximately 65 % of the river Nile's total water [105].

Lake Nasser region: The Egyptian desert starts from Khartoum, where the Nile flows northward towards Egypt through Lake Nasser (formed by the Aswan dam) and then to the Mediterranean Sea. Yates and Strzepek [105] reported a net loss of water between the joining of the Atbara river with the Nile north of Khartoum, and Lake Nasser due to evaporation and seepage.

13.5.3.1 Challenges Facing the Basin's Waters

The Nile river basin is one of the largest in the world, with an area of about 3,400,000 km^2 (almost one-tenth of Africa) and traversing some 6,500 km from south to north as it winds its way across the boundaries of eleven countries: Tanzania, Uganda, Kenya, Rwanda, Burundi, Democratic Republic of Congo (DRC), Eritrea, Ethiopia, Sudan, South Sudan, and Egypt, e.g., [80]. As it flows through these countries, it supports a livelihoods of over 200 million people.

The Nile's water resources, however, have come under threat from both anthropogenic and natural factors, e.g., [41].Anthropogenic influences have been fueled by the increasing human population that has put pressure on domestic water needs, the supply of hydroelectric power, all coupled with the need to sustain economic growth. However, not only are the demands on water increasing, but the available water supplies appear to be decreasing, with environmental degradation of the upper Blue Nile catchment having increased throughout the 1980s [100]. Whittington and McClelland [100] found that about 86 % of the annual Nile river flow into Egypt originates from Ethiopia, and warn of significant implications for Egypt and Sudan should Ethiopia undertake any potential extractions; this issue emphasizes the need for cooperation between the three Blue Nile riparian states.

Natural factors include the changing climate, which has been the subject of numerous studies, e.g., [105, and the references therein]. Therefore, a combination of human population growth, unsustainable water usage and development, and desertification are just some of the factors that threaten the Nile's ability to supply crucially needed water to the people of the basin.

The present-day state of the stored water variations and their relations to climate variability (e.g., El Niño and Southern Oscillation (ENSO) and the Indian Ocean Dipole (IOD)) in the Nile Basin are, however, also not well understood. Most studies dealing with the Nile Basin to date, however, have dealt mainly with modeling the impacts of climate change (e.g., [19, 26]), with very little being reported on how to monitor the spatial and temporal variations in the stored water (surface, groundwater and soil moisture) of the basin in a holistic manner, and linking them to climate variability. The reason for such few studies, e.g., [21, 84] has been partly attributed to its large size, as well as the lack of appropriate monitoring techniques that could cover such a vast spatial extent. For instance, the hydrological water balance involves the flow of surface water, the movement of deeper groundwater, and the coupling of the land, ocean and atmosphere through evaporation and precipitation. Monitoring these components requires an accuracy and completeness of geographical data coverage that challenges conventional measurement capabilities.

Using GRACE, Global Land Data Assimilation System (GLDAS),[8] and TRMM (Tropical Rainfall Measuring Mission)[9] data for the period 2002–2011, Independent Component Analysis (ICA) method, e.g., [36], is applied in the examples below to localize the Nile Basin's hydrological signals into their respective sources. In Fig. 13.11, it is seen that the dominant signal associated with the Lake Victoria Basin, i.e., the White Nile is localized for all the three data set. This clearly shows the contribution of the Lake Victoria basin to the Nile waters. An analysis of the correlation between these signals and climate variability indicate a strong correlation (0.85) between the GRACE's total water storage and ENSO for the period 2006–2011. This confirms the known fact that climate variability, particularly ENSO, influences the changes in stored water of Lake Victoria Basin, and should be taken into consideration in evaluating the basin's hydrology. Section. 13.5.4 discusses Lake Victoria Basin in more detail.

Figure 13.12 shows the localization of the signals within the Ethiopian highlands, thus indicating the importance of the region to the Nile basin. The Blue Nile receives its waters mainly from the heavy rainfall in the region as seen from the TRMM results (Fig 13.12; IC5). Any land use patterns that could alter the use of water within the region would be capable of significantly impacting upon the entire Nile Basin. GRACE signals show a correlation of 0.52 with ENSO while GLDAS show 0.44 with ENSO and 0.4 with the Indian Ocean Dipole (IOD) index. Compared to Lake Victoria basin, the correlation to climate variability is not so strong, nonetheless, the fact that the changes in the stored water in the Ethiopian highlands is influenced by climate variability is noticeable. Figure 13.13 shows the localization of the signals within the Bahr-El-Ghazal region, which also contributes water to the River Nile by joining the tributaries from Sobat around Malakal (see Fig. 13.10). Both GRACE and GLDAS signals show a correlation of 0.68 with ENSO respectively, thus indicating that the variability of the stored water is influenced by climate variability. Finally, Fig. 13.14 shows the dynamics around Lake Nasser along the Red Sea. More discussion on this region and the impacts of over-extraction in the Nubian aquifer for irrigation purposes is presented, e.g., in [5, 6].

Positive correlations between the changes in total water storages and IOD corresponding to cool waters in the Indian Ocean associated with large scale circulation changes that leads to above average rainfall in East Africa leading to flooding, while Indonesia and several parts of Australia experience drought have been documented

---

[8] http://disc.sci.gsfc.nasa.gov/services/grads-gds/gldas

[9] http://trmm.gsfc.nasa.gov/data_dir/data.html

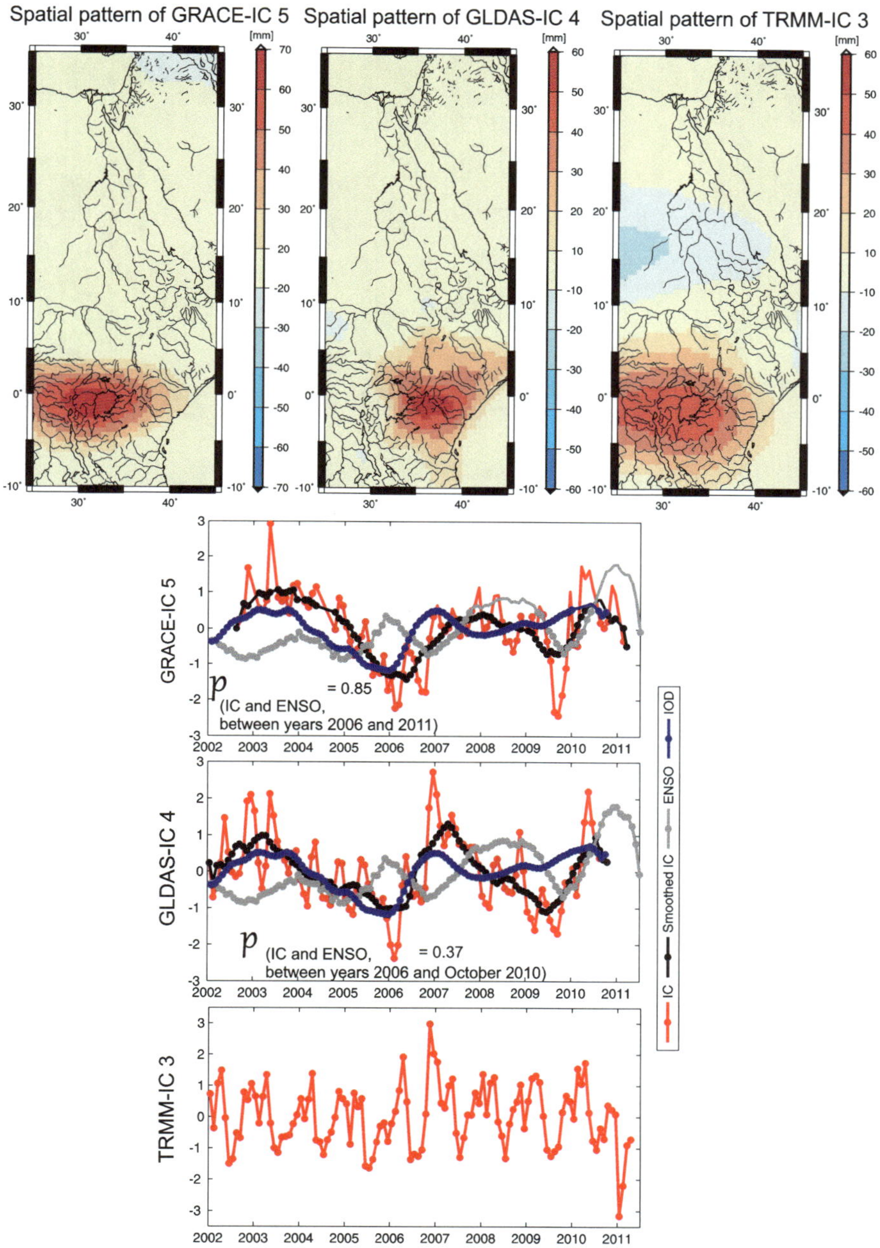

Fig. 13.11 ICA analysis of the GRACE, GLDAS, and TRMM data for the Nile basin. In this figure the signals are localized within Lake Victoria basin in all the data set

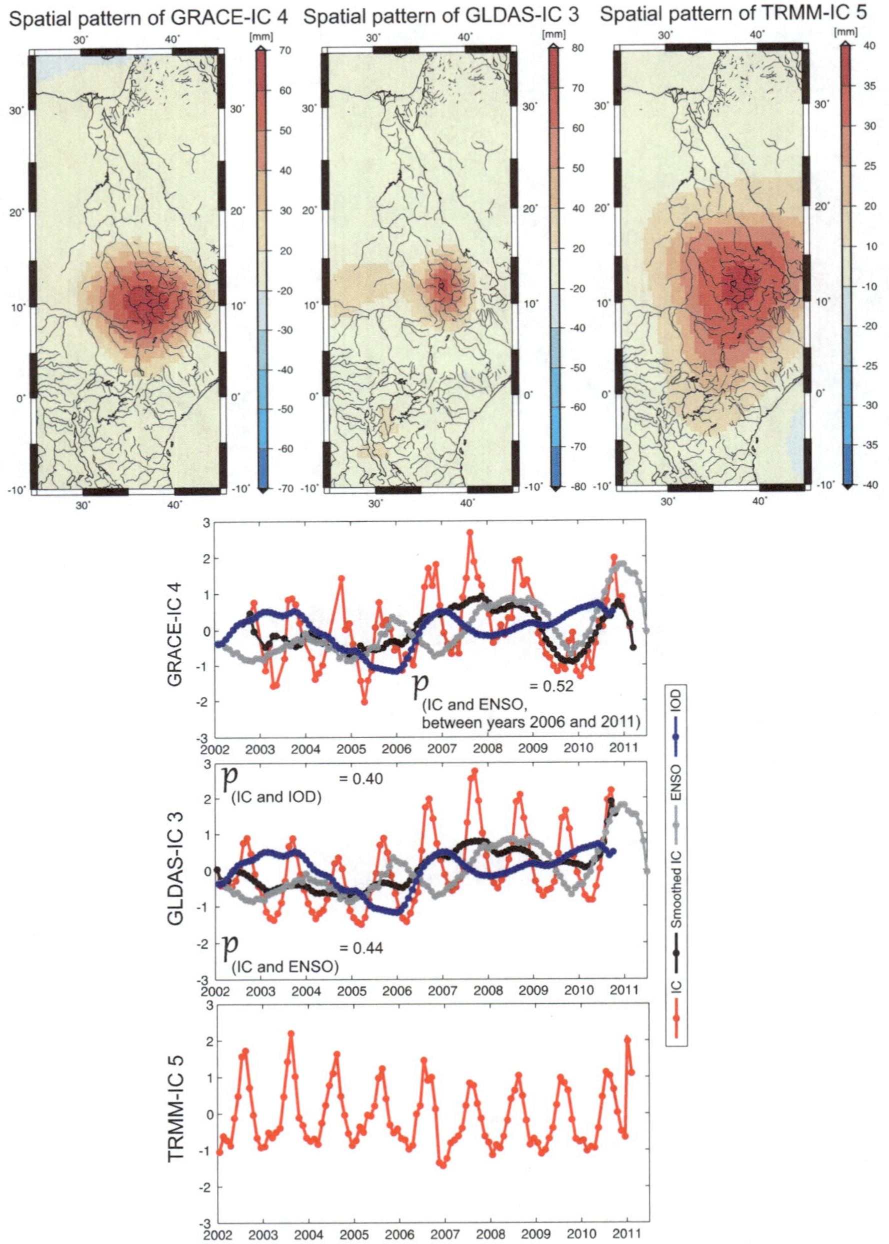

Fig. 13.12 ICA analysis of the GRACE, GLDAS, and TRMM data for the Nile basin. In this figure the signals are localized within the Ethiopian highlands in all the data set

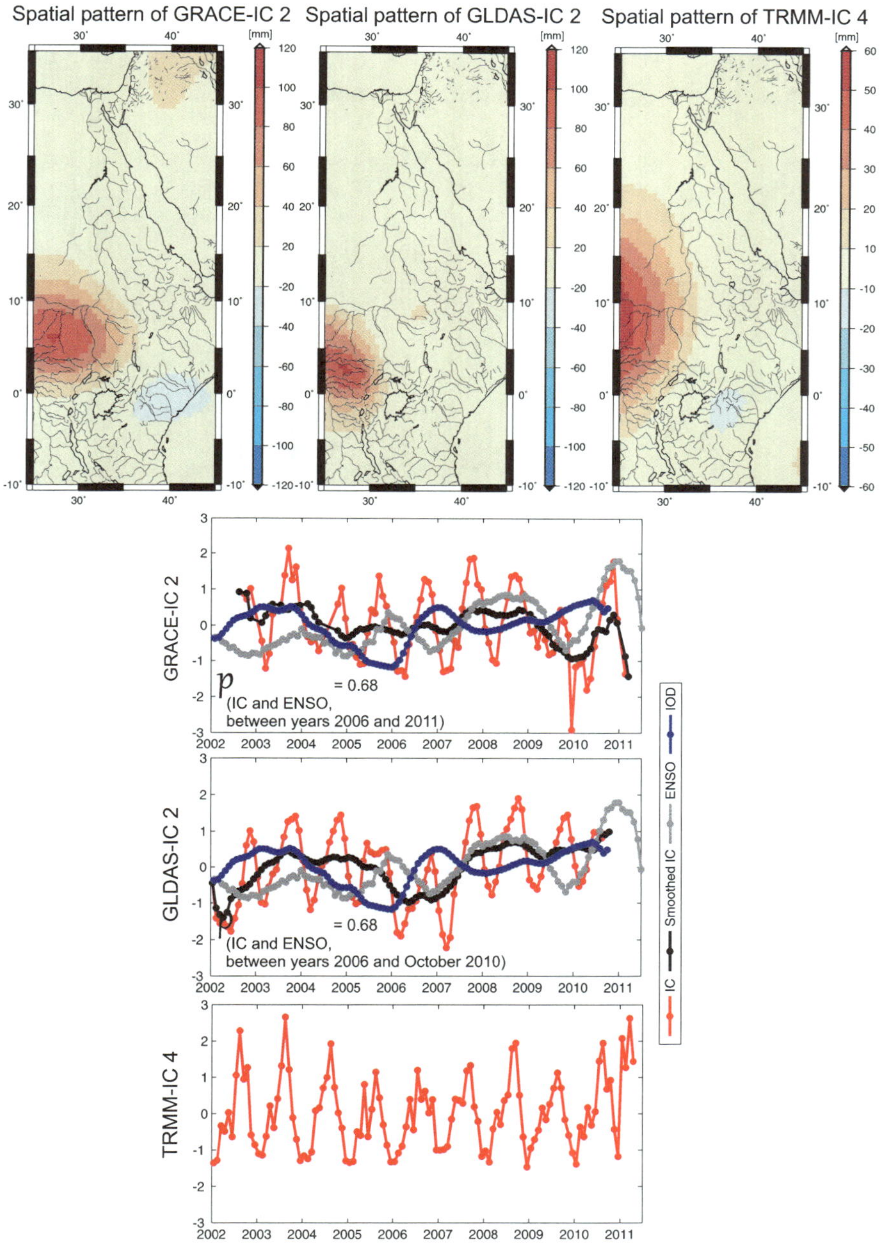

Fig. 13.13 ICA analysis of the GRACE, GLDAS, and TRMM data for the Nile basin. In this figure the signals are localized within the Bahr-el-Ghazal region in all the data set

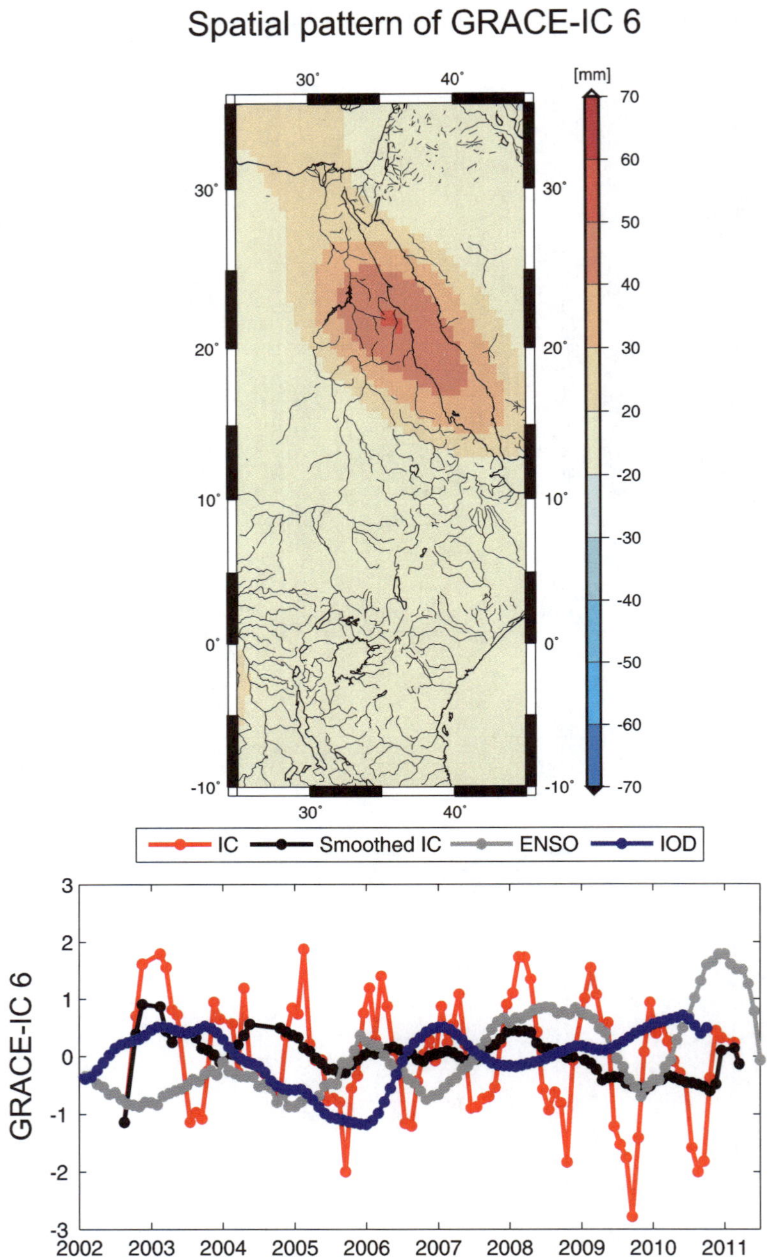

Fig. 13.14 ICA analysis of the GRACE, GLDAS, and TRMM data for the Nile basin. In this figure the signals are localized within the Lake Nasser region in GRACE data set

e.g., [17, 38]. This is true for the Lake Victoria Basin, the Ethiopian highlands and the Bar-El-Ghazal regions, which are also related to ENSO. For the Lake Nasser region, the effect of climate variability is negligible. For the definitions and measured indices of ENSO and IOD, see Sect. 16.4.5.

13.5.4 Decline in Lake Victoria Water Level

Lake Victoria (Fig. 13.15), the world's third largest lake and the largest in the developing world, is a source of water for irrigation, transport, domestic and livestock uses, and supports the livelihood of more than 30 million people who live around it [8]. Nicholson [63, 64] documents its significance as an indicator of environmental and climate change over long-term scales. Since the 1960s, the lake level had experienced significant fluctuation, see e.g., [63, 64]. From 2001 to 2006, however, Lake Victoria's water level showed a dramatic fall that alarmed water resource managers as to whether the lake was actually drying up. Kull [52] reported that the lake's levels fell by more than 1.1 m below the 10 year average.

With the receding of the lake waters, acres of land that were lost to the floods of the 1960s were fast being reclaimed, creating sources of conflicts between man and wildlife. In some beaches, e.g., Usoma in Kenya, wetlands that were once breeding

Fig. 13.15 Lake Victoria basin. *Source* Kayombo and Jorgensen [50]

places for fish were dying up, leaving areas of land as playing fields for children and farmland. Ships were now forced to dock deep inside the lake, while the landing bays needed to be extended. Those who directly depended on the lake waters for domestic use were forced to go deeper into the lake to draw water, thus exposing women and children to water-borne diseases and risks of snakes and crocodiles. Water intakes that supplied major towns and cities had to be extended deeper into the lake, thus causing more financial burden to the municipalities that were already strained financially [9].

With 80 % of Lake Victoria water coming from direct rainfall, changes in the lake level are directly related to the variation in the water stored in its basin, which contributes around 20 % in the form of river discharge. A decrease in stored basin water was therefore suspected to contribute to the drop in the lake level. An analysis of the stored water in the Lake Victoria basin in relation to rainfall and evaporation was therefore necessary as a first diagnosis. This would provide water resource managers and planners with information on the state and changing trend of the stored water within the basin. Such basin scale observations could only be achieved through the use of satellites such as GRACE. Conventional methods for studying variations in stored water such as the Artificial Neural Network, GIS and remote sensing could not diagnose the problem, see e.g., [9].

Having been motivated by the potential of the GRACE satellites, Awange et al. [9] undertook a satellite analysis of the entire lake basin in an attempt to establish the

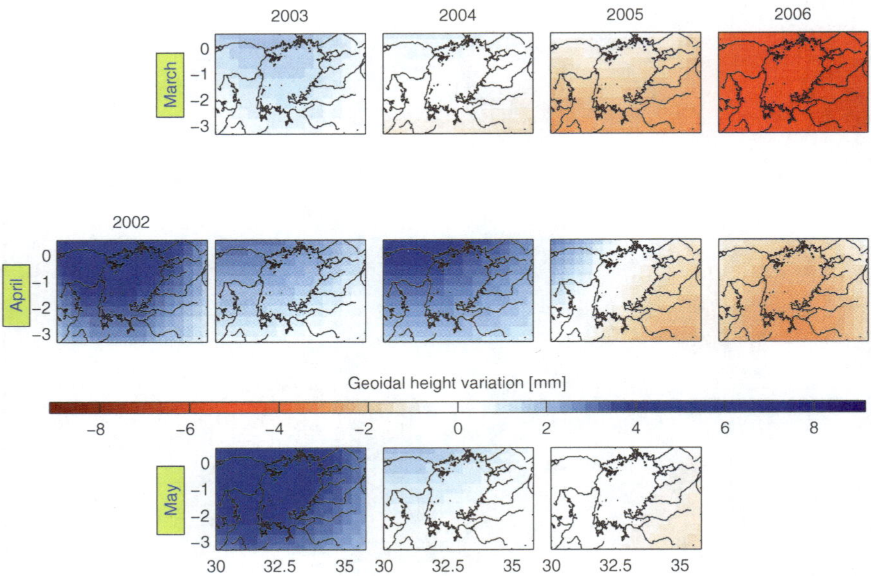

Fig. 13.16 Inter-annual comparison in the geoid during the high rainy season of MAM from 2002 to 2006. The figure indicates a decline in total water storage in the Lake Victoria basin during this period. *Source* Awange et al. [9]

cause of the decline in Lake Victoria's water levels. The GRACE and CHAMP satellites (e.g., [12, 13]) together with data from the TRMM satellite were employed in the analysis. Using 45 months of data spanning a period of 4 years (2002–2006), the GRACE satellite data were used to analyze the gravity field variation caused by changes in the stored waters within the lake basin. Figure 13.16 presents the annual variation of the geoid in the lake's basin during the high rain season months of March, April and May (MAM) for the period 2002–2006. The GRACE results indicated that the basin's total water storage dramatically decreased at a rate of 6.20 mm/month. These changes are expressed in equivalent water thickness (also known as total water storage (TWS)) in Fig. 13.17. For the period 2002–2006, the results indicate a general decline in the lake basin's water level at a rate of 1.83 km^3/month [9].

To validate the GRACE results, TRMM Level 3 monthly data for the same period of time were used to compute mean rainfall at a spatial resolution of $0.25 \times 0.25°$ (25×25 km), as shown in Fig. 13.18, from which the rainfall trends were analyzed (Fig. 13.19). To assess the effect of evaporation, GNSS remote sensing data (59 CHAMP satellite occultations) for the period 2001 to 2006 were analyzed to define if tropopause warming took place (see the approach in [12, 13]). The results indicated that the tropopause temperature fell in 2002 by about 3.9 K and increased by 2.2 K in 2003 and remained above the 189.5 K value of 2002. The tropopause heights showed a steady increase from a height of 16.72 m in 2001 and remained above that value reaching a maximum of 17.59 km in 2005, an increase in height by 0.87 m. Temperatures did not, therefore, increase drastically to cause massive evaporation. TRMM results indicated the rainfall over the basin (and directly over the lake) to have been stable during this period (see Figs. 13.18 and 13.19). Since rainfall over the period remained stable, and temperatures did not increase drastically to cause increased evaporation, the remaining major contributor during the period 2002–2006 was suspected to be discharge from the expanded Owen Falls dam. Awange et al. [9] concluded, thanks to the GRACE and GNSS satellites, that the fall in Lake Victoria's water level between 2001 and 2006, also noted in Sect. (13.5.3), was due to human impact on the basin's environment (i.e., expanded dam) as opposed to natural factors.

In a related work, Swenson and Wahr [88] used satellite gravimetric and altimetric data to study trends in water storage and lake levels of multiple lakes in the Great Rift Valley region of East Africa for the years 2003–2008. GRACE total water storage estimated by Swenson and Wahr [88] corroborated the findings of Awange et al. [9] that the lake's water level had declined by as much as 60 mm/year, while their altimetric data indicated that levels in some large lakes in the East African region dropped by as much as 1–2 m. Swenson and Wahr [88] concluded that the largest decline occurred in Lake Victoria and, like Awange et al. [9], attributed this to the role of human activities.

Both the findings of Awange et al. [9] and Swenson and Wahr [88] provide evidence that the GRACE satellites (supported by GNSS) could be used to provide independent means of assessing the relative impacts of climate and human activities on the balance of stored water that does not depend on in-situ observations, such as dam discharge values, which may not be available to the public domain.

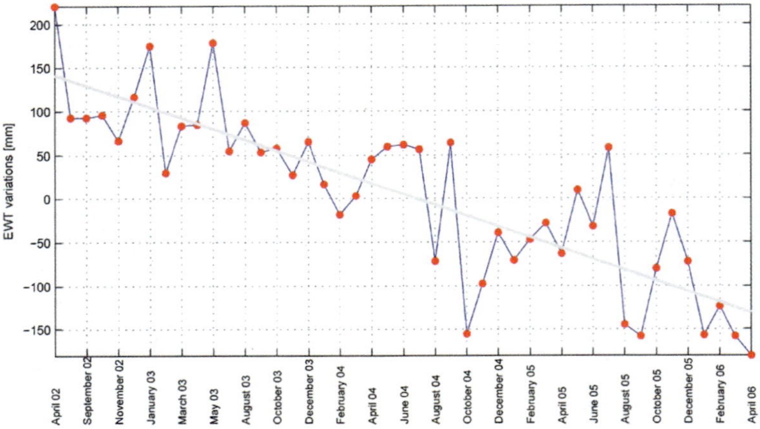

Fig. 13.17 Lake Victoria basin total water storage (equivalent water thickness) changes between 2002 and2006, as seen by the GRACE satellite. The figure indicates that the GRACE satellites observed a general decline in the lake's basin waters over this period (cf. results obtained from satellite altimetry in [12, 13]). *Source* Awange et al. [9]

13.5.4.1 Geospatial Application to LVB Water Conflict Resolution

Fishermen in the three East African countries have frequently found themselves trespassing each others water boundaries chasing the meager fish resources. The end result has often been conflict, leading to arrests and the confiscation of boats and fishing equipment. In such cases, hand-held GNSS receivers and a map could easily resolve such a dilemma. A real-case scenario is illustrated by Migingo Island in Fig. 13.20,[10] which is an island currently disputed between Kenya and Uganda due to its being home to the dwindling Nile Perch (*Lates niloticus*) fish. Owing to uncertainty about the boundary, GNSS receivers were used by a team of surveyors from both countries to mark the boundary and establish that the disputed island belongs to Kenya.

13.5.5 Water, a Critical Dwindling Australian Resource

Warning of Australia's acute water problem (see e.g., Fig. 13.21) had already been sounded by the National Land and Water Resource Audit (NLWRA [60]), which reported that Australian water resources were scarce and in high demand by agricultural, industrial and urban users. The alarming finding of the report was the fact that 26 % of the river basins and 34 % of the groundwater management units in Australia were approaching or beyond sustainable extraction limits.

---

[10] Source: http://www.nation.co.ke

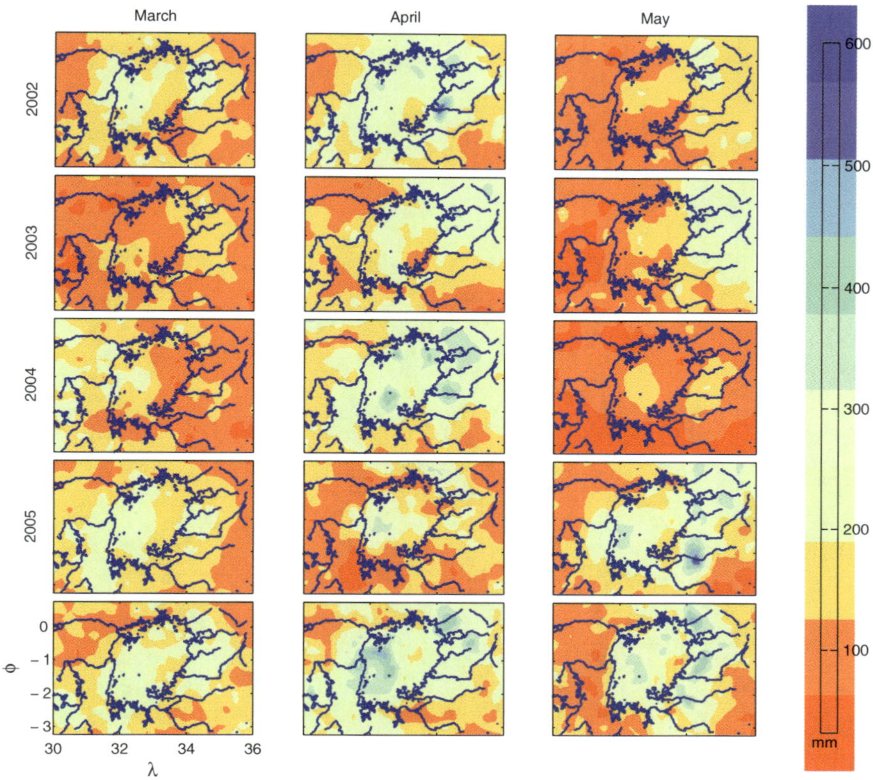

Fig. 13.18 Inter-annual comparison of rainfall over the Lake Victoria basin during the high rainy season of MAM from 2002 to 2006 using TRMM results. *Source* Awange et al. [9]

The NLWRA highlighted the need for information that could assist in improving water resource management and conservation. Steffen et al. [86] issued a further alert that Australia would be faced with the impacts of climate change on its water quantity due to decrease in precipitation over parts of Australia.

Even though the need for up-to-date information on stored water resource to support policy formulation and management issues had already been realized, e.g., NLWRA [60], and specifically with the prevailing drought conditions in Australia, the lack of appropriate techniques that offered high spatial and temporal resolution monitoring of changes in stored water remained a stumbling block [32]. The problem was further compounded by the fact that groundwater suitable for human consumption and utility is normally trapped inside aquifers (see Sect. 13.2) that are beyond reach of modern remote sensing methods. With the introduction of GRACE satellites (Sect. 13.4.2), however, the situation changed.

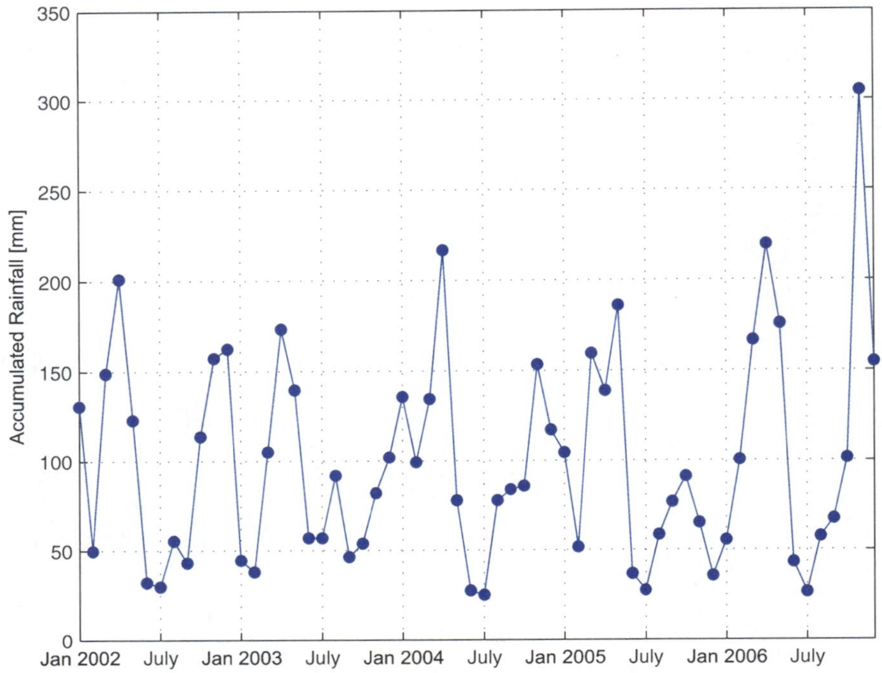

Fig. 13.19 Time series of rainfall 2002–2006 for the lake Victoria basin as observed by the TRMM satellite. *Source* Awange et al. [9]

Fig. 13.20 The disputed Migingo Island in Lake Victoria (*right*). GNSS was used to establish that the island belongs to Kenya, thus resolving a territorial dispute between Kenya and Uganda. *Source* Daily Nation, Kenya

Example 13.2 (Monitoring changes in Australia's stored water [11])

In order to use the GRACE satellite for monitoring the variation in Australia's stored water, Awange et al. [11] investigated the regional $4 \times 4°$ mascon (mass

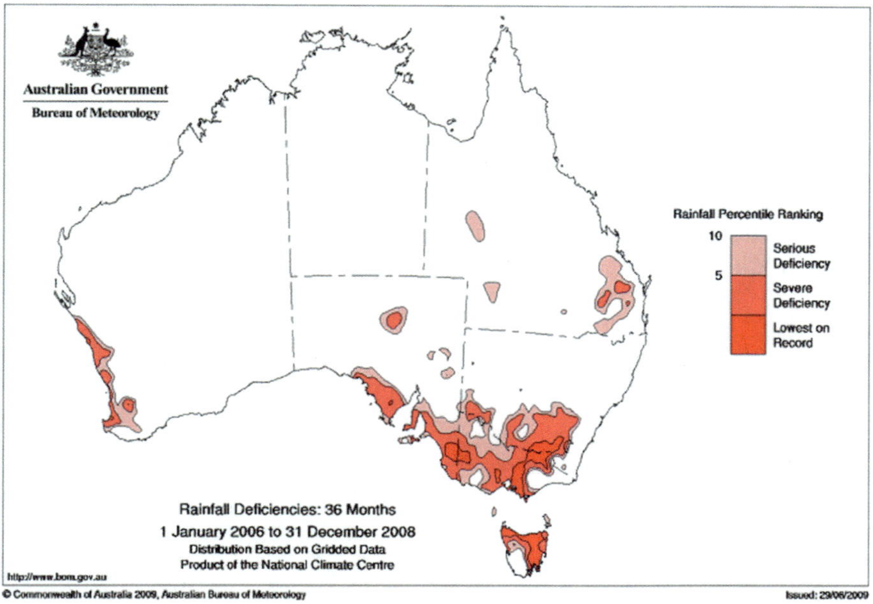

Fig. 13.21 Rainfall deficiencies for the period 1st January 2006 to 31st December 2008 (3 years, *Source* BOM)

concentration) GRACE solutions provided by GSFC (Goddard Space Flight Center, NASA) for their suitability for monitoring Australian hydrology, with a particular focus on the Murray-Darling Basin (MDB). Using Principal Component Analysis (PCA) and multi-linear regression analysis (MLRA), the main components of the spatial and temporal variability in the mascon solutions over both the Australian continent as a whole and the MDB in particular were analyzed and the results compared to those from global solutions provided by CSR (the Center for Space Research, University of Texas at Austin) and CNES/GRGS (Centre National d'Études Spatiales/Groupe de Recherche de Geodesie Spatiale, France) and validated using TRMM, water storage changes predicted by the WaterGap Global Hydrological Model (WGHM) and ground-truth (river-gauge) observations. The results of Awange et al. [11] indicated that for the challenging Australian case with weaker hydrological signals, all the solutions gave similar results.

For the PCA results in Fig. 13.22, the Australia-wide case was considered mainly to compare the different GRACE releases among themselves and with TRMM and WGHM time-series. The results of the PCA analysis for the first two modes are shown in Fig. 13.22. It was noticed that most of the variability were contained in the first mode (>50 %), while considering the first 2 modes accommodates between 63 % (for CNES/GRGS) and 81 % (for mascon) of the total variability of each signal. The 1st mode (upper panel, Fig. 13.22) shows similar behavior among all data sets, with all data displaying a general north-south varying empirical orthogonal function (EOF) pattern and strong annual signal in the principle components (PC), indicative

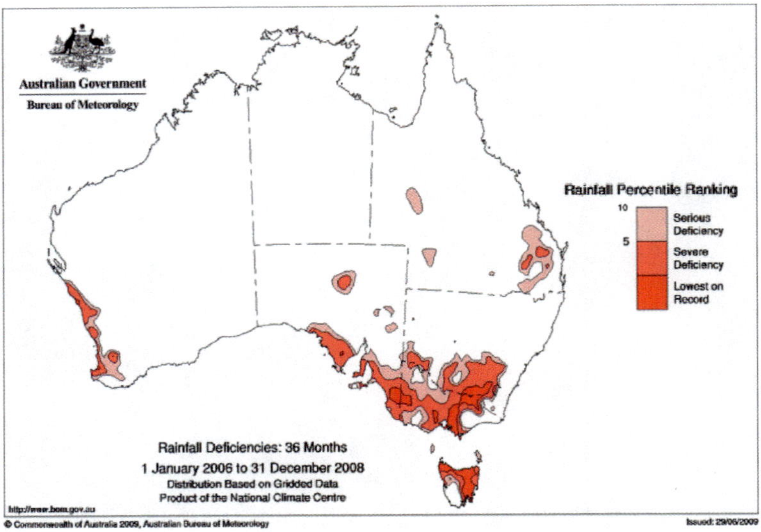

Fig. 13.22 Results of the PCA applied to the whole of Australia for the 1st and 2nd modes. *Top rows* (**a–e**) the 1st mode results, *bottom rows* (**f–j**) the 2nd mode results. (**a, f**) mascon, (**b, g**) CNES/GRGS, (**c, h**) CSR, (**d, i**) TRMM and (**e, j**) WGHM. Australiaś drainage divisions (from the Australian Bureau of Meteorology, see the Acknowledgments) are marked by the gray boundaries (Lambert conformal conic projection). *Source* Awange et al. [11]

of seasonal variations. The annual signal is also apparent in the 2nd PC mode (lower panel, Fig. 13.22), especially for the GRACE time-series, although less so for the filtered TRMM and WGHM results.

The dominant signal in northern Australia is a result of the annual monsoonal rains, and is much stronger than that in the southern part of the continent. Therefore, the 1st mode is dominated by changes in the north which may lead to smaller hydrological changes in the south being excluded from this mode. The northern signal is very obvious in the 1st mode EOF patterns for all data sets examined, and also in the 2nd mode EOF for the mascon, CNES/GRGS, CSR and TRMM time-series. The PC of the 2nd modes also appears to show strong linear trends in the time-series, especially for the CNES/GRGS and CSR results. For both the 1st and 2nd modes, Central Australia shows a relatively low signal, a consequence of the aridity of this area having small hydrological changes. The shift in the seasons that receive the high rainfall (summer in the north, winter in the south) can be seen by the opposite signs in the signals given by the EOF (time-series in the upper panels, Fig. 13.22).

The Australia-wide results therefore show that all the GRACE solutions provided similar results and were able to identify the major climatological features of Australia, in particular the dominance of the monsoonal rainfall over northern Australia, and the offset (~6 months) between the northern and southern wet seasons, as well as the areas of mass gain (northern Australia) and loss (the MDB in the southeast and

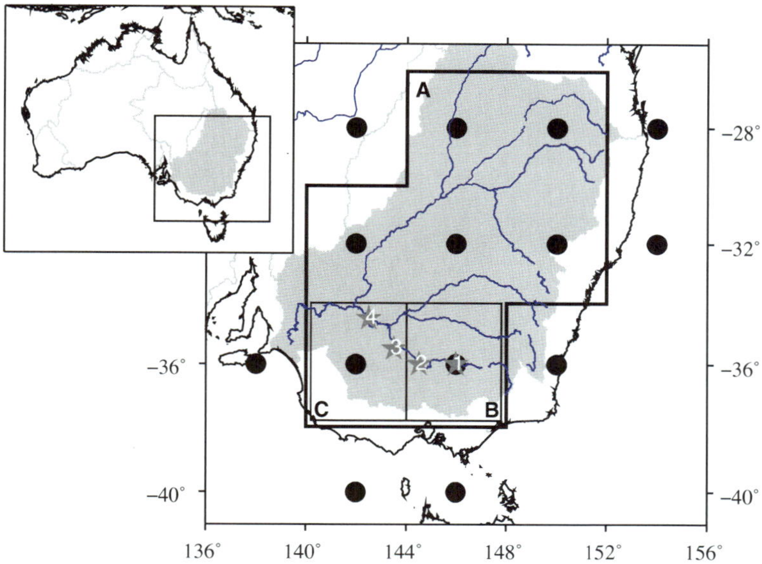

Fig. 13.23 Location map of the Murray-Darling Basin (MDB). The grey shading marks the basinś extent, with Australiaś other river basins also outlined in grey. *Black filled circles* are the centres of the mascon grid cells provided by GSFC (4 × 4° grid). The *thick black-bounded area* **a** covers most of the MDB, while the *finer black-bounded areas* **b** and **c** are examined with respect to ground-truth data in the form of river-gauge data (numbered stars). The river-gauges are at Yarrawonga (1), Swan Hill (2), Euston (3) and Torrumbarry (4). *Source* Awange et al. [11]

southwest Western Australia). In Forootan et al. [37], changes in Australia's stored water in relation to climate variability of ENSO and IOD are treated.

Example 13.3 (Localized Murray-Darling Basin. Source: Awange et al. [11])

Awange et al. [11] further examined the MDB to determine how the GRACE solutions performed for more localized areas, since the MDB is one of Australia's most important regions for agricultural production and is an area that has been severely affected by the recent drought conditions [54, 94]. First, they examined the area outlined in Fig. 13.23 denoted as A, which is defined by the mascon grid elements that cover much of the MDB. Figure 13.24 compared the inferred stored water variation from each data set.

Examining the three GRACE solutions in Fig. 13.24 over the time period covered by the mascon solutions (grey shaded area), Awange et al. [11] noted that the time-series generally follow each other reasonably well, as also shown by the resulting cross correlation values, i.e., CSR and CNE/GRGS, being global solutions, appear to be in closer agreement with each other ($R = 0.83$), than when compared to the mascon (mascon to CNES/GRGS, $R = 0.70$, and mascon to CSR, $R = 0.74$).

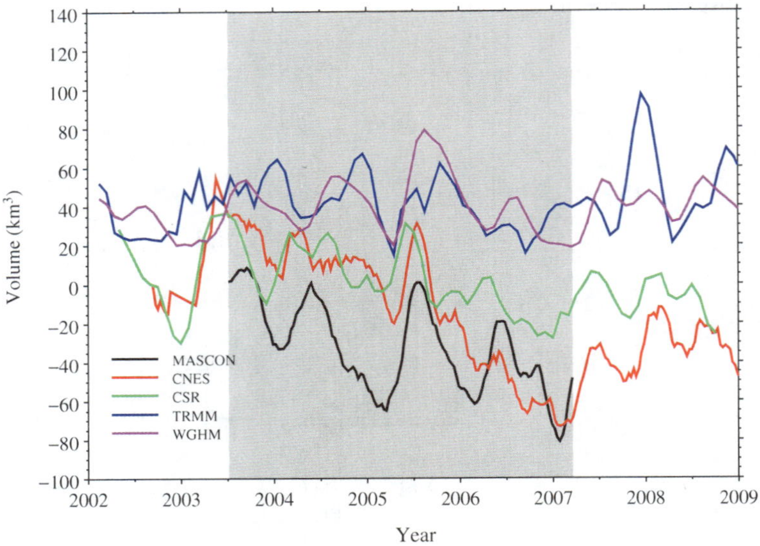

Fig. 13.24 The change in water storage over the MDB, as outlined by sector A in Fig. 13.23 for the data sets used in this work. A three-month moving average has been applied to each time-series. The gray shading marks the time span over which the mascon solutions are available. Cross-correlations between pairs of data sets are listed in Awange et al. [11]

The correlation between GRACE solutions is relatively high (>0.7), although much poorer when the GRACE solutions are compared to the TRMM and WGHM time-series (<0.5).

13.6 Concluding Remarks

As part of the solution to water scarcity problem, quick fix approach has revolved around supply management approach where infrastructures, e.g., dams have been constructed or expanded to increase the available water supply. Rijsberman [75] argue, however, that although the quick fix approach has largely succeeded in producing cheap food, water supply, and sanitation to a large number of people, many people still do not have access to safe and affordable drinking water despite huge investments. Quoting Cosgrove and Rijsberman (2000), Rijsberman [75] states:

> ...close to half the world population lacks access to sanitation, many rural poor do not have access to water for productive purposes, groundwater levels in key aquifers are falling rapidly, many rivers are no longer reaching the sea, etc.

As a shift from this school of thought, Rijsberman [75] points to the emergence of integrated water resource movement that has brought about organizations such as the World Water Council and the Global Water Partnership that is pushing for a demand management approach seeking [75]

1. to involve users more in the management of water, often through the establishment of forms of water user associations;
2. to price water and/or make it a trade-able commodity; and
3. to establish river basin authorities that integrate the usually fragmented government responsibilities for water into a single authority responsible for a hydrographically defined area, the river basin.

It is in support to (3) above that GRACE satellites have been recognized as having the potential to provide the first space-based estimate of changes in terrestrial water storage as demonstrated in the examples presented in Sect. 13.5. Such information will, in essence, assist water managers in conserving and controlling the utilization of dwindling water resources in a sustainable way. Water is arguably one of the most precious resources in the world, therefore, it is logical to try to monitor its distribution as efficiently as possible, and GRACE provides such opportunity. This is because one of the environmentally important signals detected by GRACE is the temporal gravity field variation induced by changes in the distribution of water on and below the Earth's surface, i.e., hydrology, e.g., [10]. Satellite altimetry provides the possibility of monitoring sea or lake surface heights as was demonstrated for Lake Naivasha. Global Navigation Satellite System plays a pivot role in supporting these satellites and also in providing location-based information for monitoring groundwater wells, and source of water pollutants as discussed in Chap. 14.

References

1. Abiya IO (1996) Towards sustainable utilization of lake Naivasha, Kenya. Lakes Reservoirs Res Manage 2(3–4):231–242. doi:10.1111/j.1440-1770.1996.tb00067.x
2. Alcamo J, Doll P, Kaspar F, Siebert S (1997) Global change and global scenarios of water use and availability: an application of WaterGAP 1.0. University of Kassel, CESR, Kassel
3. Alcamo J, Henrichs T, Rosch T (2000) World water in 2025: global modeling and scenario analysis. In: Rijsberman FR (ed) World water scenarios analyses. World Water Council, Marseille
4. Andersen OB, Seneviratne SI, Hinderer J, Viterbo P (2005) GRACE-derived terrestrial water storage depletion associated with the 2003 European heat wave. Geophys Res Lett 32(L18405):2–5. doi:10.1029/2005GL023574
5. Awange JL, Forootan E, Fleming K, Omondi P, Odhiambo G (2013) Remote sensing of the decadal water changes in the Nile Basin between. AGU books. Chapman conference on remote sensing of the terrestrial water cycle. (under review)
6. Awange JL, Forootan E, Kuhn M, Kusche J, Omondi P, Heck B (2013) (Submitted) Water storage changes and climate variability within the Nile Basin between 2002–2011. J Hydrol (under review)
7. Awange JL, Forootan E, Kusche J, Kiema JBK, Omondi PA, Heck B, Fleming KM, Ohanya S, Goncalves RM (2013) Understanding the decline of water storage across the Ramser-Lake Naivasha using satellite-based methods. Adv Water Resour doi:10.1016/j.advwatres.2013.07.002

8. Awange JL, Ong'ang'a O (2006) Lake victoria-ecology, Resource of the lake basin and environment. Springer, Berlin
9. Awange JL, Sharifi M, Ogonda G, Wickert J, Grafarend EW, Omulo M (2007) The Falling lake victoria water level: GRACE. TRIMM and CHAMP satellite analysis. Water Resour Manage. doi:10.1007/s11269-007-9191-y
10. Awange JL, Sharifi MA, Baur O, Keller W, Featherstone WE, Kuhn M (2009) GRACE hydrological monitoring of Australia. Current limitations and future prospects. J Spat Sci 54(1):23–36. doi:10.1080/14498596.2009.9635164
11. Awange JL, Fleming KM, Kuhn M, Featherstone WE, Heck B, Anjasmara I (2011) On the suitability of the 4 × 4° GRACE mascon solutions for remote sensing Australian hydrology. Remote Sens Environ 115:864–875. doi:10.1016/j.rse.2010.11.014
12. Awange JL (2012) Environmental monitoring using GNSS. Global navigation satellite system. Springer, Berlin
13. Awange JL, Kiema JBK (2013) Environmental geoinformatics. Monitoring and management. Springer, Berlin
14. Becht R, Haper DM (2002) Towards an understanding of human impact upon the hydrology of Lake Naivasha Kenya. Hydrobiologia 488(1–3):1–11. doi:10.1023/A:1023318007715
15. Becht R (2007) Environmental effects of the floricultural industry on the lake naivasha basin. ITC, The Netherlands
16. Becht R, Odada EO, Higgins S (2005) Lake Naivasha : experience and lessons learned brief. In: Lake basin management initiative: experience and lessons learned briefs. including the final report: managing lakes and basins for sustainable use, a report for lake basin managers and stakeholders. International lake environment committe foundation (ILEC), Kusatsu, pp 277–298
17. Becker M, Llowel W, Cazenave A, Güntner A, Crétaux J-F (2010) Recent hydrological behaviour of the East African Great Lakes region inferred from GRACE, satellite altimetry and rainfall observations. C R Geosci 342(3):223–233. doi:10.1016/j.crte.2009.12.010
18. Bettadpur S (2007) UTCSR Level-2 Processing standards document for Level-2 product release 0004. Gravity Recovery and climate experiment (GRACE) Rev 3.1, GRACE 327–742 (CSR-GR-03-03). Center for space research. The University of Texas at Austin
19. Beyene T, Lettenmaier DP, Kabat P (2010) Hydrologic impacts of climate change on the Nile River Basin: implications of the 2007 IPCC scenarios. Clim Change 100:433–461. doi: 10.1007/s10584-009-9693-0
20. Birnie, P, Boyle, A (1993) International law and the environment. Cambridge Law J 52(3):540 (Cambridge University Press)
21. Bonsor HC, Mansour MM, MacDonald AM, Hughes AG, Hipkin RG, Bedada T (2010) Interpretation of GRACE data of the Nile Basin using a groundwater recharge model. Hydrol Earth Syst Sci Discuss 7:4501–4533. doi:10.5194/hessd-7-4501-2010
22. Bower DR, Courtier N (1998) Precipitation effects on gravity measurements at the Canadian absolute gravity site. Phys Earth Planet Inter 106:353–369
23. Brutsaert W (2005) Hydrology—an introduction. 4th edn. Cambridge University Press, New York
24. Carolina BF (2002) Competition over water resources: analysis and mapping of water-related conflicts in the catchment of Lake Naivasha (Kenya). ITC. MSc Thesis
25. Casanova MT (1994) Vegetative and reproductive responses of charophytes to water-level fluctuations in permanent and temporary wetlands in Australia. Australian J Mar Freshw Res 45:1409–1419
26. Conway D (2005) From headwater tributaries to international river: observing and adapting to climate variability and change in the Nile Basin. Glob Environ Change 15:99–114. doi: 10.1016/j.gloenvcha.2005.01.003
27. Cosgrove WJ, Rijsberman FR (2000) World water vision: making water everybodys business. Earthscan Publications, London
28. Crowley JW, Mitrovica JX, Bailey RC, Tamisiea ME, Davis JL (2006) Land water storage within the Congo Basin inferred from GRACE satellite gravity data. Geophys Res Lett 33:L19402. doi:10.1029/2006GL027070

29. Damiata BN, Lee TC (2002) Gravitational attraction of solids of revolution —part 1: vertical circular cylinder with radial variation of density. J Appl Geophys 50(3):333–349. doi:10.1016/S0926-9851(02)00151-9
30. Darling W, Allen D, Armannsson H (1990) Indirect detection of subsurface outflow from a rift valley lake. World Dev 26(12):2169–2179
31. Duraiappah A (1998) Poverty and environmental degradation: a review and analysis of the nexus. World Dev 26(12): 2169–2179
32. Ellett KM, Walker JP, Rodell M, Chen JL, Western AW (2005) GRACE gravity fields as a new measure for assessing large-scale hydrological models. In: Zerger A, Argent RM (eds) International congress on modelling and simulation (MODSIM 2005). The modelling and simulation society of Australia and New Zealand, December 2005, pp 2911–2917, ISBN: 0-9758400-2-9
33. Ellett KM, Walker JP, Western AW, Rodell M (2006) A framework for assessing the potential of remote sensed gravity to provide new insight on the hydrology of the Murray-Darling Basin. Australian J Water Resour 10(2):89–101
34. Everard M, Harper DM (2002) Towards the sustainability of the Lake Naivasha Ramsar site and its catchment. Hydrobiologia 488(1–3):191–203. doi:10.1023/A:1023390430571
35. Everard M, Vale JA, Harper DM, Tarras-Wahlberg H (2002) The physical attributes of the Lake Naivasha catchment rivers. Hydrobiologia 488(1–3):13–25. doi:10.1023/A:1023349724553
36. Forootan E, Kusche J (2011) Separation of climate-driven signals in time-variable gravity using independent component analysis (ICA). J Geodesy 86:477–497. doi:10.1007/s00190-011-0532-5
37. Forootan E, Awange J, Kusche J, Heck B (2012) Independent patterns of water mass anomalies over Australia from satellite data and models. Remote Sens Environ 124:427–443. doi:10.1016/j.rse.2012.05.023
38. Garcia-Garcia D, Ummenhofer CC, Zlotnicki V (2011) Australian water mass variations from GRACE data linked to Indo-Pacific climate variability. Remote Sens Environ 115:2175–2183. doi:10.1016/j.rse.2011.04.007
39. Gleick PH (1999) The human right to water. Water Policy 1:487–503
40. Goodkind JM (1986) Continuous measurement of nontidal variations of gravity. J Geophys Res., 91(B9):9125–9134
41. Hamouda MA, Nour El-Din MN (2009) Vulnerability assessment of water resources systems in the Eastern Nile Basin. Water Resour Manage 23:2697–2725. doi:10.1007/s11269-009-9404-7
42. Hanjra MA, Qureshi ME (2010) Global water crisis and future food security in an era of climate change. Food Policy 35(5):365–377
43. Harper DM, Mavuti KM, Muchiri SM (1990) Ecology and management of Lake Naivasha, Kenya, in relation to climatic change, alien species' introductions, and agricultural development. Environ Conserv 17(04):328–336
44. Hofman AR (2004) The connection: water and energy security. http://www.iags.org/n0813043.htm Accessed 25 Aug 2010
45. Intergovernmental Panel on Climate Change (IPCC) (2007) Contribution of working group I to the fourth assessment report
46. International Lake Environment Committee (ILEC) (2005) Managing lakes and their basins for sustainable use. A report for the lake basin managers and stakeholders. In: International lakes environmental committee foundation, Kusatsu
47. IRIN humanitarian news and analysis, UN office for coordination of humanitarian affairs (2006) Global: The global water crisis: managing a dwindling resource. http://www.irinews.org Accessed 25 Sept 2011
48. Johnson LE (2009) Geographic information systems in water resources engineering. CRC Press, Taylor & Francis Group. ISBN 978-1-4200-6913-6
49. Jury WA, Vaux HJ, Jr (2007) The emerging global water crisis: managing scarcity and conflict between water users. Advances in agronomy, vol 95. Elsevier Inc., San Diego, p 77. doi:10.1016/50065-2113(07)95001-4

50. Kayombo S, Jorgensen SE (2006) Lake Victoria: experience and lessons learned brief. In: International lake environment committee, Lake Basin management initiative, Kusatsu. http://www.ilec.or.jp/eg/lbmi/reports/27_Lake_Victoria_27February2006.pdf Accessed 25 Sept 2011
51. Kenya Flower Council (KFC) (2011) Kenya flower council. http://www.kenyaflowercouncil.org Accessed 25 July 2010
52. Kull D (2006) Connections between recent water level drops in Lake Victoria, dam operations and drought. http://www.irn.org/programs/nile/pdf/060208vic.pdf. Accessed on 25 Sept 2011
53. Lambert A, Beaumont C (1997) Nano variation in gravity due to seasonal groundwater movements: implications for the gravitational detection of tectonic movements. J Geophys Res, 82(2):297–306
54. Leblanc M, Tregoning P, Ramillien G, Tweed S, Fakes A (2009) Basin-scale, integrated observations of the early 21st century multiyear drought in southeast Australia. Water Resour Res 45:(W04408). doi:10.1029/2008WR007333
55. Leick A (2004) GPS satellite surveying, 3rd edn. Wiley, New York
56. Leirião S, He X, Christiansen L, Andersen OB, Bauer-Gottwein P (2009) Calculation of the temporal gravity variation from spatially variable water storage change in soils and aquifers. J Hydrol 365:302–309
57. Mekonnen MM, Hoekstra AY (2010) Mitigating the water footprint of export cut flowers from the Lake Naivasha Basin, Kenya. Value of water research report series No. 45, UNESCO-IHE, Delft
58. Montgomery EL (1971) Determination of coefficient of storage by use of gravity measurements. Ph.D. Thesis, University of Arizona, Tucson
59. Mukai A, Higashi T, Takemoto S, Nakagawa I, Naito I (1995) Accurate estimation of atmospheric effects on gravity observations made with a superconducting gravity meter at Kyoto. Phys Earth Planet Inter 91(1–3):149–159
60. National Land and Water Resource Audit (NLWRA) (2001) Water resources in Australia. A summary of the National Land and Water Resource Audit's Australian water resources assessment 2000. Surf Water Groundwater Availability Qual
61. Neff RA, Parker CL, Kirschenmann FL, Tinch J, Lawrence RS (2011) Peak oil, food systems, and public health. Am J Public Health 101(9):1587–1597
62. Neumeyer J, Barthelmes F, Dierks O, Flechtner F, Harnisch M, Harnisch G, Hinderer J, Imanishi Y, Kroner C, Meurers B, Petrovic S, Reigber C, Schmidt R, Schwintzer P, Sun HP, Virtanen H (2006) Combination of temporal gravity variations resulting from superconducting gravimeter (SG) recordings, GRACE satellite observations and global hydrology models. J Geodesy 79(10–11):573–585. doi:10.1007/s00190-005-0014-8
63. Nicholson SE (1998) Historical fluctuations of Lake Victoria and other lakes in the Northern rift valley of East Africa. In: Lehman JT (ed) Environmental change and response in East African lakes. Kluwer, Dordrecht, pp 7–35
64. Nicholson SE (1999) Historical and modern fluctuations of lakes Tanganyika and Rukwa and their relationship to rainfall variability. Clim Change 41:53–71. doi:10.1023/A:1005424619718
65. Omondi P, Awange J, Ogallo Ininda J, Forootan E (2013) The influence of low frequency sea surface temperature modes on delineated decadal rainfall zones in Eastern Africa region. Adv Water Resour doi:doi.org/10.1016/j.advwatres.2013.01.001
66. Omondi P, Awange J, Ogallo LA, Okoola RA, Forootan E (2012) Decadal rainfall variability modes in observed rainfall records over East Africa and their relations to historical sea surface temperature changes. J Hydrol 464–465:140–156. doi:dx.doi.org/10.1016/j.jhydrol.2012.07.003
67. Owor M, Taylor RG, Tindimugaya C, Mwesigwa D (2009) Rainfall intensity and groundwater recharge: empirical evidence from the Upper Nile Basin. Environ Res Lett 4:035009. doi:10.1088/1748-9326/4/3/035009
68. Payne WA (2010) Farming systems and food Security in sub-saharan Africa. In: Lal R, Stewart BA (eds) Food security and soil quality, Taylor and Francis, New York

69. Pool DR, Eychaner JH (1995) Measurements of aquifer-storage change and specific yield using gravity surveys. Ground Water 33(3):425–432
70. Ramillien G, Cazenave A, Brunau O (2004) Global time variations of hydrological signals from GRACE satellite gravimetry. Geophys J Int 158(3):813–826. doi:10.1111/j.1365-246X.2004.02328.x
71. Ramillien G, Frappart F, Cazenave A, G?ntner A (2005) Time variations of land water storage from an inversion of two years of GRACE geoids [rapid communication]. Earth Planet Sci Lett 235(1–2):283–301. doi:10.1016/j.epsl.2005.04.005
72. Rekacewicz P (2006) Increased global water stress. Vital water graphics 2. Le monde diplomatique. http://www.grida.no/publications/vg/water2/page/3289.aspx Accessed 15 April 2012
73. Richardson JL, Richardson AE (1972) History of an African rift lake and its climatic implications. Ecol Monogr 42(4):499–534. doi:10.2307/1942169
74. Rieser D, Kuhn M, Pail1 R, Anjasmara IM, Awange J (2010) Relation between GRACE-derived surface mass variations and precipitation over Australia. Australian J Earth Sci 57(7):887–900. doi:10.1080/08120099.2010.512645
75. Rijsberman FR (2006) Water scarcity: fact or fiction. Agric Water Manag 80:522, Elsevier
76. Rodell M, Famiglietti JS (1999) Detectability of variations in continental water storage from satellite observations of the time dependent gravity field. Water Resour Res 35(9):2705–2724. doi:10.1029/1999WR900141
77. Rodell M, Famiglietti JS, Chen J, Seneviratne SI, Viterbo P, Holl S, Wilson CR (2004) Basin scale estimates of evapotranspiration using GRACE and other observations. Geophys Res Lett 31:L20504. doi:10.1029/2004GL020873
78. Rodell M, Chen J, Kato H, Famiglietti JS, Nigro J, Wilson CR (2006) Estimating groundwater storage changes in the Mississippi River basin (USA) using GRACE. Hydrogeol J 15(1):159–166. doi:10.1007/s10040-006-0103-7
79. Rodell M, Felicogna I, Famiglietti JS (2009) Satellite-based estimates of groundwater depletion in India. Nature 460:999–1003. doi:10.1038/nature08238
80. Sahin M (1985) Hydrology of the Nile Basin. Developments in water science, vol 21. Elsevier, New York, p 575
81. Sansome HW (1952) The trend of rainfall in East Africa. E. Afr. Meteorol. Dep Tech Mem 1:14
82. Schelton D (1991) Human rights, environmental rights, and the right to environment Heinonline. Stan J Int Law 28(1):103–138
83. Schmidt R, Schwintzer P, Flechtner F, Reigber C, Güntner A, Döll P, Ramillien G, Cazenave A, Petrovic S, Jochmann H, Wünsch J (2006b) GRACE observations of changes in continental water storage. Global Planet Change 50(1–2):112–126. doi:10.1016/j.gloplacha.2004.11.018
84. Senay GB, Asante K, Artan G (2009) Water balance dynamics in the Nile Basin. Hydrol Process 23:3675–3681. doi:10.1002/hyp.7364
85. Smith AB, Walker JP, Western AW, Ellett KM (2005) Using ground based measurements to monitor changes in teresstrial water storage. In: 29th Hydrology and water resources symposium (CD Rom), Institute of Engineers Australia, Sydney
86. Steffen W, Sanderson A, Tyson PD, J?ger J, Matson PA, Moore BIII, Oldfield F, Richardson K, Schellnhuber HJ, Turner BLII, Wasson RJ, (2005) Global change and the Earth system: A planet under pressure. Springer-Verlag, Berlin
87. Swenson S, Yeh PJ-F, Wahr J, Famiglietti J (2006) A comparison of terrestrial water storage variations from GRACE with in situ measurements from Illinois. Geophys Res Lett 33:L16401. doi:10.1029/2006GL026962
88. Swenson S, Wahr J (2009) Monitoring the water balance of Lake Victoria, East Africa, from space. J Hydrol 370:163–176. doi:10.1016/j.jhydrol.2009.03.008
89. Swenson S, Wahr J, Milly PCD (2003) Estimated accuracies of regional water storage variations inferred from the gravity recovery and climate experiment (GRACE). Water Resour Res 39(8):1223. doi:10.1029/2002WR001736
90. Syed T, Famiglietti J, Rodell M, Chen J, Wilson C (2008) Total basin discharge for the Amazon and Mississippi River basins from GRACE and a land-atmosphere water balance. Water Resour Res 44:W02433. doi:10.1029/2006WR005779

91. Tapley BD, Bettadpur S, Ries JC, Thompson PF, Watkins MM (2004) GRACE measurements of mass variability in the Earth system. Science 305:503–505. doi:10.1126/science.1099192
92. Taylor CJ, Alley WM (2001) Ground-water-level monitoring and the importance of long-term water-level data. U.S. Geological Survey Circular 1217, Denver
93. Tiwari VM, Wahr J, Swenson S (2009) Dwindling groundwater resources in northern India, from satellite gravity observations. Geophys Res Lett 36:L18401. doi:10.1029/2009GL039401
94. Ummenhofer C, England M, McIntosh P, Meyers G, Pook M, Risbey J, Gupta A, Taschetto A (2009) What causes southeast Australias worst droughts? Geophys Res Lett 36:L04706. doi:10.1029/2008GL036801
95. UNEP (2002) A World of Salt: Total global saltwater and freshwater estimates. http://www.unep.org/dewa/assessments/ecosystems/water/vitalwater/freshwater.htm Accessed 25 Aug 2010
96. Vincent CE, Davies TD, Beresford UC (1979) Recent changes in the level of Lake Naivasha, Kenya, as an indicator of equatorial westerlies over East Africa. Clim Change 2(2):175–189. doi:10.1007/BF00133223
97. Vorosmarty CJ, Green P, Salisbury J, Lammers RB (2000) Global water resources: vulnerability from climate change and population growth. Science 289:284288
98. Wallace JS (2000) Increasing agricultural water efficiency to meet future food production. Agric Ecosyst Environ 82:105119
99. Wallace JS, Gregory PJ (2002) Water resources and their use in food production. Aquat Sci 64:363375
100. Whittington D, McClelland E (1992) Opportunities for regional and international cooperation in the Nile Basin. Water Int 17(3):144–154. doi:10.1080/02508069208686134
101. Winsemius HC, Savenije HHG, van de Giesen NC, van den Hurk B, Zapreeva EA, Klees R (2006) Assessment of gravity recovery and climate experiment (GRACE) temporal signature over the upper Zambezi. Water Resour Res 42:W12201. doi:10.1029/2006WR005192
102. World Bank (2003) Water resource and environment. In: Davis R, Hirji R (eds). Technical Note G.2, Lake Management
103. Yan JP, Hinderer M, Einsele G (2002) Geochemical evolution of closed-basin lakes: general model and application to Lakes Qinghai and Turkana. Sed Geol 148(1–2):105–122. doi:10.1016/S0037-0738(01)00212-3
104. Yang H, Reichert P, Abbaspour K, Zehnder AJB (2003) A water resources threshold and its implications for food security. Environ Sci Technol 37:30483054
105. Yates DN, Strzepek KM (1998) Modelling the Nile Basin under climate change. J Hydrol Eng 3(2):98–108. doi:10.1061/(ASCE)1084-0699(1998)3:2(98)
106. Zehnder AJB, Yang H, Schertenleib R (2003) Water issues: the need for action at different levels. Acquatic Sci Res Across Boundaries 65(1):1–20. doi:10.1007/s000270300000

Chapter 14
Environmental Pollution

People blame their environment. There is only one person to blame—and only one—themselves.

Robert Collier

14.1 Concept of Pollution

There exist various definitions to the word *pollution* depending on one's jurisdiction and the laws of a particular country. Springer [39, see references therein] looks at the meaningful concept of defining pollution in international law by posing the questions: "What are you talking about when you are talking about pollution? What is pollution? How would you define it if you are going to remove the concept of damage from it?" These questions are not easily answerable and as Springer [39] acknowledges, the term pollution is a word whose precise meaning in law, particularly international law, is not easily discerned [39]. It has been used in a wide variety of contexts, from international conventions to pessimistic speeches about the state of the environment, to describe different levels and kinds of man-induced changes in the natural world [39].

Within a particular context, pollution assumes a meaning, either explicitly stated or implicitly developed, which may bear little resemblance to its usage elsewhere. Springer [39] discusses the basic conceptual categories within which pollution has been approached in the past such as; pollution as an alteration of existing environment, pollution as a right to territorial sovereignty, pollution as a damage (to man and his property, and the environment), pollution as interference with other uses of environment, and pollution as exceeding assimilative capacity of environment.

N. O. Agola and J. L. Awange, *Globalized Poverty and Environment*,
DOI: 10.1007/978-3-642-39733-2_14, © Springer-Verlag Berlin Heidelberg 2014

14.1.1 Pollution: Poverty-Environmental Nexus

Pollution can occur on land , in the air or in water and can be local or regional in scope. The role it plays in the poverty-environmental linkage nexus has been analysed, e.g., by Duraiappah [15] and Dasgupta et al. [12]. Impacts of pollution on land, water or air often results in health related problems (e.g., Mink [27]), with those mostly affected being the poor who have difficulties in affording medical treatments. For instance, the World Bank estimated that approximately 1.3 billion people mostly in developing countries lived in towns that did not meet world health organization (WHO) standards [15]. According to Duraiappah [15], the estimation by World Bank above only covers outdoor air pollution, and were about 400–700 million (mostly rural women and children) people exposed to unsafe levels of indoor pollution to be included, then approximately two-fifths of the world's population, most of them located in developing countries, would not enjoy the basic right to clean air (e.g., Oodit and Somonis [33]).

Although Chap. 13 highlighted the scarcity of water as a major threat to the poverty-environmental nexus, water contamination and pollution pose a much more immediate serious problem since safe water and adequate sanitation are essential for health and well being of the society. In developing countries, access to safe drinking water is still out of reach for many people (see, e.g., Mink [27]), and the most vulnerable group are often the poor, and particularly children. For instance, Dasgupta et al. [12] writes:

> Safe water and adequate sanitation are critical determinants of health status, particularly for children. Ingestion of coliform bacteria from contaminated drinking water or food is a prime cause of diarrheal disease, which is in turn a major cause of infant mortality in developing countries.

In most developing countries access by the poor to clean form of energy, e.g., electricity is out of reach. Wood, cow dung, and charcoal, which are highly polluting biomass, form the main source of energy for cooking while kerosene is often used in lamps for lighting (e.g., Mink [27]; Tolba et al. [42]). These source of energy, i.e., wood fuel, contribute not only to unsustainable deforestation activities (e.g., [15]) but also to indoor pollution that lead to significant health problems (e.g., Dasgupta et al. [12]). Outdoor air pollution is attributed, e.g., by Duraiappah [15] to industry and higher income groups that are largely motivated by profit and affluence that benefit from lack of policy instruments. By its nature, outdoor pollution can be said to be largely an urban problem (e.g., those emanating from vehicle and industry pollution, see Yang et al. [50]) although its extend can take on a regional or global scale as is the case of global warming caused by emission of carbon dioxide (CO_2). Dasgupta et al. [12] points out that the severity depends on the scale of polluting activities, their pollution intensity (or pollution per unit of output), and the characteristics of the urban air shed.

As pointed out by Leitman [22], the impact of air pollution (indoor and outdoor) are income dependent. For instance, in case of outdoor air pollution, Dixon et al. [14] finds the high income group capable of coping compared to the low income group.

Duraiappah [15] attributes the difficulties experienced by the low income group to the fact that in many cases, factories are situated in or close to the low income neighbourhood. In linking pollution to the poverty-environmental nexus, Duraiappah [15] writes on water pollution:

> In the case of water pollution, commercial agents are driven primarily by profit motives. On the other hand, the low-income groups pollute because of a lack of provision of proper sanitation and drinking water facilities by governmental agencies. The presence of endogenous poverty causing environmental degradation in this case is to be expected as the water degradation leaves the low-income groups no other option but to degrade further the existing water supply. This in turn causes the impacts, which then set into motion the case where environmental degradation causes poverty and the spiral continues.

On air pollution, Duraiappah [15] writes

> The resulting health consequences arising from outdoor air pollution are more prominent in the low income groups. The rise in respiratory diseases among the low income groups implies a drop in productivity, which in turn forces many to lose their jobs and source of income. The ensuing drop in income forces these groups to experience economic and social hardship which over time results in poverty. This is a classic example of how environmental degradation causes poverty.

14.1.2 Geospatial Applications to Pollution

Within the poverty-environmental nexus, pollution, whether on land, water, or air often take on spatial dimension. Although several studies have been undertaken to assess the health impacts that pollution poses to the poor (e.g., Mink [27]; Akbar and Lvovsky [1]; Bosch et al. [7]), those that relate pollution to the spatial dimension have been few (e.g., Dasgupta et al. [12]). Geospatial tools provide possibilities of providing the spatial dimension that could complete the poverty-environmental-pollution link. Some of the possible applications include [3, 4]:

(a) providing maps that indicate the precise *locations of the point and non-point sources of pollutants*. Global navigation satellite system (GNSS) could be very useful in mapping point sources, while non-point sources could be mapped using remote sensing.
(b) mapping the *spatial coverage* of areas that are already affected as well as those likely to be affected by pollution. Remote sensing and geographical information system (GIS) could be very useful in this endeavor. GNSS could also be useful in providing boundaries for sampling data for the purpose of pollution monitoring, e.g., air pollution monitoring that is irregular and cannot be represented by regular grids, see e.g., [16].
(c) *documenting* and *visualizing* the *changes* in areas affected by pollution. High resolution, multispectral and multi-temporal satellite imagery could be very useful in this.

(d) *modeling* various possible impact scenarios for pollution to develop strategies for curtailing or minimizing human exposure to pollution. This is the strength of GIS.

14.2 Water Pollution

That fresh water has played a key role in agriculture, industries, and municipalities, among others is highlighted for instance by Mackenzie [24, p. 314]. Whereas most of the world's fresh water lakes carry natural substances and nutrients, today, in addition to these natural materials, pollutants, e.g., chemicals and excess eroded soil also find their way into the lakes, e.g., [6].

Oceans, home to salt water marine, also face pollution from human-induced activities as witnessed by the BP oil leak at the Gulf of Mexico in 2010. There is need, therefore, to ensure that this vital source of life suffers minimum pollution as possible. Mackenzie [24, p. 314] defines water pollution as any physical or chemical change in water from both natural and anthropogenic sources that adversely affects the organisms living in it. Water resources undergoes pollution from both ***point*** and ***non-point*** sources discussed next.

14.2.1 Point and Non-point Sources

Points sources are, e.g., factory outlet pipes and sewage treatment plant outlets whose positions can be accurately mapped using GNSS receivers. *Surface pollutants* include those produced from factories and industries, and agricultural activities. During rainy seasons, storm water carries with it chemicals from fertilizers and pesticides from farms, together with other nutrients from the neighboring towns. These materials are either swept into water bodies (rivers, dams, boreholes, etc) through surface runoffs or percolate into the ground to reach groundwater flows, and finally into the rivers. The rivers eventually pour their contents into the lakes and oceans. The consequence of sewage contamination of water quality include outbreak of human diseases and eutrophication problem, which result in the proliferation of algal blooms and undesirable aquatic macrophytes, e.g., water hyacinth, see e.g., [6]. In some developing countries, where the compliance and enforcement of environmental laws is generally weak, such as Kenya, raw sewage carrying human waste and other industrial effluent are emptied untreated directly into the rivers or lakes. Regarding increased effects of water pollution resulting from human waste, Duraiappah [15] states (Fig. 14.1):

> In the past, human waste was deposited naturally in natural systems but with increasing populations, the load of human waste has far exceeded the natural systems absorption and cleansing rate. Therefore, without modern sanitation systems to help relieve the natural systems, these systems, including water, degrade.

Fig. 14.1 Mobile hand-held GNSS. *Source* ©1995–2010 ESRI. http://www.esri.com/technology-topics/mobile-gis/graphics/geoxt-web-lg.jpg

These sentiments of Duraiappah [15] is exemplified by the case of Nyalenda in Kisumu (Kenya) where pit latrines (Fig. 14.2) are in the same locality with boreholes supplying groundwater (Fig. 14.3) thus contaminating the water. Nyalenda is inhabited mainly by low income group, most of who rely on these boreholes for their daily water needs. With the pit latrines contaminating their water source, most of these poor people with limited affordability of health services are at risk of water-borne diseases.

Geospatial technique of GNSS could be useful in mapping the locations and perimeters of these sources of sewage pollution (pit latrines) and boreholes, what could simply be achieved using a low cost hand-held GNSS receiver (Fig. 14.1). This data can then be integrated into a GIS system to inform decisions by the stakeholders. Details of the application of these geospatial technique together with GIS are discussed in Awange [4] and Awange and Kiema [3].

An example of the effect of toxic chemical poisoning of water and its subsequent repercussion on human health was illustrated by the mercury poisoning of Minamata Bay in Japan [24, p. 324]. Around 1950s, mercury was discharged from a chemical plant located on a river flowing into the bay. The traces of this mercury found themselves in fish and were eventually consumed by people leading to the Mad Hatter's disease, a disease that affects the nervous system. Mackenzie [24, p. 324] report that older Japanese living around the bay today still exhibit evidence of the disease that has killed a number of people.

Non-point sources are those that are difficult to identify, e.g., pollution from transportation sector and waste from mining operations, agricultural activities, and waste carried by street runoffs and drainage. From the atmosphere, precipitation containing chemical such as *sulfur* and *nitrogen* oxides also contaminate water sources. Owing to their dispersive nature, non-point sources are difficult to control and can only be mapped using remote sensing techniques.

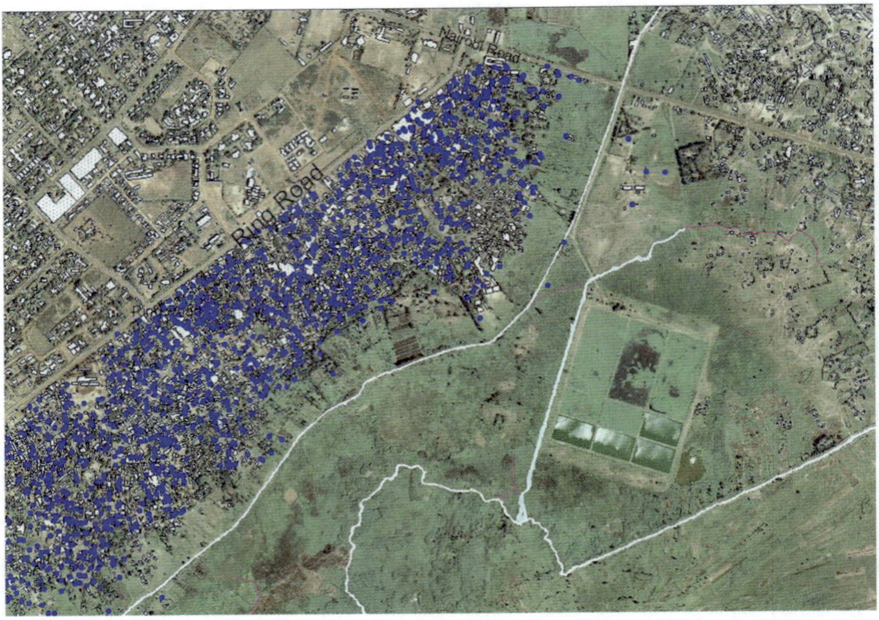

Fig. 14.2 Pit latrine (*dotted blue*) plotted on a Map of Nyalenda (Kisumu, Kenya). Nyalenda is inhabited by low income group that rely on the boreholes for water supply. GNSS could be useful in providing the perimeter/area covered by these pit latrines in addition to their actual locations. This could be mapped into a GIS system to support management issues. *Source* Department of Planning, MCK (Municipality of Kisumu, Kenya) and Regional Center for Mapping of Resources for Development (2006) in Opande [34]

For Lake Victoria in East Africa that supplies more than 30 million people with water (see e.g, Sect. 13.5.4 and also Awange and Ong'ang'a [6]), studies indicate that the high *phosphorous* and *nitrogen* loans choking it are attributed to several non-point sources. The most important of these are *agriculture*, *livestock*, *domestic* and *industrial effluents*. Most of these activities use large amounts of synthetic compounds including fertilizers and pesticides, see e.g., [20]. Although animal manure and domestic wastes contribute about 3000 tons and 132 tons of the total, respectively, their nitrogen inputs are estimated at over 40,000 tons per year [2, p. 106]. This represents more than three times the amount contributed by synthetic fertilizers. GNSS could be useful in providing sampling boundaries for measuring the contributions of these pollutants, see e.g., [16].

14.2.2 Eutrophication of Lakes

Mackenzie [24, p. 318] defines eutrophication as the process of being fed too well. Eutrophication leads to water quality deterioration, taste and odour problems, oxygen

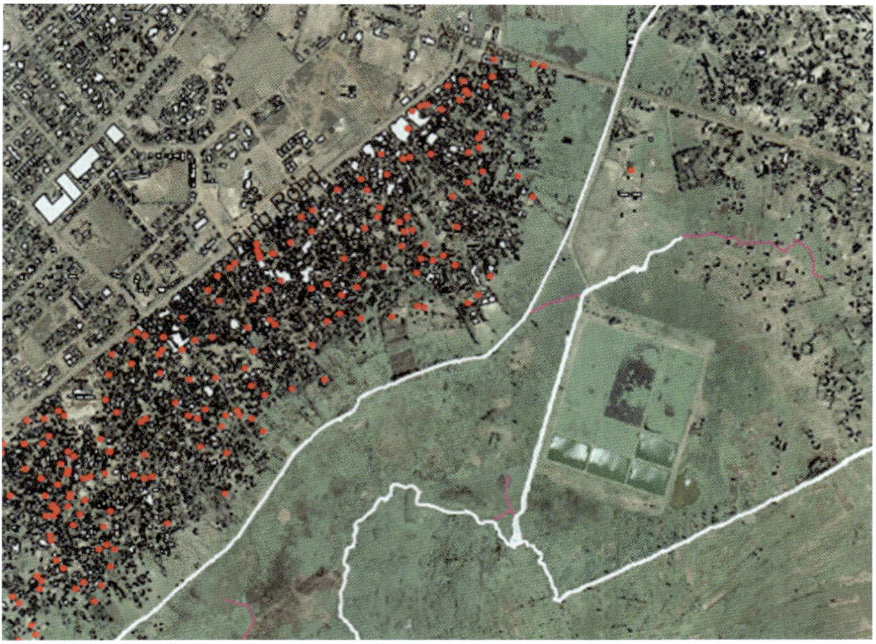

Fig. 14.3 Borehole point locations (*dotted red*) plotted on a Map of Nyalenda (Kisumu, Kenya). GNSS could be useful in providing the positions of these wells, which could be mapped into a GIS system to support management issues. *Source* Department of Planning, MCK and Regional Center for Mapping of Resources for Development (2006) in Opande [34]

depletion, increased turbidity, decline of fisheries, possible fish kills, clogging of waterways and toxic effects on animals and human beings. The phenomenon is a typical result of nutrient imbalances at several levels. The source of nutrients include agricultural activities, which produce a nutrient rich runoff resulting from the leaching of fertilizers and manure, garbage dumps, sewage, and industrial effluent. Odada et al. [30] listed three causes of eutrophication as:

(a) Enhanced effluent discharge.
(b) Runoff and storm water.
(c) Enhanced discharge of solids.

The first two are the most important causes of eutrophication. Sources of effluent discharge could be mapped using hand-held GNSS receivers (Fig. 14.1), while remote sensing and GIS could be applied to monitor effluent discharge. In the US, the *Clean Water Act* 1972 focused on eliminating point source pollutants through regulating discharges from such point-sources. This was achieved by setting effluent limits which businesses were required to adhere to. By the 1980s, better control point-source posed a minor threat to public waters and focus shifted to storm water management. The San Francisco Public Utilities Commission (SFPUC) in its attempt to reduce storm water pollution mapped the city's storm drain using a hand-held GNSS-based data collection device (e.g., Fig. 14.1) that captured the drain's precise locations and recorded digital notes on their conditions creating a GIS database [11].

14.3 Air Pollution

14.3.1 Background

Environmental monitoring is concerned with monitoring the Earth's surface as well as the atmosphere. The Earth's atmosphere is known to be sensitive to *global warming*, *ozone layer depletion*, and *air pollution*. Air pollution is the emitting of harmful substances into the air that are detrimental to the environment. These substances could include, e.g., harmful poisonous gases, volcanic eruptions, e.g., the Indonesian eruption of 2010, greenhouse gases such as CO_2 that contribute to global warming, just to list but a few. Perhaps, the mostly crucial, is the emission of greenhouse gases from transport sector.

In both developed and developing nations, urban air pollution is increasingly being recognized as a major public health and environmental concern (see, e.g., Dasgupta et al. [12] and the references therein). In most developing countries, however, air quality monitoring is not routinely conducted and in some urban areas, such information does not even exist, though signs of deteriorating air quality and health problems related to air pollution are visible [31]. In Awange [4] and Awange and Kiema [3], the role and possible contribution of geospatial to global warming monitoring is discussed. In the next section, the role and possible contributions of geospatial in monitoring local pollution are presented. The benefits of local monitoring is that it supports environmental quality management. The fact that environmental management improves by knowledge of local conditions is acknowledged by Dasgupta et al. [12] who state that an efficient regulation requires local inspection of damaged sources such as pollution, as well as more centralized information collection, storage and analysis.

Remote sensing and GIS have been proven to be most appropriate and effective techniques for handling spatial data that describe urban air pollution, urban thermal environment, and land use and land cover (LULC) patterns [49]. For example, by making use of remote sensing techniques, it is possible to obtain reasonably high-quality land surface temperature (LST) estimates [29, 32, 47]. Satellite data has been extensively used in air pollution mapping and monitoring in many urban environments, see e.g., [38, 44, 48] etc. On the other hand, as a decision support tool, GIS is effective in mapping the spatio-temporal pattern of air pollution and examining its association with diverse land use activities [28, 49]. GIS has also been employed to model air pollution variation through various statistical interpolation methods, see e.g., [9, 10, 13, 41] etc. Furthermore, GIS offers the potential for exposure assessment in support of not only air pollution epidemiology, but also air quality policy and environmental justice [8, 18, 25].

GNSS is useful in providing location data. It is advantageous that an integrated approach drawing from remote sensing, GIS, and GNSS be developed and applied when examining the fairly complex relationships among spatial variables such as urban land use, pollution, and thermal variation, especially within an urban context. At a local level, one major air pollutant is the transport sector that emits CO_2 gases to the atmosphere. Example of possible use of GNSS to monitor transport related pollution is presented in Sect. 14.3.2. Other source include industrial pollutants whose exact locations can readily be mapped using a hand-held GNSS receivers. These GNSS-based location data could then be integrated with other information such as the emission level per location in a GIS system to give a real-time air pollution monitoring capability.

For non-point sources, GNSS could play a role of mapping atmospheric quality parameters by being integrated with other devices in a geosensor network (e.g., [3, 4]), e.g., gas sensors, remote sensors, cell phones, and laptops in a mobile device such as that discussed, e.g., in Schreiner et al. [35]. In the work of Schreiner et al. [35], GNSS provided location-based data in terms of longitude, latitude, and altitude at points where the sensor recorded the pollution level. Both the sensor and GNSS transmitted their measured data to a laptop, which processed the information and relayed it to a mobile phone for remedial actions to be taken.

14.3.2 Pollution from Transportation Sector

Increased traffic volumes (see e.g., Fig. 14.4, right) in major towns such as Beijing, New York, etc., has potential for affecting ambient air. In most developing nations, such as Kenya, most of the vehicles are of the older generation (e.g., Fig. 14.4, left) and cannot sustainably run on unleaded fuel, as the conditions generated during combustion contribute to a rapid erosion of the valve seats, resulting in loss of engine performance through poor valve sealing.

The high lead content combined with the low combustion efficiency of the older type of engines lead to the emission of high volumes of exhaust gases, which contain semi-combusted fuels combined with worn out engine parts and waste oils [4, 6], hence high concentration of particulate matter in the air. From their study of air pollution in Beijing, Yang et al. [50] found that the relative particulate mass and the elemental concentrations of crustal and pollutant elements in the air particulate matter collected over the urban area was higher than in rural areas.

The possible role that could be played by geospatial tools in supporting monitoring of transportation related emission and noise pollution is exemplified in the study of Taylor et al. [40] who discuss the Transportation System Center (TSC) developed at the University of South Australia based on the integration of GIS and GNSS systems. In this system, GIS played the role of data management, i.e., data entry and integration, data analysis and display, while GNSS had the role of determining the locations of both static observations and dynamic recordings of the vehicle positions over time. The GIS-GNSS system is further integrated with an engine management

Fig. 14.4 Exhaust pollution and vehicle congestion

Fig. 14.5 Informal waste dumping locations (*dotted red*) plotted on Map of Nyalenda (Kisumu, Kenya). In Fig. 14.3 note that boreholes provide source of water for Nyalenda. Solid waste disposal indicated in this figure could thus be a source of pollution to this very water source that sustains livelihood to the inhabitants of Nyalenda. GNSS could be useful in providing these locations. *Source* Opande [34]

system of the vehicle to provide time-tagged data on GNSS position and speed, distance traveled, acceleration, fuel consumption, engine performance, and air pollution on a second-by-second basis [40].

Such data is vital not only for air pollution monitoring but also for *energy conservation* through monitoring of engine performance and fuel consumption. Acceleration data are useful in monitoring *noise pollution* level coming from the engine. An essential component of transportation that contributes to large fuel consumption and higher emission of greenhouse CO_2 gases is *traffic congestion*. In this GIS-GNSS system of [40], GNSS played a role of monitoring traffic congestion by measuring some components of total delay time, e.g., stopped time and acceleration noise (indicated by speed variation depending on the road condition). GNSS could also be useful in providing speed data, see e.g., [51, 52]. Further, congestion can be identified by the presence of queues, and as such, the knowledge of the incidence of queuing in any congestion monitoring system is essential [40]. As already stated, GNSS could help in measuring the proportion of stopped time (PST), which indicate the amount of time spent in queues during a journey [40].

Example 14.1 (Air pollution in Guangzhou, China. Source: [49])

This study investigated the air pollution pattern in this fast growing political, economic, and cultural center in South China using geospatial. Findings from the study indicate that high-quality data and reliable information regarding Land use and land cover (LULC) patterns and Land surface temperature (LST) can be derived with satellite remote sensing and GIS technologies and that the data so derived closely correspond to ambient air-quality measurements. Furthermore, the spatial patterns of air pollution depend on many factors tied to land-use activities, e.g., division of functional districts, distribution of land-use types, water bodies and parks, building and population densities, layout of the transportation network, and air flushing rates.

In Guangzhou, ambient air quality was monitored between 1980 and 2000 and concentrations of sulfur dioxide (SO_2), nitrogen oxides (NO_x), carbon monoxide (CO), and total suspended particles (TSPs), as well as dust level spatially documented using geographic distribution maps. These maps were digitized and converted into GIS data layers for interpretation of the spatial patterns of air pollution. Results indicate that the impact of various factors varied with different pollutants, and the street canyon effect was particularly evident owing to Guangzhous closely spaced highrise buildings and low-wind environment. Because of the locations of industrial plants, high population density, clustering of the catering industry, and low air-flushing rates, two urban localities (namely the Liwan and Yuanchun Districts) became the pollution hubs for SO_2, dust, and other pollutants [49]. Positive correlation identified between the concentrations of the pollutants probed and satellite-derived LST values confirms that both ambient air quality and LSTs were related to land use activities. The study demonstrates that GIS is effective geospatial tool in examining the spatial pattern of air pollution and its association with urban built-up density.

End of Example 14.1

14.4 Land Pollution

14.4.1 Solid Waste Collection and Management

Natural resources on the fringes of urban areas always suffer severe depletion and degradation as urban centers create demand for goods and services thus exerting tremendous pressure on fragile ecosystems. The high rate of urbanization therefore rapidly increases the pressure on these ecosystems. The increasing population and economic growth of cities associated with urbanization create externalities due to the demand for resources and waste disposal [22].

Solid waste includes refuse from households, non-hazardous solid waste from industrial, commercial and institutional establishments (including hospitals and other health care facilities), market waste, yard waste and street sweepings. Solid Waste Management (SWM) encompasses the functions of storage, collection, transfer, treatment, recycling, resource recovery and disposal. Its first goal is to protect the health of the population. Other goals include promotion of *environmental quality* and *sustainability*, support of economic productivity and employment generation. Achievement of SWM goals requires sustainable systems, which are adapted to and carried out by the municipalities and their local partners including the communities within their jurisdictions.

In many cities of developing countries, private or public systems of SWM are inadequate, only able to achieve collection rates of between 30 and 50 %. The uncollected wastes are often disposed off in ways detrimental to the environment such as open burning, burying, or dumping in rivers [17]. This problem is further compounded in low income areas where the authorities normally prioritize SWM lowly as compared to water supply, electricity, roads, drains, and sanitary services [36]. SWM is a major responsibility of local governments in developing countries consuming between 20 and 50 % of their total budgets [46]. It is a complex supply driven task, which depends as much upon organization and cooperation between households, communities, private enterprises, and government authorities, as it does upon the selection and application of appropriate technical solutions for collection, transfer, recycling and disposal [43]. It should therefore, be approached from the perspective of the entire cycle of material use with a broad scope encompassing planning and management, *waste generation* and waste handling processes [36].

14.4.2 Role of Geospatial in Solid Waste Management

The functioning of solid waste management systems is influenced by the waste handling patterns and underlying attitudes of the urban population who are the waste generators; these factors are themselves, conditioned by the people's social and cultural context [36]. Fast growing low-income residential communities often comprise of a considerable social, religious and ethnic diversity, which strongly influences the waste characteristics and socio-economic patterns thereby influencing the choice of SWM techniques to be applied.

In urban areas, the physical characteristics of a settlement such as the road conditions, topography, waste characteristics, etc., need to be considered when selecting and/or designing waste collection and transportation procedures and equipment. Use of geospatial technology could be valuable in visualizing this physical characteristic to improve on collection efficiency. More importantly, geospatial tools (GNSS and GIS) could be vital tools for urban planners in that they could enable them harmonize solid waste management and other services such as water, electricity, access roads, physical planning, and sanitation, all of which are associated with inadequate service delivery within a municipality. This technology would help in getting the exact location and properties of the containers, understanding the attributes of the infrastructure, and routing of the secondary collection vehicles.

At the level of natural systems, the interaction between waste handling procedures and public health conditions is influenced by climatic conditions such as seasonal weather variations and other ecological factors, e.g., the nature of animals reared within the homestead. In practical terms, climate determines the frequency by which waste collection points must be serviced in order to limit negative environmental consequences. The contribution of geospatial to weather and climate monitoring are presented in Awange [4] and Awange and Kiema [3].

The management methods and techniques employed in SWM should pay more attention to integrated approaches based on adequate information systems, e.g., GIS among other tools. With regard to operational planning and appropriate management methods, the approach should include among others, data collection techniques, analysis of waste composition, projecting waste amounts, scenario techniques and formulation of equipment specifications [37], all of which could be supported by geospatial as discussed above.

14.4.3 Solid Waste from Transportation Sector

Although pollution from industry, agrochemical factories and farmlands have been well documented (see, e.g., Sect. 14.2), the contribution of the transportation sector to the pollution of environment in terms of solid waste also deserve to be mentioned. Contribution of pollution from the vehicles, particularly in developing countries, takes the form of;

Fig. 14.6 Waste accumulation at bus termini in Kisumu (Kenya)

Fig. 14.7 Wastes from vehicle repair known as Jua Kali in Kisumu (Kenya)

- litter generated at the termini and along the roads (e.g., Fig. 14.6),
- vehicle repair, service and maintenance yards and garages commonly known in Kenya as "Jua Kali" (i.e., open garages indicated by Figs. 14.7 and 14.8),
- exhaust fumes and oil spills on to the roads, which are washed into the water sources (Figs. 14.9 and 14.10).

A hand-held GNSS receiver (e.g., Fig. 14.1) could be useful in providing locations of these sources, which could then be mapped and integrated with other data in a GIS to develop a real-time monitoring system for decision makers in municipalities.

Car Washing. In East Africa, an increasingly popular practice at the beaches today is the car washing (Fig. 14.11) whose history stretches back to the late 1950s. The beach records very high turn over of vehicles of different classes that are washed at the Lake every day, see e.g., [5, 6]. Casual observations reveal that the shore waters are now dirtier and greasier. The waste oils from these vehicles, and dirt accumulated in transit are all washed into the Lake. Some of the vehicles have leaking systems hence the oils or the fuels drip into the Lake while they are being washed, further compounding the problem.

As can be seen in Fig. 14.12, the shores are no longer supporting the original vegetation or other aquatic life forms, due to the conscious removal by the car washers or possibly due to the decrease in the oxygen levels in the water occasioned by the oil covering the water surface.

In Fig. 14.12, the floral and faunal composition has also changed as can be detected by the lack of shells at the car washing area. There has been a change in the trophic

Fig. 14.8 The spray painting process and wastes in Kisumu (Kenya)

Fig. 14.9 Oil spills in Kisumu (Kenya)

Fig. 14.10 Petrol station operations and the Caltex petrol depot at the Lake side in Kisumu (Kenya)

relations at the car washing area, which affects the entire Lake. It should be pointed out that Kichinjio beach in Kisumu (Kenya) is one of the breeding sites for some fish species and other aquatic life forms within the lake, a characteristic that is fast disappearing. The disturbance accompanied by oils in the water could have adversely affected them. A hand-held GNSS receiver as well as high resolution satellite imagery

Fig. 14.11 Washing of heavy petroleum vehicle in Lake Victoria in Kisumu (Kenya)

Fig. 14.12 Environmental impacts of washing of heavy petroleum vehicles in Lake Victoria in Kisumu (Kenya). A hand-held GNSS could provide the location, perimeter, and area information of such impacted areas. This information could then be incorporated into a GIS system to support habitat conservation and management

could play a vital role in providing locations, perimeters, and areas of such car washing impacted areas for the purposes of habitat conservation and management.

14.4.4 Acid Mine Deposit Sites

Sulfide minerals occurring naturally are common minor constituents of the Earth's crust but occur in large quantities in some metallic ore deposits (e.g., Cu, Pb, Zn, Ni, U, Fe), phosphate ores, coal seams, oil shales and mineral sands [23, p. 29]. When not exposed to oxygen, sulphide are harmless. However, when exposed to oxygen through, e.g., mining, excavation, waste rock dams, etc., they become oxidized to produce sulphuric acid and high concentration of heavy metals such as Cu and Zn, see e.g., [21, p. 33]. Lottermoser [23, p. 29] point to the fact that the mineral pyrite (FeS_2) tend to be the most common sulphide mineral present whose weathering at mine sites cause the largest, and most testing, environmental problem facing the industry today—the acid mine drainage (AMD).

AMD generation can lead to the contamination of surface and ground water resulting in expensive treatment. Johnson et al. [19] indicate that the cost of remediating impacts of AMD in Canada for example was in excess of $3 billion. In Australia, the environmental impact of AMD is most significant in the abandoned mines, e.g., Rum Jungle (NT), Mt. Lyell (Tas) and Mt. Morgan (Qld) [21, p. 34]. The problem with AMD is that where it enters streams, there is a drop in pH leading to the disappearance of *aquatic ecosystems* and backside *plant communities*, soil contamination, the degradation of the food chain and contaminated groundwater [23, p. 69] and [45]. According to Lottermoser [23, p. 29], long-term exposure to contaminants in farmed food products may have the possibility of increasing health problems for humans. The extent to which extreme conditions of AMD develop depends on [21]:

- Water availability for oxidation and transportation.
- Oxygen availability.
- Physical characteristic of materials.
- Temperature, ferrous/ferric iron equilibrium, and microbial activity.

An effective approach for handling AMD entails managing of one of the factors above. Encapsulations of acid generating materials to prevent water leaching through the sulfidic materials and to limit oxygen contact with it are the two most important principles to apply [21]. The two steps above can only be applied once the materials on the mine sites, which have acid generating potential, have been identified [21].

GNSS could be useful in providing the locations of the materials once they have been identified and also in mapping sample sites for which tests such as net acid generation (NAG) are to be undertaken. When integrated in a GIS, these location-based data could support management measures by showing decision makers the exact places where the factors listed above have been implemented.

14.5 Concluding Remarks

This chapter has outlined some examples of areas where geospatial could be employed to support in the monitoring and management of pollution. The materials presented in the chapter are by no means exhaustive, but are intended to provide an impetus and motivation for further research on how geospatial application could be developed to enhance comprehensive pollution monitoring and management. This is demonstrated, e.g., in the case of monitoring air pollution and storm water pollution, wherein GNSS is integrated with gas sensors, remote sensors and GIS respectively. It is also shown that GNSS will play a crucial role in transport pollution related monitoring through measurements, e.g., of stoppage time, and provision of location-based data. On the other end of the divide, away from technical matters, it is imperative to educate people on the need for citizens to take responsibility for their own environment, while at the same time enforcing full compliance of environmental rules and legislation. Such education could be enhanced by the use of GIS, which could be used to provide visual maps that are simple enough to be understood by the local community.

References

1. Akbar S, Lvovsky K (2000) Indoor air pollution: energy and health for the poor. ESMAP Newsletter, No. 1., World Bank, Washington, DC
2. Aseto O, Ong'ang'a O (2003) Lake Victoria (Kenya) and its environs: resource, opportunites and challenges. Africa Herald Publishing House, Kendu Bay
3. Awange JL, Kiema JBK (2013) Environmental geoinformatics. Monitoring and management. Springer, Berlin
4. Awange JL (2012) Environmental monitoring using GNSS. Global navigation satellite system. Springer, Berlin
5. Awange JL, Obera B (2007) Motor vehicles: are they emerging threats to Lake Victoria and its environment? Water Air Soil Pollut 182(1–4):43–56. doi:10.1007/s11270-006-9319-3
6. Awange JL, Ong'ang'a O(2006) Lake Victoria-ecology, resource of the Lake Basin and environment, Springer, Berlin
7. Bosch C, Hommann K, Rubio GM, Sadoff C, Travers L (2001) Water, sanitation and poverty chapter, poverty reduction strategy papers source book. World Bank, Washington, DC
8. Briggs D (2005) The role of GIS: coping with space (and time) in air pollution exposure assessment. J Toxicol Environ Health 68(13–14):1243–1261. doi:10.1080/15287390590936094
9. Carletti R, Picci M, Romano D (2000) Kriging and bilinear methods forestimating spatial pattern of atmospheric pollutants. Environ Monit Assess 63:341–359
10. Collins S, Smallbone K Briggs, D (1995) A GIS approach to modelling small area variations in air pollution within a complex urban environment. In: Fisher P (ed) Innovations in GIS 2. Taylor & Francis, London, pp 245–253
11. Corbley KP, Stauffer R (2006) Curbing water pollution with mobile GIS: mobile data collection with handheld GPS units facilitates a San Francisco utility's quest for cleaner water, enabling a two-pronged effort to build a storm drain GIS layer and educate citizens about pollution. Geospatial Solutions. http://www.gpsworld.com/gis/local-government/curbing-water-pollution-with-mobile-gis-5361. Accessed on 20 Jan 2011
12. Dasgupta S, Deichmann U, Meisner C, Wheeler D (2005) Where is the poverty-environment nexus? Evidence from Cambodia, Lao PDR, and Vietnam. World Dev 33(4):617638. doi:10.1016/j.worlddev.2004.10.003
13. Denby B, Walker SE, Horálek OJ, Eben K, Fiala J (2005) Interpolation and assimilation methods for European scale air quality assessment and mapping. Review and recommendations. European Topic Centre on Air and Climate Change Technical Paper, vol, Part I, p 7
14. Dixon J et al (1995) Economic analysis of environmental impacts. Earthscan Publications, London
15. Duraiappah A (1998) Poverty and environmental degradation: a review and analysis of the nexus. World Dev 26(12):2169–2179
16. Gibson J, MacKenzie D (2007) Using global positioning systems in household surveys for better economics and better policy. World Bank Res Obs 22(2):217–241. doi:10.1093/wbro/lkm009

17. Hoornweg D, Thomas L (1999) What a waste: solid waste management in Asia. Urban development sector unit East Asia and Pacific region. The International Bank for Reconstruction and Development/THE WORLD BANK, Washington
18. Jerrett M, Burnett RT, Kanaroglou P, Eyles J, Finkelstein N, Giovis C, Brook JR (2001) A GIS: environmental justice analysis of particulate air pollution in Hamilton, Canada. Environ Plan 33(6):955–973
19. Johnson R, Wichern D (2007) Applied multivariate statistical analysis, 6th edn. Prentice Hall, Upper Saddle River
20. Kairu JK (2001) Wetland-use and impact on Lake Victoria. Lakes Reserv Resour Manag 6(2):117–125. doi:10.1046/j.1440-1770.2001.00135.x
21. Manual Laboratory (2008) ENV 310 land management practical manual. Learning Materials: School of Environmental Science, Murdoch University
22. Leitman J (2000) Rapid environmental assessment: lessons from cities in the developing world. Methodology and preliminary findings, urban management programme. vol 1, UNHCS, UNDP and the World Bank
23. Lottermoser B (2003) Mine wastes. Characterization, treatment and environmental impacts. Springer, Berlin
24. Mackenzie FT (2003) Our changing planet; an introduction to earth system science and global environmental change, 3rd edn. Prentice Hall, New Jersey
25. Maantay J (2005) Asthma and airpollution in the Bronx: methodological and data considerations in using GIS for environmental justice and health research. Health & place: environmental justice, population health, critical theory and GIS, vol 13, issue no 1. Elsevier, pp 32–56
26. Municipal Council of Kisumu (2004) Kisumu City Environmental Profile. UN Habitat
27. Mink SD (1993) Poverty, population, and the environment. World Bank Discussion Paper 189. World Bank, Washington DC
28. Mulaku GC, Kariuki LW (2002) Mapping air pollution in Nairobi, Kenya. Afr J Environ Assess Manag 4(1):29–37 [ISSN 1436 7890]
29. Nichol JE (2005) Remote sensing of urban heat islands by day and night. Photogramm Eng Remote Sens 71:613–623
30. Odada EO, Olago DO, Kulindwa K, Ntiba M, Wandiga S (2004) Mitigation of environmental problems in Lake Victoria, East Africa: causal chain and policy options analyzes. Roy Swed Acad Sci 33(1):13–17
31. Odhiambo GO, Kinyua AM, Gatebe CK, Awange JL (2010) Motor vehicles air pollution in Nairobi, Kenya. Res J Environ Earth Sci 2(4):178–187
32. Oke TR (1988) The urban energy balance. Prog Phys Geogr 12:471–508
33. Oodit D, Somonis UE (1992) Poverty and sustainable development. In: Ditetz F, Somonis UE, van der Straaten J (eds) Sustainability and environmental policy. Sigma, Berlin
34. Opande T (2008) Planning solid waste collection in low income settlements of Nyalenda and Ondiek in Kisumu town, Kenya. MSc Thesis, Maseno University, Kenya
35. Schreiner C, Branzila M, Trandabat A, Ciobanu RC (2006) Air quality and pollution mapping system, using remote measurements and gps technology. Glob NEST J 8(3):315–323
36. Schubeler P, Wehrle K, Christen J (1996) Conceptual Framework for municipal solid waste management in low income countries. UMP working papers series no.9, UNDP/SDC/UNCHS/World Bank, UMP/SKAT
37. Senkwe B, Mwale A (2001) Solid Waste in Kitwe, Zambia. A solid waste characterization study for the city of Kitwe, Zambia. Phase I. A consultancy report for SINPA, Zambia; Institute for Housing and Urban Development Studies, Netherlands; and the Copper Belt University, Kitwe, Zambia
38. Sifakis NL, Soulakellis NA, Paronis DK (1998) Quantitative mapping of air pollution density using earth observations: a new processing method and application to an urban area. Int J Remote Sens 19(17):3289–3300
39. Springer AL (1977) Towards a meaningful concept of pollution in international law. Int Compar Law Quart 26:531–557

40. Taylor MAP, Woolley JE, Zito R (2000) Integration of the global positioning system and geographical information systems for traffic congestion studies. Transp Res Part C 8(1–6): 257–285. doi:10.1016/S0968-090X(00)00015-2
41. Tayanc M (2000) An assessment of spatial and temporal variation of sulphur dioxide levels over Istanbul. Turkey. Environ Pollut 107(1):61–69
42. Tolba MK, El-Kholy OA, El-Hinnawi E, Holdgate MW, McMichael DF, Munn RE (1992) The world environment, 1972–1992: two decades of challenge. Chapman and Hall, London
43. United Nations Centre for Human Settlement (UNCHS) and United Nations Environmental Programme (UNEP) (1996) International source book on environmentally sound technologies for municipal solid waste management. IECT Technical Publication Series, New York
44. Ung A, Wald L, Ranchin T, Weber C, Hirsch J, Perron G, Kleinpeter J (2001) Satellite data for air pollution mapping over a city- virtual stations. Proceeding of the 21th EARSeL symposium. New solutions for a new millenium, Paris, Observing Our Environment From Space, pp 14–16
45. U.S. Environmental Protection Agency (US-EPA) (1994) Technical document. Acid mine drainage prediction. EPA 530-R-94-036; NTIS PB94-201829
46. Van Beukering P, Sekher M, Gerlagh R, Kumar V (1999) Analyzing urban solid waste in developing countries; A perspective on Bangalore, India. International Institute for Environment and Development (IIED), Working paper No. 24. CREED, London
47. Voogt JA, Oke TR (2003) Thermal remote sensing of urban climate. Remote Sens Environ 86:370–384
48. Wald L, Baleynaud J-M (1999) Observing air quality over the city of Nantes by means of LANDSAT thermal infrared data. Int J Remote Sens 20(5):947–959
49. Weng Q (2010) Remote sensing and GIS integration: theories, methods, and applications. McGraw-Hill, New York, p 416
50. Yang S, Dong J, Cheng B (2000) Characteristics of air particulate matter and their sources in urban and rural area of Beijing. China J Environ Sci 12:402–409
51. Zito R, D'Este GM, Taylor MAP (1995) Global positioning systems in the time domain: how useful a tool for intelligent vehicle-highway systems. Transp Res C3(4):193–209
52. Zito R, Taylor MAP (1994) The use of GPS in travel time surveys. Traffic Eng Control 35(12):685–690

Chapter 15
Poverty–Environment–Agriculture Interactions

> *"The Earth provides enough to satisfy every man's needs, but not every man's greed."*
>
> Mahatma Gandhi

15.1 Agriculture and Land Degradation

In most developing countries, agriculture accounts for most land use and is probably the single most powerful influence on environmental quality [69]. With the world's population having surpassed 7 billion by 2011, pressure is being exerted on the available natural resources such as land, leading to the degradation of the vary resource that is supposed to support life. Land, for example, provides the base upon which *social*, *cultural* and *economic* activities are undertaken and as such is of significant importance to the poverty-environmental link due to the fact that it is the principal livelihood of the rural poor (see, e.g., [55, 69]). When these resources are not utilized in a sustainable way by humans (both poor and rich), degradation occurs, with the most affected being the poor (see, e.g., the findings of Duraiappah [22]).

Scherr [69] list environmental concerns associated with agriculture as:

- Sustainability of the resource base for agricultural production (e.g. soil quality),
- protection of biodiversity and habitats,
- and environmental services of resources influenced by agricultural land use (e.g. carbon sequestration),

and attributes the blame to both the wealthy and the poor alike. On the cause of environmental degradation, Scherr [69] writes:

> Wealthier farmers, developers and multinational corporations typically control greater total land area and play a prominent role in many types of environmental degradation. However, the poor play a significant role in unsustainable agricultural intensification, expansion of

DOI: 10.1007/978-3-642-39733-2_15,

> farming into marginal lands and vegetative over exploitation and the consequences for their livelihood can be more serious because they lack assets to cushion the effects.

In many developing countries, lack of proper farming technologies such as crop rotation has seen the once fertile soil degrade on the one hand, and the impacts of soil erosion, salinity, and nutrient depletion increase on the other hand. Vegetation degradation has not been spared either. Increased population growth of urban cities lead to vegetation destruction to pave way to settlement or infrastructure. In this scenario, forests are cleared to provide timber for building and fuelwood. As a result, topsoil erosion and sedimentation are enhanced leading to land degradation and increased suspended solid loads in the rivers that feed the lakes. Monitoring of vegetation, therefore, provides information on the impacts on the environment as a result of human-induced activities or natural causes.

Bell [10] classifies land degradation as both a local and global problem, both of which have a bearing on poverty measures. Locally, the individual land owners are faced with the problem of declining land productivity, which increase the problem of the rural poor who dependent on land for food. This is supported by Scherr [69] who points out that in most regions, the rural poor depend more for their livelihoods on agricultural production and employment, and on common lands, than do the rural non-poor. The rural poor practise agriculture for subsistence purposes rather than for economic purposes and are faced with the difficulty of resource mobilization to achieve sustainable agriculture. To this end, Scherr [69] writes:

> Poverty is recognised as a significant constraint on agricultural growth because of poor peoples need to concentrate resources on lower value food crops to ensure subsistence security and their difficulties in mobilising production and investment resources.

Globally, the increase in population poses a threat to the fertile land and decreasing resources, again, worsening the problem of the poor, most of whom rely on farming for livelihood. The demand for increased food production for feeding the increasing population must increasingly be met from increased production per unit area [10]. Indeed, identifying the serious effects of land degradation on the environment leads to increased efforts in quantifying the problem, developing possible solutions, and engaging with the community to resolve these problems [10]. All these measures contribute towards addressing food poverty alleviation.

Land degradation can be seen through the effect of salinity, water erosion, wind erosion, waterlogging, remnant vegetation decline, sub-soil compaction, soil structure decline, structural decline, fertility decline, eutrophication, acidification and water repellence [49]. Duraiappah [22] put the figure of total world arable land lost annually due to land degradation to about 0.3–0.5 % (5–7 million hectares). The main cause of these forms of land degradation has been attributed to water, which is often an off-farm impact as a result of water recharge, water discharge, water runoff and flooding, e.g., [10].

Again, similar to the deductions of Scherr [69] on both the wealthy and the poor causing environmental degradation, Duraiappah [22] noted that for soil salinization and desertification, both commercial farmers (wealthy) and low-income farmers were involved. The government policies encouraging the use of water-dependent green revolution were considered to be the primary cause of salinization (see, e.g., Duraiappah [22] and the references therein). From literature analysis, Duraiappah [22] concludes that (i) power, wealth and greed causes environmental degradation, (ii) institutional failures cause environmental degradation, (iii) market failure cause environmental degradation, (iv) the resulting environmental degradation from (i), (ii), and (iii) above in turn causes poverty and subsequently endogenous poverty causes environmental degradation.

This chapter presents the degradation caused on land resources and their impacts on poverty and environmental quality. Geospatial-based methods for monitoring such degradation and indicators to support management efforts that would lead to sustainable utilization of resources are presented.

15.2 Geospatial Monitoring of Land Resources

Monitoring of *changes* in land through indicators could help in policy formulation and management issues for the betterment of the environment and also enhanced poverty eradication measures. Some of the vital indicators for land management include vegetation, soil quality and health, biosolids and waste disposed on land, land evaluation, land use planning, contaminated land, integrity of the food supply chain, mine closure completion criteria, and catchment management, in particular water balance, salinity, eutrophication, and riparian/wetland vegetation.

Environmental monitoring of the *extent* and *quality* of vegetation plays a crucial role of indicating land condition. By monitoring the changes in vegetation colour, often as a results of stress, environmental impacts such as those arising from the exposure of plants to pollutants (e.g., air, acid rains, heavy metal contamination of the soil), insect infestation and disease can be deciphered [36]. Besides the stress factors mentioned above, increase in population growth of urban cities lead to vegetation destruction to pave way to settlement or infrastructure. In this scenario, forests are cleared to provide timber for building and fuelwood. As a result, topsoil erosion and sedimentation are enhanced leading to increased suspended solid loads in the rivers that feed the lakes. Monitoring of vegetation, therefore, provides information on the impacts on the environment as a result of both human-induced activities or natural causes.

Remote sensing satellite provide means for detecting vegetation changes through monitoring of the changing vegetation colour, which occurs as a result of the stress stated above. This is achieved through monitoring of vegetation greeness/vigour quantified through the normalized difference vegetation index (NDVI) of Tucker [73]. Through integration with of remote sensing satellites, GNSS (global navigation satellite systems) provides location-based data and controls for vegetation mapping (e.g., [5]). GNSS could also find use in monitoring of rangelands, and can also be indirectly used with altimetry satellites such as ICESat-2 to map vegetation.

James et al. [40] defines rangelands as grasslands, shrublands, and open woodlands managed as natural ecosystems that are traditionally used by grazing animals. In this context, GNSS could be used in digital cameras to provide observations on amount and locations of poisonous, invading, or noxious plants; and threatened or endangered species [40]. Next, we present geospatial methods relevant to monitoring changes in land.

15.3 Reconnaissance and Validation

Before any field measuring campaign is undertaken, pre-field preparations are essential. At this initial stage, all available data sets and imagery required for the project need to be acquired and a desk study undertaken in preparation for the field work. *Topographic maps*, *aerial photographs*, *satellite imagery* and any *existing reports* normally provide the main sources of information for determining land use capability. Maps, photographs and coarse resolution satellite imagery such as *Landsat* from varying time span can be visually examined and inspected (e.g., [3]) to identify any potential land use changes that may have occurred during the time interval under consideration. During the desk study, existing reports need to be examined to extract any relevant information.

Field reconnaissance entails the pre-visitation of a survey area before the actual survey. Hand-held GNSS receivers (see, e.g., Fig. 14.1) are useful for georeferencing as well as in providing navigation services. For example, if the crew members require to visit a specific location, they could easily use the navigation function of a hand-held GNSS receiver to reach the desired location or to identify the derived feature. In addition, *still* or *video images* may be taken to capture important spatial information. It is possible today to employ *digital cameras* that integrate both GNSS and GIS, thus allowing images acquired in the field to not only be directly georeferenced, but also uploaded directly into existing GIS databases. This may even allow elementary GIS spatial analysis [3] to be undertaken in *real-time* during the field reconnaissance.

Terrain evaluation to identify *relief changes* within the study area and any landmark features, e.g., remnant vegetation (Fig. 15.1), is important as this may be useful in georeferencing. Hand-held GNSS receivers that position to within 3–15 m are useful for georeferencing tasks. This is because the spatial dimension of most land features under investigation are normally larger than the 15 m accuracy typical for positioning using single frequency hand-held receivers.

15.4 Monitoring of Land Conditions

15.4.1 Soil Landscape Mapping

Soil maps define mapping units where a particular soil type is believed to be located. On the other hand, *soil landscape mapping* is a survey of land resources, which

Fig. 15.1 Ground-truthing using aerial photos: example of remnant vegetation at Mt. Kokeby, Australia. This could be located using a hand-held GNSS receiver similar to that shown in Fig. 14.1

delineates repeating patterns of landscapes and associated soil types and differs from soil mapping in that the landscape component is an explicit part of the mapping [66]. Uses of land resource data collected during a soil-landscape mapping program include [66]; reducing the risk in decision-making (ensuring that changes in land management are the result of land resource information that reduces the uncertainty about the impact of different strategies), improving our understanding of biophysical processes, designing large-scale land use changes (e.g., targeted re-vegetation to control dry land salinity), environmental regulation and trading systems, mapping and monitoring land conditions to support national policies on natural resource management, support of international agreements (e.g., Kyoto Protocol demands for predictions of the distribution and dynamics of soil carbon over a range of scales), and support of environmental management systems. For example, Pieri [62] used GNSS to collect spatial data needed for planning sustainable land management. Using GNSS information, photographs and photo mosaics, simple information such as field boundaries, farmers' appreciation of soil productivity or constraints (using, e.g., indigenous classification of soils), and land use rights were collected and registered. The report proposed the use of GNSS to spatially reference all physical and social information in land management planning.

Schoknecht et al. [66] divided the information about the soils and landscapes of Western Australia into point data stored in a soil profile database, which is comprised of site specific information such as soil profile descriptions, the results of laboratory analysis and photographs, map unit polygons stored as digital lines, boundaries drawn around areas containing similar soil and landscape patterns, and a map unit database that contained descriptions of the map unit polygons and related them to broad areas rather than a specific point. In both point and polygon data, GNSS satellites played an important role during the data collection process, with GIS used to provide the spatial database.

15.4.2 Provision of Point Data

Field surveys for soil landscape mapping involves occupying selected sites identified during a preliminary survey and reconnaissance. Factors influencing the selection of these sites include ease of access and the changing landscape. Representative soils from these selected sites are examined in soil pits with samples taken for chemical and physical analysis. Besides the examination of the soil, other physical descriptions of the site, such as slope and vegetation, are also recorded.

The important contribution of GNSS satellites in this process would be the provision of the site locations, which can be achieved using a hand-held receiver as shown in Fig. 15.2. This allows point data information on which all the landscape information relating to a specific site, such as; site *description*, *soil profile* description, *soil classification*, *soil profile's chemical*, and *physical properties* are related. The site description consists of information such as slopes measured using clinometers, landform patterns and elements of the survey area, the surrounding vegetation (e.g., Fig. 15.1), land surface and use. All these attributes can be input into a GIS database. An example of mapping soil characteristics that significantly affects crop production, e.g., clay content, organic matter content, nutrient content, and plant available water, is illustrated, e.g., in the work of Brubaker et al. [14].

15.4.3 Provision of Polygon Data

Since the aim is to produce the soil landscape maps, the boundary of the maps are traced on aerial photographs or satellite imagery and then captured digitally using computer-aided mapping systems (e.g., in a GIS). This is done by either scanning the photographs or by keying in directly the two dimensional x, y coordinates. The boundary coordinates $\{x_i, y_i\}$ that form the map unit polygon are provided through satellites or ground surveying methods and are then mapped onto those of adjoining surveys. GNSS satellites provides control and tie points needed to match these boundaries. These points are then captured by aerial photographs and the GNSS

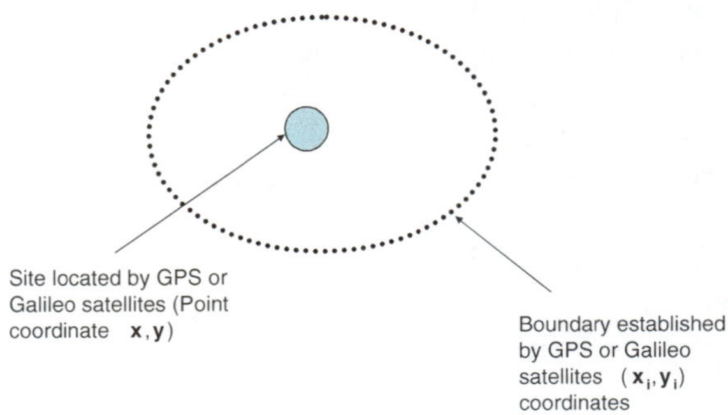

Fig. 15.2 Schematic description of how GNSS can provide site and boundary locations. These values can easily be integrated in a GIS system to enhance analysis

positions compared with those of the photograph to control the quality of the maps derived from the photographs.

Alternatively, points could be located in older photographs and their positions established using GNSS receivers to control the quality of the digital maps derived from these photographs. Tie points on the other hand are common points that appear on different overlapping photographs. They too could be located using GNSS satellites, and together with the control points provide the base upon which landscape maps could be prepared. Point data and map unit polygons form map unit databases that relate to the entire landscape. Use of such digital maps include precision farming discussed in Sect. 15.6.1. Details of the actual geospatial procedures needed to generate maps are discussed, e.g., in Awange and Kiema [3].

15.5 Monitoring of Land Degradation

Besides the challenges listed in Sect. 15.1, another emerging global challenge to agriculture is the increasing conversion of productive agricultural land to other usage such as biofuel production leading to reduced food production and increased food prices. For instance, the contribution of biofuel production has been associated, e.g., by Kgathi et al. [45] to influence increase in food prices. This is due to the fact that food availability is affected if food crops or productive resources such as land, labour, water are diverted from food production to that of biofuel (e.g., [45]).

15.5.1 Soil Erosion Monitoring

Soil erosion occurs through the process of detachment, transportation and the deposition of soil. The net effect is the loss of fertile top soil for agriculture, siltation of streams and lakes, eutrophication of surface water bodies, and the loss of aquatic biodiversity [59]. Mackenzie [53, p. 322] points out that an estimated 480 billion tons of topsoil have been lost to the world's farmers by erosion in the past few decades. Of this, 18 billion tons are said to be transported by rivers to oceans annually, while the rest moves to other terrestrial ecosystem.

In order to minimize the impact of soil erosion, it is essential that management practices take into consideration the magnitude and *spatial distribution* of soil erosion. Soil erosion models such as the Universal Soil Loss Equation (USLE) [38] and its modified version (Revised Universal Soil Loss Equation RUSLE, e.g., [59]) have been developed to estimate the rate of soil erosion. Determination of soil erosion, however, poses a challenge due to the contribution of biophysical factors (e.g., soil and climate) and the interactions between them. The limitation of these models is their inability to cope with the large amount of data that describes the heterogeneity of the natural system [61]. Within the scope of these models, remote sensing finds use in mapping and assessing the landscape attribute that control soil erosion such as land use/land cover, soil type, relief, slope, drainage etc.

Multi-temporal satellite images are useful in studying seasonal land use dynamics that contribute to soil erosion. Satellite imagery also provide information on erosion features such as gullies and vegetation cover, and also contribute in the generation of Digital Elevation Models (DEM), useful for soil erosion models. This is achieved through the analysis of stereoscopic and microwave (SAR) data. Lufafa et al. [52] combined aerial photo interpretation and interpretation of Landsat TM images of 1994 to generate land use/land cover maps using the normalized digital vegetation index (NDVI). Parameters of the land use/land cover maps were then inserted in the USLE model to predict the rate of soil erosion.

In Onyando et al. [59], multi-spectral image processing is used to extract thematic information from Landsat TM (bands 2, 3 and 4) data to generate land cover maps from which parameters for the USLE model are derived. Supervised classification using maximum likelihood is adopted. Other studies that have generated land cover maps useful for soil erosion include [39, 51, 61]. In [51], geometric rectification of the images was achieved using ground reference data collected from topographic maps.

15.5.2 Salinity Monitoring: The Catchment Approach

Salinity can be classified into primary and secondary salinity. Over millions of years, salt from the ocean has been collected and deposited by wind and rain across landscape, with few rivers returning this salt to the sea. This led to a gradual accumulation of salt, which eventually became part of the natural landscape termed *primary salinity*.[1] *Secondary salinity* of soil, or 'salinization', results from rise in water table, which is often caused by changing land use actions such as land clearing or irrigation, where the term 'water table' refers to underground water (or groundwater), which may be very deep, but can be quite close to (or at) the surface.

The major origin of salt causing secondary salinity is rainfall which has deposited large amount of salt over ten thousands of years. Other contributors are weathering of minerals in the underlying bedrock and marine sediments. The processes leading to *secondary salinity* are traced to the altered water balance due to the fact that deeply rooted perennial native vegetation, which helped maintain the groundwater balance (e.g., Figs. 15.3 and 15.4) has been replaced by pastures and crops that do not use enough rainfall to lower water tables and significantly reduce salinity.

The clearing of such vegetation has seen increased recharge at a much faster rate than would otherwise occur naturally. The excess water not used by vegetation seeps into the soil, past the root zone, and contributes to increase in stored groundwater. With time, water table rises and in the process dissolves the solid salt stored deep in

[1] http://www.environment.gov.au/water/publications/quality/pubs/salinity-and-water-qualityfs.pdf [Accessed on 4-9-2013].

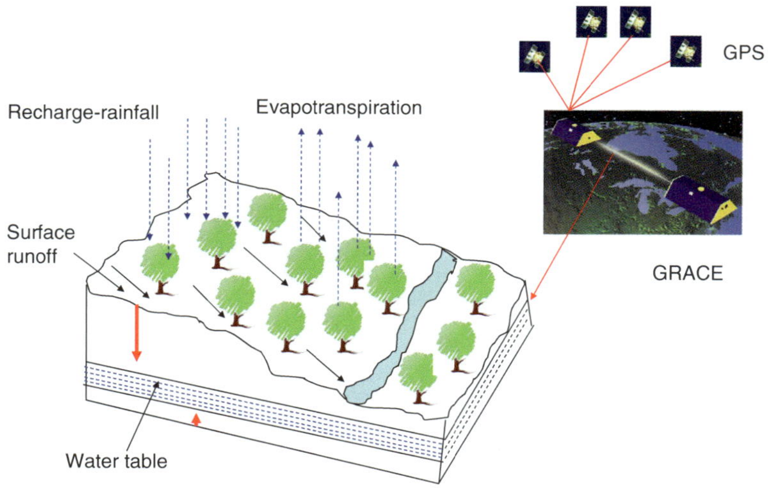

Fig. 15.3 With natural vegetation, the water balance is maintained, groundwater levels are low, less recharge, less runoff and salinity is curbed

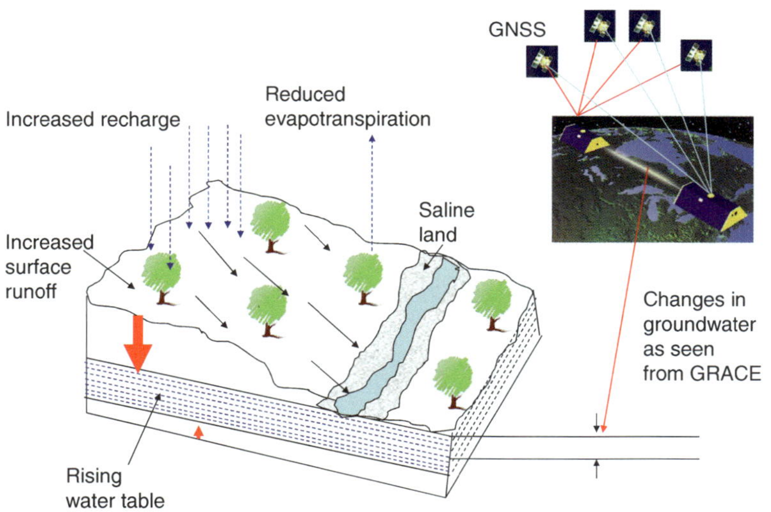

Fig. 15.4 With vegetation cleared, the water balance is disturbed. Groundwater levels rise, recharge increases with increased runoff and salinity increases. GRACE satellites discussed in Awange [5] and Awange and Kiema [3] could provide a possible means of measuring changes in water level at catchments having sufficient spatial resolution for its use

the soil. When the water table reaches within 1–2 m of the soil surface, salt enters the plant root zone and growth is affected. This process is called *dryland salinisation* and has been reported, e.g., in [57] to affect a significant proportion of agricultural soils in south-western Australia.

Fig. 15.5 Impacts of salinity at Mt. Kokeby, Australia

Apart from the dryland salinity, there is also *irrigation salinity*, which is caused by the recharge of groundwater by excessive irrigation and leakage from irrigation channels. This again causes water tables to rise and in so doing brings salt to the root zones. Besides vegetation clearing and irrigation, the local landscape (i.e., geology and topography) and groundwater aquifer characteristics also play an important role in the susceptibility of land to salinization. In both dryland and irrigation salinity, saline groundwater near the soil surface kills native vegetation and reduces biodiversity, thus threatening the overall health of natural ecosystems (see Fig. 15.5). It also results in the loss of agricultural productivity, deteriorates the quality of drinking water, and damages infrastructure such as roads and buildings.

In the Murray-Darling Basin (MDB) in Australia for example, agricultural production is reported to account for approximately 40 % of the total production in Australia [25]. This high level of agricultural production comes despite MDB's high aridity and low runoff compared to other major river basins [26]. This comes at the expense of irrigation, which uses about 75 % of the total water used in Australia, and extensive clearing of native vegetation for cropping and pastoral land use [25].

As already illustrated in Fig. 15.4, vegetation clearing leads to salinity as does extensive irrigation. Since both irrigation and clearing of the MDB has occurred to sustain agricultural productivity, the landscape and river systems have been dramatically altered leading to several undesirable effects, including increased salinity, diminished biodiversity and ecosystem degradation [26]. The number of land parcels falling victims to salinity seems to be on the rise, calling for integrated catchment-based remedies. For instance in New South Wales (Australia), over 500,000 ha of land were estimated to be impacted by salinity in the year 2000, with a projection of 2–4 million hectares by 2050 if changes in land use are not enacted [26].

Management Options: Management of salt-affected land (or potentially salt-affected land) must be based at least on a whole catchment approach, e.g., [64]. Catchment management involves all those activities related to the *capacity to produce* and *maintain* runoff and groundwater with desirable quality and quantity characteristics, and by reducing the undesirable consequences of land use on stream behavior [10]. Catchment management should, therefore, be holistic in outlook, i.e., it should involve the consideration of all aspects of the physical and socio-economic environments that impinge on the catchment, its resources, and their use. Management of *discharge* areas will differ from the *recharge* areas. Discharge areas are part of the landscape where groundwater approaches or reaches the soil surface. Recharge areas on the other hand are part of the landscape where water passes below the root zone and adds to the groundwater (see Fig. 15.4).

Management of discharge areas aims at obtaining production (i.e., of salt tolerant shrubs or pasture) from the area, while at the same time reducing the salt-affected land. Actions taken on, or immediately around, discharge areas can take the form of engineering remedies such as digging drains or banks, to biological methods such as planting shrubs, grasses or trees or a combination of both engineering and biological. Management of salt-affected areas to alleviate the problem uses four principles [10]; *increase water use*, *decrease surface evaporation*, *decrease water-logging*, and *increase leaching* of salt.

Management of recharge areas essentially leads to the prevention of salinity with the objective being to reduce recharge water (Fig. 15.4) reaching groundwater. This can be achieved by increasing plant water use and by using engineering means to divert excess water from the saline areas to avert ponding and water-logging. One approach for achieving this is to plant deep rooted plants with the ability to use water throughout the year. This approach has been found to yield fruit in some parts of Western Australia (WA). For instance, Greenwood [31] found that at North Bannister, the annual evaporation from trees (*Eucalyptus cladocalyx*, *E. globulus* and *E. maculata*) was some seven times the evaporation from grazed pasture [64]. Planting trees and shrubs on recharge areas of agricultural catchments is also being advocated by the Department of Agriculture in WA. While it is not economic to indulge in large scale plantings, there are often areas in catchments which can be identified as "specific recharge areas" [58], which are small, make a major contribution to recharge, and do not produce economic crops or pastures, but can be used for planting trees or shrubs. Encouraging extensive planting, especially in high recharge zones, of deep-rooted perennial plants (trees, shrubs and pastures) is, therefore, one of the clearest mechanism to fight salination.

Plants such as barley and lupins are also known to have high water uptake owing to their deep roots and are encouraged for prevention. The difference in plant water usage is attributed to *rooting depth*, *leaf area index*, and *biomass*. For example, barley, and lupin have root penetration down to 2.5 m. Barley also has been shown to attain the highest leaf area index, followed by lupins, wheat and clover. Lupins have been shown to have higher transpiration rates than wheat and pasture in late winter due to greater biomass. The disadvantage with the plants approach is that they are not annual crops and are unable to utilize water throughout the year, thus rain which fall outside of the growing season still seep into the soil [10].

Example 15.1 (Geospatial support of salinity management)

This can possibly be achieved in two basic ways:

(1) In the first instance, remote sensing and photogrammetry could be employed to map the spatial extent of affected areas. Given their penetration abilities, microwave remote sensing sensors would be better suited for this. GNSS handheld receivers could then be used in monitoring salt-tolerant pasture areas through the mapping of the boundary of salt-tolerant pasture species (e.g. barley). The GNSS data could then be input into a GIS database for monitoring purposes.
(2) The second approach would be the monitoring of changes in groundwater as illustrated in Fig. 15.4. For large spatial coverage, use of piezometers may not suffice and hence the requirement of space techniques. The monitored variations in groundwater could then be correlated with the re-vegetation measures to assess the impacts of replanting of trees or retained native vegetation to curb salinity. In this particular scenario, geospatial techniques of GRACE could be applied. It should be mentioned, however, that the resolution of the catchment area must be suitable for use of GRACE. Other components of the total water storage such as surface water and soil moisture need to be isolated to leave only the groundwater. Soil moisture could be measured using microwave scatterometers (see e.g., [3]). This overall approach of using GRACE satellites, however, remains an ambitious proposal to be tested.

♣ End of Example 15.1

15.6 Role of Geospatial in Precision Farming

Scherr [69] promotes the use of agricultural technologies with environmental benefits as part of the policies to jointly address poverty. In particular, desirable technologies are those that lower the unit output costs of variables, e.g., nutrients, lower costs of conservation, and investment, leading to management system that integrate agriculture and environmental objectives. Arguably, geospatial applications to precise farming and topographic map generation discussed below fits into such technologies.

15.6.1 Precise Farming

Traditionally, farming normally treats the "*whole field*" as a single (homogeneous) entity upon which decisions (based on field averages) have been made. Inputs (e.g., fertilizers) are then applied uniformly across the field. Modern farming methods (also known as precision farming), however, divide large pieces of land into smaller segments that are managed separately to optimize productivity (see, e.g., Fig. 15.6) through the use of GNSS.

Precision farming is the gathering of information dealing with *spatial* and *temporal variation* within a field with the aim of using the information to manage inputs and practices. This is made possible by linking *computers*, *mobile sensors*, *GNSS*, *GIS* and other devices [32]. It is a comprehensive approach to farm management and has the following goals and outcomes; *increased profitability and sustainability*, *improved product quality*, *effective and efficient pest management*, *energy, water and soil conservation*, *surface and groundwater protection*, *yield rates*, *pest infestation* management, and *other factors that affect crop production* [33]. To achieve these outcomes, farm managers base their decisions on the requirements of each zone to control input resources. *GNSS and GIS can be used to control the input of these zones* as discussed, e.g., in [70, Chap. 9].

Precision farming allows the management of various zones, with the zone having a higher potential for economic return receiving more inputs, if needed, than less productive zones so as to achieve maximum economic return for each input [33]. Dobermann and White [21] refer to precision farming as site specific crop management and discuss how GNSS satellites are applicable to nutrient management.

Precise farming attempts to increase food production per unit area and as such, requires proper land utilization and management of the input resources, e.g., *fertilizers* to maximize the yield. When this is done in a manner such that the environment is protected and conserved, and social values also promoted, it can lead to *sustainable agriculture*, whose indicators are *state of resource*, *biophysical* and *economic* trends.

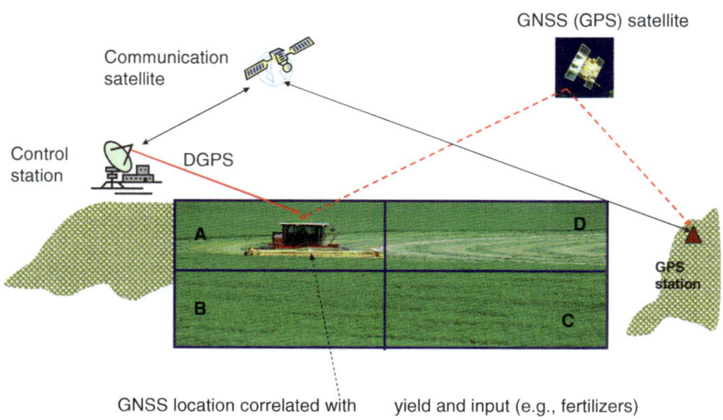

Fig. 15.6 Application of geospatial to precision farming. *Source* Awange [5]

For instance, the application of GNSS to *precision farming* have been shown to include soil sample collection, chemical application controls, and harvest yield monitoring [24, p. 142]. The applications of GNSS for soil sampling was discussed in Sect. 15.4.1. Spraying of the field from airplanes could be integrated using aerial guidance systems in such a way that the field sprayer is guided using a moving map display. Based on the sprayer's location, the system will apply the chemical at the correct locations, with minimal overlap, and automatically adjust the chemical's use rate, thus increasing the efficiency of chemical and fuel usage [24, p. 142]. The position of these field spots could be provided by GNSS as discussed in Sect. 15.4.3.

Example 15.2 (Trimble's AgGPS).

Trimble[2] has developed the AgGPS system, which combines in-field guidance using GPS satellite receivers and intelligent farm management. The AgGPS field positioning systems enable field guidance at a higher accuracy, allowing farmers the possibility of collecting information on tillage, cultivation, irrigation, topography and infestations, which can then be mapped and analyzed in a GIS to provide a better understanding of the elements affecting farm operations and management [15]. The system can be tailored for site specific applications such as yield mapping and variable rate management. Buick [15] lists the primary reasons why many farmers are considering (or reconsidering) numerous site-specific precision agriculture practices as the net gain in sustainable agriculture, i.e., *economic*, *environmental* and *sociological* paybacks from GNSS-guidance and automated steering systems, and the decrease in the cost of GNSS machine control systems. With more GNSS satellites on the way, this cost is expected to be even lower.

♣ End of Example 15.2.

15.6.2 Farm Topographic Maps

A farm topographic map is an example of maps produced using either Differential Global Positioning System (DGPS) or Real Time Kinematic (RTK) methods of GNSS positioning respectively, see, e.g., Awange [4]. They are useful in depicting topographical features such as elevation, landscape, drainage, soil attributes, slopes etc. Since spatial yield variability is usually related to topographical features, see e.g., [16, 43, 47, 65, 71, 78] etc., the availability of high accurate GNSS *farm topographic maps* should enhance *operational efficiency* and lead to better, improved

[2] http://www.trimble.com

agricultural and environmental management. For instance, such maps should lead to improved management of agricultural inputs such as fertilizers, and in water deficient countries such as Australia, contribute to better management of surface and groundwater.

Example 15.3 (Controlled Traffic Farming (CTF) Solutions)

Controlled Traffic Farming (CTF) Solutions is an Australian-based organization that seeks to provide solutions on focusing on efficient and effective use of *farm resources* to maximize production and minimize *environmental impacts* such as *soil erosion* and *water-logging*.[3] CTF Solutions collects topographic data using a 2 cm auto steer operations during the period when the tractor is in the field undertaking farm operation or by using a vehicle with an antenna mounted on it. In order to collect gridded data for the purpose of generating elevations, the vehicle is driven at regular swath intervals (width of 20–40 m) with the GNSS receiver logging data at 5–10 m intervals. Besides these regular grids, data for other features of environmental interest, such as flood runners and erosion washers, are also collected. The data is then analyzed using a GIS system to generate topographic maps.

An illustration of the CTF Solution's application of GNSS (RTK approach) is presented in Fig. 15.7, where CTF's RTK collected topographical data at 25 cm contour interval are overlayed on a high resolution IKONOS imagery.[4] The figure essentially shows that low lying areas of production have suffered significant crop loss due to water-logging (Tim, pers. comm).

Following the results outlined in Fig. 15.7, drainage was undertaken on the paddock in 2005 and GNSS together with high resolution imagery once again applied to monitor the environmental impact of water-logging in 2006 (Fig. 15.8). In 2004, this water-logging cost the farmers in the order of Australian $50,000 in lost production (Tim, pers. comm). The combination of GNSS and high resolution imagery was the key to the success, otherwise each monitoring method on their own would not have told the whole story (Fig. 15.8). This example, therefore, provides a good illustration of the role played by geospatial in monitoring *environmental impacts* on farms.

♣ End of Example 15.3

[3] see http://www.ctfsolutions.com.au/ [Accessed on 8-09-2009].

[4] © Geoeye 2004 NIR band.

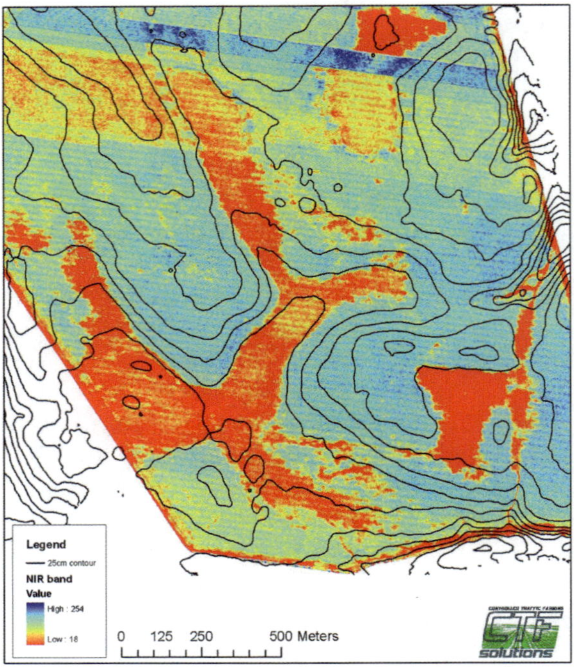

Fig. 15.7 GNSS topographical map taken in 2004 by CTF overlayed on a high resolution IKONOS imagery. *Dark red colours* indicate areas of poor drainage affected by waterlogging. *Source* http://www.ctfsolutions.com.au

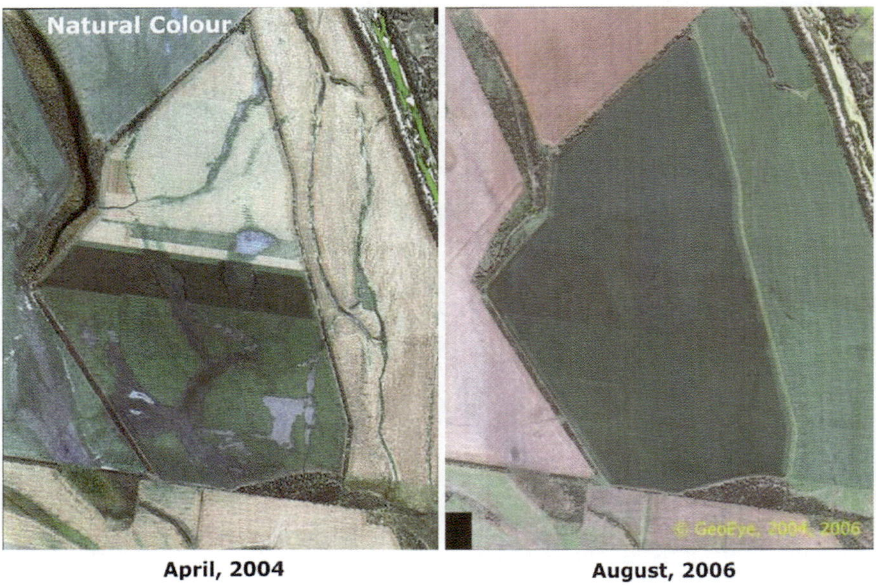

Fig. 15.8 GNSS topographical map taken by CTF overlayed on a high resolution IKONOS imagery. The *left figure* was taken in April 2004 (see Fig. 15.7). The *right figure* indicates the same area in August 2006 after drainage was undertaken in 2005. *Source* http://www.ctfsolutions.com.au

15.7 Vegetation

Environmental monitoring of the *extent* and *quality* of vegetation plays a crucial role of indicating land condition. By monitoring the changes in vegetation color, often as a result of stress, environmental impacts such as those arising from the exposure of plants to pollutants (e.g., air, acid rains, heavy metal contamination of the soil), insect infestation and disease can be deciphered [36].

15.7.1 Forests

Geospatial monitoring of deforestation can help identify the loss of critical ecosystem and biodiversity, as well as increased risk of soil erosion in steepy sloped areas, see e.g., Dasgupta et al. [20] who mapped forested areas and rates of deforestation in Cambodia, Lac PDR and Vietnam. Althlough deforestation is always viewed as having negative impacts, Durraiappah [22] argues that in itself, it is not a problem and may even result in economic growth, though caution is added that unsustainable deforestation can result in environmental degradation.

Due to population growth and the need to cultivate land for farming, forests are normally the first casualty since since the conception among the people living nearby is that the forests belong to the community leading to increase in deforestation (e.g., Fig. 15.9). An example is the case of the Mau forest in Kenya (Fig. 15.10), which was

Fig. 15.9 Deforestation of the once densely forested Gwasi hills in Kisumu (Kenya) to pave way for farming. Satellite-based methods could be useful in mapping of the deforested region

Fig. 15.10 Destruction of the Mau forest in Kenya to pave way for cultivation. *Source* UNEP [74]

invaded for cultivation purpose and severely degraded such that the four main water catchments; Sondu River, Mara River, Lake Nakuru, and Molo River were impacted [74]. The impact was felt as far as Nakuru town with a population of over 1 million as per the 2010 census.

Another pressure exerted on forests is the need for fuelwood. In most developing nations, the main source of energy for cooking and warming is derived from fuelwood. The impact of clearing the forest, therefore, could be grouped under direct, e.g., land degradation and indirect, e.g., water shortage in major cities due to destruction of catchment.

Both direct and indirect consequences of forest destruction contribute to suffering by the poor who need need the land to produce food for their daily livelihoods as well as water for their daily utilities. In Honduras for example, a correlation between deforestation from slash-and-burn agriculture and persistent poverty was reported in Bojö and Reddy [11], where deforestation affected 61 % of the forest area in the western and southern regions and 55 % in the Atlantic region. The annual economic loss from deforestation, as seen in damage to timber and non-timber products and in losses to biodiversity, water resources, and ecotourism, was estimated at US$112 million [11].

In order to manage the forests in a sustainable manner that contributes to poverty alleviation, a concept that entails community management of forests is fast gaining momentum. Rather than adopt a one sided approach where the authorities try to achieve forest protection through eviction orders as was witness in the case of Mau forest in Kenya in 2009, in many parts of the world, communities living on forest fringes or within forests have regained some control over the management of these resources [8]. This is exemplified by the cases of Nepal where the community forest management was instituted in 1993 and has helped stem forest degradation [8, 9, 23], Mexico where co-management of forests has led to increased profits to local communities [8, 46], and Africa where an estimate of at least 4,500 rural communities are involved in some form of forest management [8, 77]. Since most of those engaged in forest clearing for farming are often the rural poor, such an integrated use of forest resource, where the community is involved in its usage and management could lead to a win-win situation between the environment and fighting poverty.

Yet deforestation is not only caused by the poor whose motive is subsistence. Commercial agents pursuing logging and agricultural/pastural expansion activities have been identified, e.g., by Duraiappah [22, and the references therein] to cause deforestation by taking advantage of market and institutional failures. In general, Duraiappah [22] identify the impacts of deforestation as productivity drop, fuelwood shortage, destruction of safety buffer, soil erosion, and destruction of watershed protection. These impacts in-turn lead to water shortage (see Sect. 13.22), rainfall disruption, increased flooding potential (see Sect. 16.4), which affects both the wealthy and the poor alike, with the poor being the hardest hit. In summary, Duraiappah [22] finds that environmental degradation leads to poverty.

15.7.2 Geospatial-based Aid to Forest Management

The distinct role that forests play towards ensuring global carbon balance through carbon sequestration, discussed in Awange and Kiema [3] and elaborated in several works (see e.g., [28, 37, 54] etc.), is worth mention. Furthermore, the initiative of Reductions in Emissions from Deforestation in Developing (REDD) countries designed to motivate developing countries to voluntarily reduce national deforestation rates and associated carbon emissions through financial incentives (see e.g., [29, 48] etc.) is equally worth recognition.

Detailed and timely information on forests are required for traditional forest management, forest certification, and in monitoring forest health and biodiversity on the one hand, while on the other hand, there is an increasing need of having the information delivered at a lower cost [36]. Population increase has also put pressure on the available resources, e.g., land leading to increase in deforestation (e.g., Fig. 15.10).

Whereas remote sensing satellites detect damage on trees by monitoring color changes from the leaves, which are indicative of the related stress, Holopainen et al. [36] discuss the role of GNSS as that of enhancing fieldwork to achieve the required efficiency of acquiring spatial forest resource information at a lower cost.

GNSS applications to forest management span almost two decades. For example, Sirait et al. [68] pointed out that effective forest management requires balancing conservation and local economic-development objectives and demonstrated the capability of mapping customary land use systems using GNSS and GIS methods among other techniques. The derived maps by Sirait et al. [68] were then used in *forest protection* and *resource management*, with the limitations being accuracy of the base maps, ability of social scientists and map makers to accurately capture the complex relationships of traditional resource-management systems on maps, and the political will of the parties involved to recognize different forms of land rights [68]. Part of the forest protection is in managing forest fires and vegetation health, an area where GIS has found wide use, thanks to the capability of remote sensing to deliver timely and comprehensive coverage data and GNSS satellites to provide accurate and reliable positioning. For example, GNSS has been in use at the Ventura County Sheriff's Office since the early 1990s to map wildfire perimeters [70]. James et al. [40] point to the use of video cameras with GNSS interface to document occurrences and locations.

Raven Environmental Services [63] indicate the possibility of integrating GIS with GPS to produce maps that describe both the property and any underlying inventory data such as timber inventory for forest management units, the location and type of unique plant communities or any other type of special-use zone. From such maps, areas, lengths, distances to other features, and many other useful measures can also be calculated. The resulting underlying data associated with any management unit can then be exported as an Excel file, and used to periodically update a written management plan [63].

Integration of GNSS, remote sensing and GIS is exemplified in the work of Brondizio et al. [12] who produces a georeferenced map of land cover and land use for an area of the Amazon estuary inhabited by three populations of caboclos with distinct patterns of land use. The maps produced by Brondizio et al. [12] permitted measurements and differentiation of land uses and change detection between small areas of managed floodplain forests and unmanaged forests, and between three distinct age/growth classes of secondary succession following deforestation. Brondizio et al. [12] suggested means of balancing between use and conservation in Amazonian floodplain and estuarine areas, and the effectiveness of monitoring these types of land cover from space-borne platforms.

The Ice, Cloud, and land Elevation Satellite-2 (ICESat-2) will contribute to mapping of forest productivity by tracking the growth of individual forest stands, observations of tree phenology, forest diseases, and pest outbreaks through associated changes in canopy structures, and the mapping of forest heights and above ground biomass at a scale that approaches one that is appropriate for forest carbon management, which would enable the global ecosystem community to further constrain the sources and sinks of carbon at regional to continental scales [1]. An illustration of ICESat-2's capability to map forest productivity is presented by [35], e.g., Fig. 15.11 that shows a clear ground return that is spread by the slope of the ground, the top of the canopy, and returns from within the canopy that are indicative of structure within the crown. Abdalati et al. [1] then provide a global estimate of mean canopy heights derived from ICESat in Fig. 15.12.

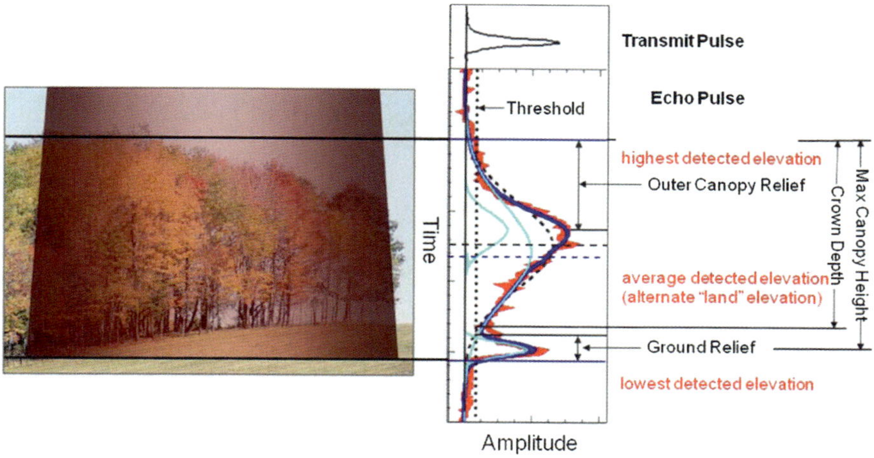

Fig. 15.11 Application of ICESat or ICESat-2 to measure tree heights as presented by [35]

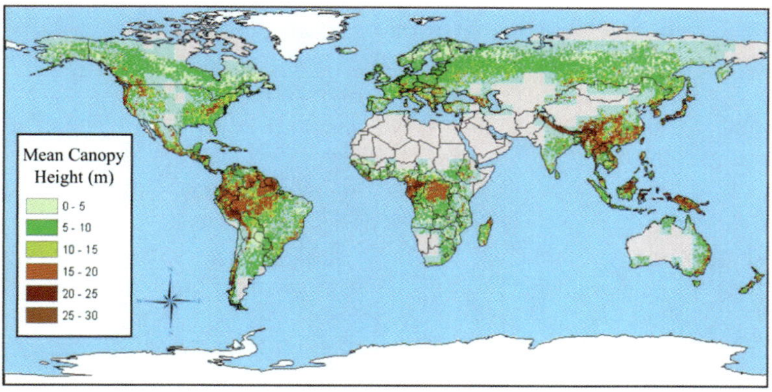

Fig. 15.12 Global estimates of mean canopy heights derived from ICESat as presented by [1]. The capability of retrieving tree height with ICESat-2 will contribute to the large-scale biomass assessments

15.7.3 Wetlands

Wetlands have been defined as a wide variety of habitats such as *bogs*, *fens*, *swamps*, *wet meadows*, *marshes*, *peatlands*, *flood plains*, rivers and lakes and coastal areas such as *salt marshes*, *mangroves* and *seagrass beds*, *coral reefs* and other marine areas no deeper than six meters at low tide, as well as man-made wetlands such as *waste water* treatment ponds and reservoirs. A simplified definition of wetland is that it is an area where the land is saturated with water long enough to support, and that do support poorly drained soils, plants and animals, which have been adopted to such environment, and biological processes suited to wet areas. Of the habitats mentioned above, bogs refer to wetlands which are formed where organic matter have accumulated for a long time and water input is entirely through precipitation. Fens are peat producing wetlands which are influenced by soil nutrients flowing through the system grasses and sledges. Swamps are mostly depicted as water logged areas while marshes are vegetation dominated by sledges and reeds.

The Ramsar International Convention on Wetlands of 1971 advocates for the wise use of wetlands by national actions and international co-operation as a means of achieving sustainable development throughout the world. Article 3.1 of the convention states that contracting parties shall formulate and implement their planning so as to promote the conservation of the wetlands, and as far as possible, the wise use of the wetlands within their territory.

Wetlands play a pivotal environmental role of providing sanctuary to biodiversity, store flood water, and improve water quality. It provides habitat for fish and wildlife, supporting a rich biodiversity that include many endangered and threatened species, see e.g., [6, 60]. Conservation of wetlands is necessary as they have the ability to sequester carbon and also tend to shift to net sources of greenhouse gases when perturbed by land use change such as drainage for agriculture or forestry. In this way, they help in preventing *global warming*, a consequence of *climate change* and its ecological effects on the environment. They have an inherent capacity for storage of carbon. Carbon cycle, conservation, and multinational conventions have a range of information requirements relating to wetland inventory, mapping and monitoring as shown in Table 15.1.

Ozesmi and Bauer [60] suggest the need for inventory and monitoring of wetlands and their adjacent uplands to assist in its conservation and management. Such inventory and monitoring can be achieved using remote sensing techniques, see e.g., [13, 27, 60]. Optical sensors that have been applied to wetlands include; Landsat Multi-

Table 15.1 Components and products of wetlands theme

Component	Product
I. Global wetland extent and properties	1. Tropical wetland extent and properties
	2. Boreal wetland extent and properties
II. Seasonal monitoring of major wetland	3. Seasonal monitoring of major regions tropical/sub-tropical wetlands
	4. Wetland extent, flood inundation patterns and vegetation change in the Greater Mekong River Basin
	5. Seasonal dynamics of the Pantanal ecosystem
	6. Seasonal monitoring of major boreal wetlands
III. Mapping and monitoring of key wetland types	7. Global mangrove extent and properties
	8. Tropical peat lands extent and properties
	9. Pan-Asian mapping and monitoring of rice paddies
	10. Global lake census

Source Lowry et al. [50]

spectral Scanner (MSS), Landsat Thematic Mapper (TM), SPOT (Système Probatoire d'Observation de la Terre), NOAA AVHRR (Advanced Very High Resolution Radiometer), and radar system (JERS-1, ERS-1, RADARSAT, and ALOS PALSAR). Landsat MSS (bands 4 (0.5–0.6 m), 5 (0.6–0.7 m), 6 (0.7–0.8 m) and 7 (0.8–1.1 m)) provide data that are useful in discriminating large vegetation wetlands [60]. Jensen et al. [41] indicate its possible use in change detection through multi-temporal analysis. Landsat TM (particularly band 5; 1.55–1.75 m) is useful in identification of wetlands and other land covers types (see, e.g., [34]) and change [56]. In practice, a window of 9 pixels would be required to consistently identify an object [60]. Band 5 has the capability to discriminate between vegetation and soil moisture levels. Jensen et al. [41] add that the separability between wetlands types could be achieved using this band.

For microwave imagery, Lowry et al. [50] conclude that L-band SAR systems are the single best option for fine spatial-resolution remote sensing of wetland extent and characteristics over large regions because they operate regardless of cloud cover, can distinguish basic vegetation structure, and provide superior canopy penetration and water surface discrimination relative to C-band. Furthermore, a dual-polarization L-band system such as that provided in the sensor ALOS PALSAR will result in improved accuracy in the discrimination between rough water surfaces and bare ground, and improved mapping of vegetation structural characteristics [50].

SPOT panchromatic (PAN) images have been used to provide reference data for registration and calibration purposes [2, 75], while [44] applied ERS-1 to monitor the presence or absence of water in wetlands in south western Florida, USA. Similarly, studies have been done to map zonation of vegetation communities and analyze biophysical properties of acquatic vegetation in the Amazon floodplain using JERS-1 and RADARSAT [18, 19]. A combination of optical and radar sensors, where the two complement one another, is presented in the work of Townsend and Walsh [72] who combine Landsat TM and SAR to detect flooding in a forested wetland on the lower Roanoke River floodplain in North Carolina. Another sensor that has been used for wetland studies is the Indian Remote Sensing (IRS) satellite whose four bands are similar to those of Landsat TM (see, e.g., [17]).

Spectral reflections of wetlands provide the possibility of separating various types of wetland vegetation. Classification techniques used for wetlands include visual interpretation, unsupervised or clustering, and supervised (see [3]). The importance of visual analysis in reconnaissance mapping of wetlands is demonstrated by Johnson and Wichern [42]. Unsupervised classification involve grouping together of pixels of similar spectral values, having the advantage of preserving spatial resolution of the images, see e.g., [41]. Supervised classification adopts statistical techniques such as the maximum likelihood. A hybrid system adopts a mixture of the two systems.

Vegetation indices find use in highlighting wetlands during classification methods above. Normalized Difference Vegetation Index (NDVI) and Transformed Vegetation Index (TVI) are some of the techniques used to enhance image identification. In addition, radar information is useful in wetland studies due to its capability to distinguish between flooded and non flooded areas. In all the remote sensing applications above, GNSS could be useful in georeferencing remotely sensed satellite data, pro-

Fig. 15.13 Virgin wetland along Lake Victoria

viding orientation for aerial photographs, and providing sampled location-based and perimeter/area data.

15.7.4 Lake Victoria's Wetlands: Overexploitation

Figure 15.13 indicate the green wetland along the shores of Lake Victoria which has not been tempered with. Figure 15.13 is a contrast showing a wetland that has been cleared to pave way for sand harvesting and agricultural activities. Naturally, wetlands will thrive on their own. Their development is however normally hampered by human activities as depicted in Fig. 15.14. These interference by humans normally reduce the sizes and bio-diversity as well as the aesthetic value of the wetlands. These activities include:

- Agriculture activities, e.g., clearing wetlands for cultivation, application of pesticides and other agro-chemicals and overgrazing. These have the negative impact on the remaining wetland through nutrient loading associated with runoff of fertilizers, pesticides and soil erosion. In addition to nutrient loading on the wetlands, shallower ones have been put under intensive cultivation for crops such as sugar cane and yams leading to the depleting of the remaining few wetlands. Some of the wetlands have become unproductive after 2 years of drainage for agriculture and are abandoned. Reclamation has been observed to lead to a reduction in the number of permanent springs and a low ground water yield in the wells.

Fig. 15.14 Destroyed wetland along Lake Victoria. This figure is a pale shadow of Fig. 15.13

- Urbanization which brings with it constructions, e.g., for roads, dams etc., leads to demand of space causing deforestation to obtain cultivable land, wood fuel and other craft products.
- Planting of exotic plants.
- Harvesting of wetland plants for construction and production of furniture, fuel etc. At Mukona, Mpigi and Amsha districts and Sango Bay in Rakai district of Uganda, deforestation of wetlands to obtain fuel has been reported.
- Fish farming and smoking activities.
- Sand, gravel and clay harvesting for housing construction and the excavation for brick making. In Fig. 15.15, the wetland has been cleared to pave way for sand harvesting. It can clearly be seen how sand harvesting has completely affected the wetlands and the environment where the pits left open attract vector borne disease causing mosquitoes and snails. The situation is similar in Kyetinda wetlands in Kampala, Uganda.
- Dumping of waste water and illegal dumping of garbage as seen in places close to Luzira prison, Masese swamp, and Walugogo valley in Iganga town in Uganda.
- During dry seasons, many people do burn the wetlands to pave way for agricultural activities (see, e.g., Fig. 15.16). The consequences of swamp burning are not known but it poses a big threat to biodiversity some of which might not be fire tolerant and also leads to replacement of natural wetland vegetation as depicted in Fig. 15.17.

In Kenya, the wetlands of Lake Victoria constitute about 37 % of the total surface area of wetlands (2,737,790 ha) in the country. The largest in Kenya is the Yala swamp which covers an area approximately 21,765 ha inclusive of water surface of Lake Kanyaboli, Lake Nyamboyo and Lake Sare.[5] Those that are adequately protected

[5] The standard, 14th December 2005.

are found within the national parks area. Various swamp forests especially outside gazetted areas are unprotected. These particular wetlands are threatened with total destruction due to degradation by the high population. Although Uganda's wetlands are protected, 75 % of the wetland area has been significantly affected by human activity and about 13 % severely degraded.

The Yala, Nyando and Sondu-Miriu swamps on the Kano plains are being drained for agriculture. It is estimated that at least 14,000 ha of the Yala swamp can be made productive. By 1980, 380, ha had been converted for rice production. On the Kano plains, 900 ha had been converted for rice and sugar cane production. Recent data indicate that 2,300 ha of Yala swamp in Siaya district were reclaimed by Dominion Farm Ltd, an affiliate of Dominion Group of companies based in Olkahoma, USA. This area of the reclaimed land is used to support the cultivation of rice. In a more recent study (see also Sect. 13.5.2), Awange et al. [4] found that excessive utilization of Lake Naivasha water for floriculture and horticulture has the potential of leading to the extinction of the Ramser-Lake Naivasha, which supports a rich wetland.

Wetlands associated with River Nyando are also rapidly shrinking because of human encroachment while the Sondu-Miriu delta is likely to experience severe ecological changes due to damming of the river to generate hydro-electric power through the support of the Japanese government. In Uganda, Munyonyo beach resort was built at the expense of wetlands for the purpose of eco-tourism.

A range of plants such as the common reed (*Phragmites australis*) and the reed-mace (*Typha latifolia*) has shown the property of breaking waste water, removing disease causing micro-organisms and pollutants. They have a large biomass both

Fig. 15.15 Wetland cleared along Lake Victoria to pave way for sand harvesting

Fig. 15.16 Burnt wetland along the shores of Lake Victoria

above and below the surface of the soil. The substrate plant tissues grow horizontally and vertically and create an extensive matrix, which binds the soil particles, and create a large surface area for the uptake of plant nutrients [67]. Wetlands act as filters of nutrient and silt loaded water. For example, papyrus swamps have been observed to take up and accumulate considerable amounts of ions in the effluent. Silt is adsorbed with nutrients. They thus play a crucial role in preventing eutrophication in the receiving waters. Degradation and drainage of wetlands will increase carbon dioxide emissions, burning will add phosphorus to the system, and agriculture usage will results in up to 50 % soil organic carbon.

15.7.4.1 Management of the Lake's Wetlands

As already mentioned, the wetlands of Lake Victoria comprise about 37 % of the total surface area of wetlands in Kenya and about 13 % of the same in Uganda. The Lake is fringed by an extensive system of swamps and wetlands, which provide habitat to a host of wildlife species. Many of the rivers and streams are fringed by swamps and often empty into the Lake through vast swamp ecosystems. The Lake is also known to contain unique plants, algae, zooplankton and invertebrates. Wetlands around Lake Victoria are increasingly threatened by agricultural activities and grazing, leading to the loss of biodiversity and reducing the capacity of wetlands to filter and reduce the amounts of pollutants reaching the Lake [7]. Lake Victoria's Wetlands filter nutrients and silt loaded water by incorporating them into the tissues

Fig. 15.17 Destroyed ecology following the burning of wetland along the shores of Lake Victoria

of wetland plants, accumulating them as sediments or by releasing Nitrogen into the atmosphere through denitrification in the root zone of the aquatic plants. The loss of wetlands also influences the hydrology of the catchment area thereby increasing the risk of flooding and erosion.

15.7.4.2 Economic Values of the Lake's Wetlands

The freshwater wetlands in Lake Victoria basin constitute an important natural resource base upon which the riparian communities depend. They are important in terms of food production, hydrological stability and ecological productivity. While Lake Victoria's wetlands support significant components of biodiversity and are important for socio-economic reasons, they have continued to be under increasing pressure from human activities, such as *conversion for agricultural purposes*, *water pollution* and *destruction of vegetation*. This led the Global Nature Fund to declare it the "*threatened Lake of* 2005" at the world wetlands day on 2nd February 2005. Although research has been undertaken to address gaps in knowledge of wetlands and develop suitable ways of monitoring and managing them in the Lake Victoria basin, a lot more still is unknown.

The wetlands in the Lake basin have found various uses which include:

1. Rainfall formation through evapotranspiration. Water lost through evapotranspiration contributes towards cloud formation and subsequently falls as rain. Wetlands therefore contribute towards balancing of the hydrological circle maintaining both surface and underground water supply.
2. Contributes to weather and climate. Other than contributing to rain formation as stated above, wetlands' contribution to water vapour leads to humidity, increase in greenhouse gases and also in breaking the speed of wind. Their use of CO_2 on the other hand ensures reduction of the greenhouse gases and thus regulating temperature.
3. Being source of goods and services for the riparian communities, i.e., sources of raw materials, handicrafts and fuel. Forested wetlands for example are a source of commercially valuable timber.
4. Supporting fisheries, grazing and agriculture. The Lake's wetlands provide breeding places for mud fish, lung fish and Tilapia. Tilapia fish normally comes in the cool wetlands to bread.
5. Providing outdoor recreation and education for human society, i.e., fishing, hunting, boating, plant identification, scientific study, and wildlife observation etc.
6. Providing habitat for wildlife, especially waterfowl, shorebirds, and other birds which depend on wetlands. This is achieved through the provision of essential water, food, cover, and reproductive areas for wildlife. In areas that are semi-arid, riparian wetlands are crucial to the survival of many wildlife species.
7. Being a source of water and food production during the dry season. Because of their high productivity, wetlands also provide essential food chain support. During dry periods, vegetable and other food crop can be planted in the wetlands without necessarily destroying them.
8. Through its effective removal of Nitrogen, Phosphorus, certain chemicals, and heavy metals from water, wetlands contribute to purification of water which subsequently leads to improved water quality through the filtration process.
9. Some parts of aquatic plants found in the wetlands are edible. These include for example, *Commelina berghalensis* whose leaves and young shoots are edible, rhizomes from *Nymphaea caerula*, leaves from *Ipomea aquatica* and leaves, stem and seeds from *Portulaca oleraceae*.
10. Some plants are of medicinal values. These include, for example, the roots of *Polygonum pulcheria* which are used to treat tropical ulcers, leaves of *Pentas longiflora* which are used to treat fever and *Adenia umicifolia* (whole plant) which is used to treat neurotic illness.
11. Wetlands reduce flood velocities, erosion erosions, and trap water-borne sediments and in so doing control erosion and sedimentation.

15.7.4.3 Sustainable Management of Lake Victoria's Wetlands

The communities living along the Lake basin combine both fishing and farming activities. Awana beach in Kenya has proved a good example of how management prowess of wetlands can yield good fruits. In dry or wet seasons, the inhabitants of Awana have plenty to take to the market, thanks to the good management of their wetlands. Indeed, the many functions that wetlands provide make *protection*, *restoration*, and *wise management* of wetlands important to the Lake inhabitants. While managing the wetlands, decision makers (e.g., Lake authorities, municipalities, fisheries departments etc.) should ensure that the Lake's wetlands are not converted into upland. The goal in managing wetlands, therefore, is to avoid activities that may convert the wetland to upland and to manage site alterations to retain the benefits provided by the wetland [76]. This therefore calls for incorporating the inhabitants surrounding the wetlands in all wetlands management programmes. Community based organizations can be set up and environmental awareness taught. Regular seminars can also be organized where people meet and exchange ideas.

Due to that fact that wetlands change in a variety of ways, where some changes are inevitable, as landscapes are changing all the time, the key to managing wetlands and woodlands is to plan carefully and thoughtfully, and to be aware of the affects that the planning activities will have on the landscape [76]. It is further noted in [76] that while some changes are natural, this may not be the case for others, or the changes may be accelerated by the activities of people. Some wetland sites may be so sensitive, and so difficult to reforest, that one may consider excluding them from one's harvest plans. As was pointed out at the beginning of this chapter, wetlands are defined by the amount of saturation of the water in the soil. The amount and duration of this water at a location affects soil development which in turn, affects the plant community characteristics. Just as specific water, soil, and vegetation characteristics define a wetland, they are the same characteristics that must be managed with care in order to maintain the functions and values of wetlands [76]. Some potential forest practices disturbances to wetlands have been pointed out by [76] to be:

- Alteration of hydrologic functions.
- Soil disturbance.
- Introduction of invasive plant species.
- Changes in the microclimate due to vegetation removal in and around the wetland.
- Alteration of habitat through the removal of vegetation, snags, and downed wood.
- Reduction of woody species or woody debris that simplify the composition, structure, and stability of wetlands.

For management techniques, [76] list forestry activities which demand careful planning to minimize damage to wetlands as; road and landing constructions, heavy equipment use in and around wetlands, reforestation, and harvesting.

15.7.4.4 Conservation Measures

Conservation positive policies and programs are required that recognize the full spectrum of intrinsic values of wetlands. The objectives of conservation and management of wetlands therefore encompass:

- Strengthening institutional and administrative structures for wetland management and conservation.
- Putting in place policy and legislation for wetland conservation and management.
- Enhancing knowledge and understanding of wetlands and associated ecological process, developing and disseminating methodologies for conservation and sustainable use of wetlands. This as already stated can be achieved through the use of community based organizations, seminars, etc.
- Improving awareness and knowledge of and support for wetlands conservation at all levels.
- A multidisciplinary unit to provide technical services to the central government, districts and local communities.
- Outline guidelines for conservation of wetlands.

Inline with the community based wetlands management and conservation, pilot projects can be carried out, e.g., on fish farming, craft making, bee keeping, and eco-tourism ventures to ascertain whether they conform to the recommended standard. There is need to promote production of booklets, posters, calendars, brochures to promote education and awareness of stakeholders and the general public. Use should also be made of radio (e.g., Radio Lake Victoria) and TV to further propagate wetlands conservation measures. Training should be carried out for participants at various levels (e.g., NGOs and local resource users).

Research need to be carried out on the wetlands in general. Efforts have been made in this direction as seen by the Research on Lake Victoria (VicRes)'s 2005 theme which was on wetlands. Through such research, various aspects e.g., level of carbon sink, rates of nitrification/denitrification, net ecosystem of biomass etc., will be unraveled. Long-term monitoring of wetlands is essential to all aspects of their conservation and sustainable use. It is an imperative prerequisite to the understanding, protection and enhancement of wetland ecosystems as carbon sinks.

International organizations have played a major role in research, conservation and socio-economic development. Their primary role has been to implement government policies and obligations to international conventions and agreements in their areas of interest. They are involved in areas of research, conservation, use and management of wetland resources. Studies by LVEMP show that the larger wetland fringes of Lake Victoria have significant buffering capacity against land based pollution. The wetland component of the programme has been carrying out research on wetland biodiversity, resource use patterns and buffering capacity of the riparian wetlands.

The studies on community utilization of papyrus and grass reeds for handicraft in Busia and Nyando districts also demonstrated that small business enterprizes based on the exploitation of wetland products are viable and can help create employment and alleviate poverty [30]. ICRAF has studied the role of Lake Victoria wetlands as sinks of sediment and other pollutants from the Lake's watershed. Preliminary results indicate that the Nyando River catchment has high erosivity and that the river transports large silt load to Nyakach bay. ICRAF has also been promoting agro-forestry and soil erosion control activities in the Lake basin.

The American Zoos and Aquaria Association (AZA) have conducted research on the assessment and conservation of indigenous fisheries. Their findings indicate that wetlands contain a more diverse fauna of indigenous fish species than Lake Victoria. Some of them like the New England Aquarium have conserved some of the haplochomines from Lake Victoria. The International Union for Conservation of Nature (IUCN) instituted studies on the compilation of biodiversity and socio-economic data on fisheries and conducted a series of training workshops. Despite all these studies, there is a wide range of knowledge gaps, which needs further actions. These are[6]:

- The significance of wetlands as an integral and vital part of Lake Victoria ecosystem is not fully appreciated. Little attention has been paid to wetland research. There have been delays in incorporating wetlands in national legislations.
- Existing open access system, lack of official resource use guidelines and the government policies of reclaiming wetlands for agricultural production negate conservation.
- Researchers have paid little attention to assessing and documenting ecosystem values of wetlands such as filtering water and processing pollutants, source of valuable products such as fish, waterfowl and medicine.
- Raw or semi analyzed data, e.g., reports, dissertations, manuscripts and technical reports.
- Limited publication of work in journals.
- Fragmented socio-economic studies not linked to ecosystem values of wetlands. Sustainability of human activities not linked to wetland health.
- Findings of biomedical research in wetlands do not filter into the local communities to help them prevent infection.
- Few wetland studies included gender and cultural values.

The following actions are essential for proper management of wetlands.

1. Pilot investments in the sustainable management of wetland products. This will go a long way in alleviating poverty among poor fishermen.
2. Identify and demonstrate practical, self sustaining remedies on wetlands.

[6] Getabu, Pers. Comm.

3. Establish mechanisms for co-operative management of resources.
4. Maximize the sustainable benefits to riparian communities from using resources within the basin to generate food, employment and income, supply safe water, and sustain a disease free environment.
5. Build capacity for ecosystem management among institutions and riparian communities.
6. Conservation and development of wetlands.
7. Provide the necessary information to improve management of the Lake ecosystem.

15.8 Concluding Remarks

This chapter has presented the possibility of using geospatial techniques to support land management measures. Discussions have been made on pertinent land management issues including reconnaissance and validation, monitoring of land conditions, land degradation, precision farming, and sustainable management and conservation of wetlands as a way of alleviating poverty and enhancing environmental quality. It is clear from the discussion that the area of precision farming has received a greater share of GNSS application compared to other areas of land management. In general, however, the uses of GNSS satellites are more vital in areas of land management that requires information on positions and spatial coverage. With more GNSS satellites being launched, they are expected to become more useful to land management, particularly when coupled with GIS and/or remote sensing methods.

References

1. Abdalati W, Zwally HJ, Bindschadler B, Csatho B, Farrell SL, Fricker HA, Harding D, Kwok R, Lefsky M, Markus T, Marshak A, Neumann T, Palm S, Schutz B, Smith B, Spinhirne J, Webb C (2010) The ICESat-2 laser altimetry mission. Proc IEEE 98(5):735–751. doi:10.1109/JPROC.2009.2034765
2. Ausseil AE, Dymond JR, Shephard JD (2007) Rapid mapping and prioritisation of wetland sites in the Manawatu-Wanganui region, New Zealand. Environ Manage 39:316–325
3. Awange JL, Kiema JBK (2013) Environmental geoinformatics: monitoring and management. Springer, Berlin
4. Awange JL, Forootan E, Kusche J, Kiema JBK, Omondi PA, Heck B, Fleming KM, Ohanya S, Goncalves RM (2013) Understanding the decline of water storage across the Ramser-Lake Naivasha using satellite-based methods. Adv Water Resour 60:7–23. doi:10.1016/j.advwatres.2013.07.002
5. Awange JL (2012) Environmental monitoring using GNSS. Global navigation satellite system. Springer, Berlin
6. Awange JL, Ong'ang'a O(2006) Lake Victoria—ecology, resource of the Lake Basin and environment. Springer, Berlin

7. Balirwa JS (2001) From vegetation to fish: structural aspects and related components of lakeshore wetlands in Lake Victoria. Paper presented at LVEMP conference, Kisumu, Kenya
8. Bandyopadhyay S, Shyamsundar P (2004) Fuelwood consumption and participation in community forestry in India. World Bank Policy Research Working Paper 3331
9. Bardhan PJ, Baland, S, Das S, Mookherjee D, Sarkar R (2002) The environmental impact of poverty: evidence from firewood collection in rural Nepal, mimeo
10. Bell RW (1997) Introduction. In: Bell RW (ed) Land management unit reader. Murdoch University, Murdoch, pp 3–7
11. Bojo J, Reddy RC (2002) Environment strategies notes: mainstreaming environment in poverty reduction strategies. No. 4, World Bank, Washington, DC
12. Brondizio ES, Moran EF, Mausel P, Wu Y (2005) Land use change in the Amazon estuary: patterns of caboclo settlement and landscape management. Hum Ecol 22:249–278
13. Brooks RP, Wardrop DH, Cole CA (2006) Inventorying and monitoring wetland condition and restoration on a watershed basin with examples from Spring Creek Watershed, Pennsylvania, USA. Environ Manage 38:673–687
14. Brubaker SC, Jones AJ, Lewis DT, Frank K (1993) Soil properties associated with landscape position. Soil Sci Soc Am J 57(1):235–239
15. Buick R (2006) GPS guidance and automated steering renew interest in precision farming techniques. White Paper, Trimble Agriculture Division, 10355 Westmoor Drive, Suite #100, Westminster, CO 80021, USA
16. Cassel DK, Kamprath EJ, Simmons FW (1996) Nitrogen–sulfur relationships in corn as affected by landscape attributes and tillage. Soil Sci Soc Am J 88(2):133–140
17. Chopra R, Verma VK, Sharma PK (2001) Mapping, monitoring and conservation of Harike wetland ecosystem, Punjab India through remote sensing. Int J Remote Sens 22:89–98
18. Costa M (2004) Use of SAR satellites for mapping zonation of vegetation communities in the Amazon floodplain. Int J Remote Sens 25(10):1817–1835
19. Costa M, Niemann O, Novo E, Ahern F, Mantovani J (2002) Biophysical Properties and mapping of aquatic vegetation during the hydrological cycle of the Amazon floodplain using JERS-1 and RADARSAT. Int J Remote Sens 23(7):1401–1426
20. Dasgupta S, Deichmann U, Meisner C, Wheeler D (2005) Where is the Poverty-Environment Nexus? Evidence from Cambodia, Lao PDR, and Vietnam. World Dev 33(4):617638. doi: 10.1016/j.worlddev.2004.10.003
21. Dobermann A, White PF (1999) Strategies for nutrient management in irrigated and rainfed lowland rice systems. In: Ladha JK, Denning GL, Balasubramanaian V (eds) Resource managements in rice systems: nutrients. Kluwer Academic Publishers, New York
22. Duraiappah A (1998) Poverty and environmental degradation: a review and analysis of the nexus. World Dev 26(12):2169–2179
23. Edmonds E (2002) Government initiated community resource management and local resource extraction from Nepal forests. J Dev Econ 68(2):89–115
24. El-Rabbany A (2006) Introduction to GPS global positioning system, 2nd edn. Artech House, London
25. Ellett KM, Walker JP, Rodell M, Chen JL, Western AW (2005) GRACE gravity fields as a new measure for assessing large-scale hydrological models. In: Zerger A, Argent RM (eds) MODSIM 2005 international congress on modelling and simulation, Modell. and Simul. Soc. of Aust. and N. Z., December 2005, pp 2911–2917. ISBN: 0-9758400-2-9
26. Ellett KM, Walker JP, Western AW, Rodell M (2006) A framework for assessing the potential of remote sensed gravity to provide new insight on the hydrology of the Murray-Darling Basin. Aust J Water Resour 10(2):89–101
27. Emerton L, Kekulandala LDCB (2003) Assessment of the economic value of Muthurajawela wetland. Occasional Papers of IUCN Sri Lanka, No. 4, January 2003

28. Fearnside PM, Laurance WF (2003) Determination of deforestation rates of the world's humid tropical forests. Science 299:10–15
29. Gibbs HK, Brown S, Niles JO, Foley JA (2007) Monitoring and estimating tropical forest carbon stocks: making REDD a reality. Environ Res Lett 2:1–13
30. Gichuki NN (2003) Review of wetland research activities in Lake Victoria basin, Kenya: analysis and synthesis report. SIDA/Inter-University Report. Council of East Africa, Kampala, Uganda, 39 pp
31. Greenwood EAN, Klein JL, Beresford JD, Watson GD (1985) Differences in annual evapouration between grazed pasture and Eucalyptus species in plantations on a saline farm catchment. J Hydrol 78:261–278
32. Grisso RB, Oderwald R, Alley M, Heatwole C (2003) Precision farming tools: global positioning system (GPS). Virginia Cooperative Extension (VCE) Publication, Virginia, pp 442–503
33. Grisso1 RB, Alley M, McClellan P, Brann D, Donohue S (2002) Precision farming: a comprehensive approach. Virginia Cooperative Extension (VCE) Publication, Virginia, pp 442–500
34. Han M, Sun Y, Xu S (2007) Characteristics and driving factors of marsh changes in Zhalong wetland of China. Environ Monit Assess 127:363–381
35. Harding DJ, Carabajal CC (2005) ICESat waveform measurements of within-footprint topographic relief and vegetation vertical structure. Geophys Res Lett 32:L21S10. doi:10.1029/2005GL023471
36. Holopainen M, Leino O, Kämäri H, Talvitie M (2006) Drought damage in the park forests of the city of Helsinki. Urban For Urban Greening 4:75–83. doi:10.1016/j.ufug.2005.11.002
37. Houghton RA (2005) Above ground forest biomass and the global carbon balance. Global Change Biol 11:945–958
38. Hudson N (1985) Soil conservation. Batsford Academic and Educational, London
39. Ismail J, Ravichandran S (2007) RUSLE2 model application for soil erosion assessment using remote sensing and GIS. Water Resour Manage. doi:10.1007/s11269-006-9145-9
40. James LF, Young JA, Sanders K (2003) A new approach to monitoring rangeland. Arid Land Res Manage 17:319–328. doi:10.1080/15324980390225467
41. Jensen JR, Christensen EJ, Sharitz R (1984) Nontidal wetland mapping in south Carolina using airborne multi-spectral scanner data. Remote Sens Enviro 16:1–12
42. Johnson RA, Wichern DW (2007) Applied multivariate statistical analysis, 6th edn. Pearson (Prentice Hall), Upper Saddle River, New Jersey
43. Jones AJ, Mielke LN, Bartles CA, Miller CA (1989) Relationship of landscape position and properties to crop production. J Soil Water Cons 44(4):328–332
44. Kasischke ES, Bourgeau-Chavez LL (1997) Monitoring south Florida wetlands using ERS-1 SAR imagery. Photogram Eng Remote Sens 63:281–291
45. Kgathi DL, Mfundisi KB, Mmopelwa G, Mosepele K (2012) Potential impacts of biofuel development on food security in Botswana: a contribution to energy policy. Energy Policy 43:70–79. doi:10.1016/j.enpol.2011.12.027
46. Klooster D (2000) Community forestry and tree theft in Mexico: resistance or complicity in conservation? Dev Change 31:281–305
47. Kravchenko AN, Bullock DG (2000) Correlation of corn and soybean grain yield with topography and soil properties. Agron J 92(1):75–83
48. Kumi-Boateng B (2012) A spatio-temporal based estimation of vegetation changes in the Tarkwa mining area of Ghana. Doctor of Philosophy, Dissertation, University of Mines and Technology, Ghana, 165 pp
49. Lantzke N, Fulton I (1993) Soils of the Northam advisory district. Volume 3-The Darling Range and West Kokeby Zone. Agriculture WA, Bulletin, pp 42–57
50. Lowry, J, Hess, L, Rosenqvist, A (2009) Mapping and monitoring wetlands around the world using ALOS PALSAR: the ALOS Kyoto and carbon initiative wetlands products. In: Jones, S, Reinke, K (eds) Innovations in remote sensing and photogrammetry. Lecture notes in geoinformation and cartography. Springer, Berlin, pp 129–144
51. Lu D, Li G, Valladares GS, Batistella M (2004) Mapping soil erosion risk in Rondonia, Brazilian Amazonia: using RUSLE, remote sensing and GIS. Land Degradation Dev 15:499–512

52. Lufafa A, Tenywa MM, Isabirye M, Majaliwa MJG, Woomer PL (2003) Prediction of soil erosion in a Lake Victoria basin catchment using GIS-based Universal Soil Loss Model. Agric Syst 76:883–894
53. Mackenzie FT (2003) Our changing planet: an introduction to Earth system science and global environmental change, 3rd edn. Prentice Hall, New Jersey
54. Malhi Y, Grace J (2000) Tropical forests and atmospheric carbon dioxide. Trends Ecol Evol 15:332–337
55. Malik SJ (1999) Rural poverty and land degradation: what does the available literature suggest for priority setting for the Consultative Group on International Agricultural Research? Report prepared for the Technical Advisory Committee of the CGIAR, Vienna, Virginia
56. Munyati C (2000) Wetland change detection on the Kafue Flats, Zambia, by classification of a multitemporal remote sensing image dataset. Int J Remote Sens 21:1787–1806
57. NLWRA (National Land and Water Resource Audit) (2001) Water resources in Australia. A summary of the National Land and Water Resource Audit's Australian water resources assessment 2000. Surface water and groundwater—availability and quality
58. Nulsen RA (1984) Saltland management—the catchment approach. Western Australia Department of Agriculture. Farmnote 133/84
59. Onyando JO, Kisoyan P, Chemelil MC (2005) Estimating the potential of soil erosion for river Perkerra catchment in Kenya. Water Resour Manage 19:133–143
60. Ozesmi SL, Bauer ME (2002) Satellite remote sensing of wetlands. Wetlands Ecol Manage 10:381–402
61. Pandey A, Chowdary VM, Mal BC (2007) Identification of critical erosion prone areas in the small agricultural watershed using USLE, GIS and remote sensing. Water Resour Manage 21:729–746
62. Pieri C (1997) Planning a sustainable land management: the hierarchy of user needs. ITC J 3(4):223–228
63. Raven Environmental Services (2008) Mapping and GIS/GPS Technology Services. http://www.ravenenvironmental.com/mapping.aspx (Accessed on 06/03/2008)
64. Read V (2001) Salinity in Western Australia a situation statement. Resource Management Technical Report No. 81. ISSN 0729–3135
65. Schmidt JP, Taylor RK, Gehl RJ (2003) Developing topographic maps using a sub-meter accuracy global positioning system. Appl Eng Agric 19(3):291–300
66. Schoknecht N, Tille P, Purdie B (2004) Soil landscape mapping in south-western Australia. Overview of methods and outputs. Resource Management Technical Report 280, Department of Agriculture, Government of Western Australia
67. Shutes RBE (2001) Artificial wetlands and water quality improvement. Environ Int 26:441–447
68. Sirait M, Prasodjo S, Podger N, Flavelle A, Fox J (1994) Mapping customary land in east Kalimantan, Indonesia: a tool for forest management. Ambio Stockholm AMBIO 23(7): 411–417
69. Scherr S (2000) A downward spiral? Research evidence on the relationship between poverty and natural resource degradation. Food Policy 25:479–498
70. Steede-Terry K (2000) Integrating GIS and the global positioning system. ESRI Press, Redlands
71. Timlin DJ, Pachepsky Y, Snyder VA, Bryant RB (1998) Spatial and temporal variability of corn grain yield on a hillslope. Soil Sci Soc Am J 62(3):764–773
72. Townsend PA, Walsh SJ (1998) Modeling floodplain inundation using an integrated GIS with radar and optical remote sensing. Geomorphology 21:295–312
73. Tucker CJ (1979) Red and photographic infrared linear combinations for monitoring vegetation. Remote Sens Environ 8(2):127150. doi:10.1016/0034-4257(79)90013-0
74. UNEP (United Nations Environment Programme), Programme kenya Wildlife Service And Kenya Forests Working Group (2005) Mau Complex Under Siege. Continuous destruction of Kenya's forest. http://www.iapad.org/publications/mau_crisis_2005f.pdf (Accessed on 4/11/2011)
75. Welch R, Remillard M, Doran RF (1995) GIS database development for South Florida's national parks and preserves. Photogram Eng Remote Sens 61:1371–1381

76. Woodland Fish and Wildlife Project Publication (1994) Wetlands as varied as our region, 1 June.
77. Wily L (2002) The political economy of community forestry in Africa—Getting the power relations right. Forest Tress People Newslett 46:4–12
78. Yang C, Peterson CL, Shropshire GJ, Otawa T (1998) Spatial variability of field topography and wheat yield in the Palouse region of the Pacific Northwest. Trans ASAE 41(1):17–27

Chapter 16
Geospatial Monitoring and Management of Disasters

> "The greatest exploiter for all of us are floods today, droughts tomorrow, earthquake some times and all of these multiply our trauma of deprivation, pains of poverty and hunger. These disasters take away not only our crops, shelters, lives of our families, friends tattles, but also destroy our hopes and dreams of the future. Is there any event comparable to these, which causes so much human sufferings and injustice?" This is the cry in bewilderment of a common farmer of Koshi River basin, Bihar (India) in the midst of recurrent floods and droughts [52].

16.1 Introductory Remarks

Since time immemorial, natural disasters have continued to plague the history of mankind. They have varied in type, frequency, coverage and severity ranging from earthquakes, landslides, droughts, floods, tornadoes, hurricanes, tsunamis, volcanic eruptions, etc. Over the last century, the frequency, severity and impact of natural disasters has increased substantially. This has been partly attributable to the destruction of the environment and the concomitant climate change paradigm discussed, e.g., in Awange [8] and Awange and Kiema [7]. To enhance the sustainability of Earth, and thereby ensure the very survival of humanity, it has become imperative that natural disasters be continually and systematically monitored. This needs to be undertaken for different natural disasters at various levels ranging from local, regional to global monitoring scales.

That space technology has made a significant contribution in disaster monitoring and management over the years is not in dispute. For example, Earth observation remote sensing satellites have continued to provide comprehensive, synoptic and multi-temporal coverage of large areas in real time and at frequent intervals. Data acquired from geospatial approaches of remote sensing and Global Navigation Satellite Systems (GNSS) have become valuable for continuous monitoring of atmospheric as well as surface parameters that are critical in disaster monitoring. In general, advancements in remote sensing, Geographical Information System (GIS),

N. O. Agola and J. L. Awange, *Globalized Poverty and Environment*,
DOI: 10.1007/978-3-642-39733-2_16,

GNSS and communication technologies have assisted in providing a framework for real time monitoring, early warning and quick damage assessment of many natural disasters.

The rest of this chapter is organized as follows; In Sect. 16.2, for purposes of putting this presentation in context and avoiding ambiguity, the definition and scope of natural disasters is outlined. This is followed by Sect. 16.3, which presents geosensor networks employed in disaster monitoring. In Sect. 16.4, floods are addressed, while Sect. 16.5 discusses droughts. Section 16.6 discusses vector-borne diseases and outbreak, with Sect. 16.7 focusing on earthquakes. Section 16.8 discusses changing sea level, while Sect. 16.9 presents Tsunami early warning systems. The chapter closes with Sect. 16.10 which discusses land subsidence and slides.

16.2 Definition and Scope

A natural disaster is the effect of a natural hazard and may be triggered by natural events such as earthquakes, landslides, droughts, floods, tornadoes, hurricanes, tsunamis, volcanic eruptions, etc. Natural disasters are often characterized by substantial financial, environmental, human and/or animal losses. The magnitude of the resulting loss, and ultimately, the impact of the disaster, depends on the *vulnerability* of the affected population to resist the *hazard*. Bankoff [16] refers to this resistance as the *resilience* of the population/community/nation. This interpretation is premised in the formulation that disasters occur when hazards meet vulnerability as argued in Wisner et al. [109]. Furthermore, a disaster will only be judged to have occurred under these situations if the affected community is unable to cope with the impending hazard. Consequently, what may be interpreted as a disaster in one location may not necessarily be seen as a disaster in another part of the world.

A hazard will therefore never transform and mature into a natural disaster in areas without vulnerability, or in situations where the community can handle the posed hazard challenge. Therefore, a strong earthquake in uninhabited areas may not necessary be flagged as a natural disaster as articulated by Ballesteros [13]. The term *natural* has consequently been disputed because the events simply are not hazards or disasters without human involvement. This has led to the global phenomenon that distinguishes such human induced disasters as *man-made natural disasters*. An example of the division between a natural hazard and a natural disaster is perhaps exemplified by the fact that the 1906 San Francisco earthquake was a disaster, whereas earthquakes are in general hazards [3].

Natural disasters, whether of meteorological origin such as cyclones, floods, tornadoes and droughts, or those having geological nature like earthquakes and volcanoes, are well known to have devastating impacts on human life, economy and environment, and are also formidable physical constraints in our overall efforts to develop and utilize natural resources on a sustainable basis [52]. Indeed, disasters have been known to hit hard as seen from the floods of 2010–2011 in Pakistan and Australia, the sludge flow in Hungary in 2010, the landslide in Brazil in 2011, and Hurricane Sandy in the US,[1] events which led to environmental catastrophe.

Disaster trends reveal that the most vulnerable and hardest hit are normally the poorest people, most of who live in developing countries. With tropical climate and unstable land forms, coupled with high population density, poverty, illiteracy and lack of infrastructure development, developing countries are more vulnerable to suffer from the damaging potential of such disasters. For example, the year 2004 witnessed one of the greatest tragedies of humankind—the great tsunami that wiped out civilizations in many parts of south-east Asia. Millions were rendered homeless, and hundreds of thousands lost their loved ones.

Though it is almost impossible to completely neutralize the damage due to these disasters, according to Jayaraman [52] it is, however, possible to;

(1) minimize the potential risks by developing disaster early warning strategies;
(2) prepare developmental plans to provide resilience to such disasters;
(3) mobilize resources including communication and tele-medicinal services; and
(4) to help in rehabilitation and post-disaster reconstruction.

Disasters such as earthquakes, tsunamis and storm events are of short duration and require a fixed amount of consequence management. Others like outbreak of contagious diseases (e.g., bird flu) and wild fires are more complex and unfold in a non-linear fashion over an extended period necessitating ongoing and adaptive consequence management as argued by Terhorst et al. [96]. Generally speaking, management plans should be put in place comprising three phases, which would address the issues related to

(a) preparedness;
(b) response and recovery; and
(c) mitigation.

The preparedness activities such as prediction and risk zone identification are usually taken up long before the event occurs, whereas the prevention activities such as early warning/forecasting, monitoring and preparation of contingency plans are taken up just before or during the event. Finally, the response/mitigation phase is where activities are undertaken just after the event, which include damage assessment and relief management. With regard to response, the concepts documented by Terhorst et al. [96] are pertinent. These include *level of preparedness*, *response times*, *sustaining the response* and *coordinating the response*. Due to the fact that time is critical, the three primary challenges in the race against time are *uncertainty*, *complexity* and *variability*. As an example, within the context of vegetation fires, Terhorst et al. [96] propose a *sensor web* (see Sect. 16.3 for discussion on geosensors) that can be used to enhance the tempo of disaster response.

To support these efforts, relevant information needed to undertake risk appraisal including vulnerability analysis of the terrain, prediction, warning, and prevention of a disastrous event need to be identified *a priori*. Disaster management measures also

---

[1] http://edition.cnn.com/2012/10/29/travel/hurricane-sandy-flight-cancellations/index.html

entail examination of the probability of occurrence and the consequences, understanding of the total processes in relation to the cause and effect, identification of preventive measures and implementation of rescue strategies [52].

Besides information communication technology (ICT), positioning and mapping technologies play a critical and central role in disaster monitoring and management in general. Consequently, geospatial technologies like GNSS, remote sensing and GIS are becoming more and more important, not just in the wake of disaster, when, for instance, relief efforts might call for quicker generated maps of flooded areas, for helicopters to navigate through thick smoke, or for the exact location of people buried alive, but in planning and preparatory phases of emergency management [94]. To support disaster management programs, GNSS could be used to provide support in pre-disaster preparedness programs, in disaster response and monitoring activities, and post-disaster reconstruction, specifically in the provision of location-based information. Video cameras fitted with GNSS interface and operating with infra-red sensors, for instance, could help in documenting the occurrences and locations of fire, insects, drought, flood, hail, and frost [51].

Other activities that could also be supported by positioning and mapping technologies include [52];

- identification of hazard zones;
- risk assessment;
- creation of awareness at various levels;
- evolving systems for monitoring, prediction and warning;
- designing long-term preventive measures (structural and non-structural) and short-term protective measures and preparedness;
- early intervention measures;
- education, training, public information;
- transfer of technology; and
- research on improved technology and disaster management.

16.3 Geosensor Networks in Disaster Monitoring

A geosensor network is a distributed ad-hoc wireless network of sensor-enabled miniature computing platforms (a sensor network) that monitors phenomena in geographic space [75, 110], and in which the *geospatial content* of the information collected, aggregated, analyzed and monitored is of fundamental importance [95]. Geosensor networks consist of thousands of sensors (also called nodes, and maybe fixed or mobile) that monitor and process data, and are also capable of communicating to each other, and to transfer data to a central station via wireless communication, hence power consumption becomes a crucial issue [21]. Since these thousands of sensors are randomly deployed, GNSS plays a crucial role in providing their locations, to which the monitored environmental parameters (e.g., temperature and humidity) can be related.

Their environmental applications include the monitoring of drinking water quality [2], disasters (earthquakes, forest fires, volcanic eruptions, oil spill, movement of glaciers, deformations in wells and bridges, flood detection and prevention) [21, 25], wildlife habitat monitoring, watershed management, environmental pollution monitoring, deep sea explorations, to monitoring food safety, e.g., in South Africa and precision agriculture for large vineyards (e.g., in Southern Australia) [74]. They also offer the advantage of real-time environmental monitoring such as the progress of oil spill as was witnessed in the gulf of Mexico or real-time detection of toxic gas plumes in open public spaces [74].

The advantage of space-based positioning and mapping technologies is that they are unaffected by the actual impact since the sensors are in space and in supporting the emerging geosensor environmental monitoring, particularly in disaster management, they contribute towards a shift from post-event, estimation based, historic data analysis to real-time, sensor-rich event detection and monitoring [74]. They are therefore useful in monitoring environmental disasters discussed in the coming sections and compliment ground based geodetic methods. In this regard, the main advantage of RTGPS (real-time GPS) data discussed e.g., in Awange [8] would be the improved temporal resolution in observations of natural processes achieved through high-rate information. RTGPS will likely demonstrate an impact similar to that of other high-rate geophysical observations (e.g., from seismological and meteorological networks) for monitoring and understanding earthquakes, seismic wave propagation, volcanic eruptions, magmatic intrusions, movements of ice, landslides, and structure and dynamics of the atmosphere [41].

In the event of a disaster on a large scale, the *International Charter on Space and Major Disasters*[2] requires its members to provide satellite remote sensing data, free of charge, once an activation of the charter is triggered. The charter aims at providing a unified framework for space data acquisition and delivery to support emergency relief efforts. Each member has devoted resources to support the provisions of the charter and thus help to mitigate the effects of disasters on human life and property. Table 16.1 shows charter members and their diverse space resources. Details on the operation of the charter are outlined in the following example.

Example 16.1 (International Charter on Space and Major Disasters. Source: http://www.disasterscharter/org)

The international charter on space and major disasters was signed in October 2000 by the founding members, the European and French space agencies (ESA and CNES), as a follow-up of the UNISPACE III[3] conference held in Austria in 1999. Since then, the charter has grown to a membership of about 20 (see Table 16.1), with the latest member the European Organization for the Exploitation of Meteorological Satellites

[2] http://www.disasterscharter.org/

[3] http://www.un.org/events/unispace3/

Table 16.1 Charter members and their space resources (up to April 2012) *Source* http://www.disasterscharter.org/web/charter/members

Member	Space resources
European Space Agency (ESA)	ERS, ENVISAT
Centre National dtudes Spatiales (CNES)	
Spotimage—France	SPOT
NSPO—Taiwan	Formosat
Canadian Space Agency (CSA)	RADARSAT
Indian Space Research Organization (ISRO)	IRS
National Oceanic and Atmospheric Administration (NOAA)	POES, GOES
Comisión Nacional de Actividades Espaciales (CONAE)—Argentina	SAC-C
Japan Aerospace Exploration Agency (JAXA)	ALOS
United States Geological Survey (USGS)	Landsat
Digital globe	Quickbird
GeoEye	GeoEye-1
Disaster Monitoring Constellation International Imaging (DMC):	
Centre National des Techniques Spatiales (Algeria)	ALSAT-1
National Space Research and Development (Nigeria)	NigeriaSat
Tübitak-BILTEN (Turkey)	BILSAT-1
UK Space Agency (UK)	UK-DMC
China National Space Administration (CNSA)	FY, SJ, ZY Satellite series
German Aerospace Center (DLR)	TerraSAR-X, TanDEM-X
Korea Aerospace Research Institute (KARI)	Kompsat-2
National Institute for Space Research (INPE)—Brazil/China	CBERS
European Organization for the Exploitation of Meteorological Satellites (EUMETSAT)	Meteosat

(EUMETSAT) joining in July 2012. The aim of the charter, underscored by the international agreement among its members, requires members to support relief efforts in the event of emergencies caused by major disasters with space-based data and information. As currently structured, however, the charter provides data to support only in immediate emergency response cases. Consequently, it does not support activities geared towards rehabilitation, reconstruction, prevention, prepardness or scientific research. Moreover, it does not provide maps suitable for use in the field.

Once a request for data support from an authorized user is received the following process is set into motion to activate the charter[4]:

(1) Identification and verification that a disaster has indeed happened and that the charter can help;
(2) Operations activity round the clock (24/7) are triggered to respond to the request for support and to quickly task satellites from members;
(3) Once the data is acquired, it is quickly processed into images; and

[4] http://www.disasterscharter.org/web/charter/activate

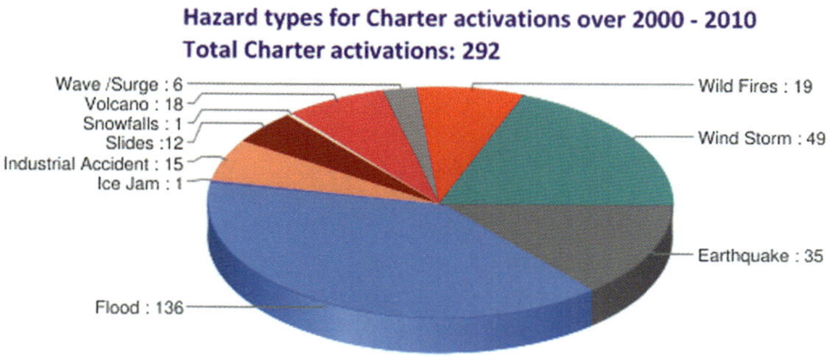

Fig. 16.1 Hazard types for charter activations between 2000 and 2010 (*Source* http://www.disasterscharter.org/web/charter/emdat)

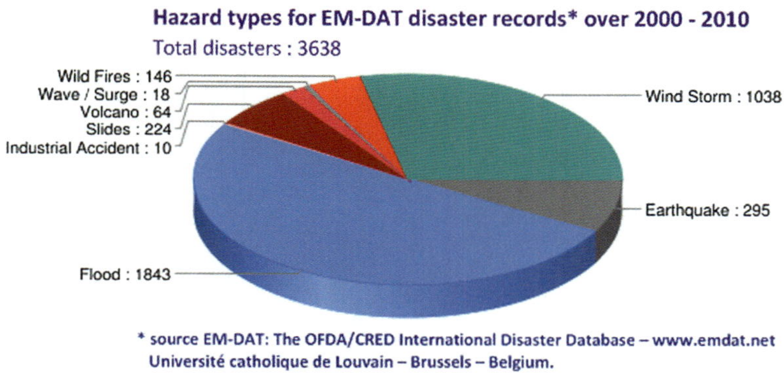

Fig. 16.2 Hazard types for EM-DAT records between 2000 and 2010 (*Source* http://www.disasterscharter.org/web/charter/emdat)

(4) The data is further processed and interpreted to extract specific data for the area affected by the disaster which is then delivered to the end user. Throughout this exercise a direct link is maintained with the user/response community who will use the information.

Up until 2012, the charter had covered a total of 314 disasters in approximately 100 countries worldwide. Figure 16.1 illustrates the hazard types for charter activations between 2000 and 2010, while Fig. 16.2 shows the total number of disasters recorded by EM-DAT[5] over the same duration. From these statistics, out of a total of 292 charter activations made, floods constituted the largest fraction with 46.5 %. Similarly, out of 3,638 disasters recorded in the EM-DAT records, floods again were the most dominant contributing about 50.7 %.

Evidently, the charter has served as a useful international disaster rapid response instrument [4, 104]. However, despite the wonderful work realized by the charter thus far, there are certain critical elements that need to be addressed including; availability

---

[5] http://www.emdat.net

of essential datasets, processing time, information extraction and usability of the products by end-users [26]. To address some of these issues, the charter intends to in future fully embrace the concept of *universal access*. This means that any national disaster management authority will be able to submit requests for emergency response support, regardless of whether or not they are within the bracket of authorized users or charter members.

End of Example 16.1

16.4 Floods

In most countries that experience frequent natural disasters, e.g., floods, droughts, earthquakes, forest fires, snow, and typhoons, floods could arguably be defined as among the most devastating natural hazards in the world, claiming more lives and causing more property damage than any other natural phenomena [53]. Moreover, in most cases, floods seem to follow droughts and vice-versa as both disaster types are subject to similar extreme weather and climate conditions. The recurrent nature of these natural disasters is also alarming. For example, almost every year, Pakistan experiences severe floods caused by monsoon rains that often result in considerable economic loss and serious damage to property. Another country that is perennially prone to floods is China, where the frequency of occurrence is higher than the world average, with historical records documenting more than 1000 floods [116]. Similarly, flooding in the City of New Orleans, USA in 2005 that was triggered by Hurricane Katrina causing levee failure set in motion an unanticipated failure of multiple infrastructure systems [60]. Indeed, floods pose one of the greatest challenges in weather prediction.

A flood can be defined as any relatively high water flow that over-tops a natural or artificial bank in any portion of a river or stream. When such a bank is overrun, water spreads over the flood plain and generally becomes a hazard. In general, floods can be classified into the following types [53]:

- river floods formed from winter and spring rains, coupled with snow melt, and torrential rains from decaying tropical storms and monsoons;
- coastal floods generated by winds from intense off-shore storms and tsunamis;
- urban floods, as urbanization increases runoff two to six times what would occur on natural terrain; and
- flash floods that can occur within minutes or hours of excessive rainfall or a dam or levee failure, or a sudden release of water.

Similar to the discussion advanced in Sect. 16.1, one can distinguish three (3) major phases in flood management namely;

(1) *Preparedness phase*: This involves activities such as prediction and risk zone identification that are taken up long before the event occurs.
(2) *Prevention phase*: Where activities such as early warning/forecasting, monitoring and preparation of contingency plans are taken up just before or during the event and
(3) *Response/mitigation phase*: This encompasses activities taken just after the event and includes damage assessment and relief management.

Evidently, the degree of disaster triggered by floods will vary depending on the type of flood. For instance, in most practical situations, flash floods may pose more disaster risk than river floods. Consequently, they require more urgent response as articulated in Scofield and Achutuni [89] in their presentation of a satellite-based funnel approach for predicting and forecasting flash floods. In the following sections, we discuss the different phases of flood management as well as present various applications of geospatial in flood management.

16.4.1 Geospatial Mapping of Flood Risk Zone Mapping

Data, information and experience gathered from previous flood events may help to inform and better understand future flooding events. Active remote sensing sensors can be used to procure time-series data of flood events regardless of the flooding weather conditions. These can be documented, mapped and/or archived within a GIS environment. Besides remote sensing data, LiDAR can also be used at this stage as demonstrated by [107]. Simulations developed using virtual reality models, as well as other animation types, can then be employed to visualize potential flooding scenarios given different hydrological conditions e.g., precipitation.

Ultimately, a flood risk zone map needs to be produced. Two types of these can be distinguished, namely [53];

(a) A detailed mapping approach, that is required for the production of hazard assessment for updating (and sometimes creating) risk maps. Such maps contribute to the hazard and vulnerability aspects of flooding.
(b) A larger scale approach that explores the general flood situation within a river catchment or coastal belt, with the aim of identifying areas that pose the greatest risk.

16.4.2 Geospatial Monitoring and Forecasting of Floods

Geospatial monitoring of floods can be carried out through remote sensing from global scale to storm scale. However, it is mostly applied at the storm scale using hydrodynamic models by monitoring the intensity, movement, and propagation of the precipitation system to determine how much, when, and where the heavy precipitation is going to move during the next zero to three hours [53]. Meteorological satellites (such as GOES and POES) help detect various aspects of the hydrological cycle including precipitation (rate and accumulation), moisture transport, and surface/soil wetness [89]. Synthetic Aperture Radar (SAR) can achieve regular observation of the earth's surface, even in the presence of thick cloud cover. NOAA AHVRR represents a family of satellites upon which flood monitoring and mapping can almost always be done in near real time. Remote sensing data such as IRS, SAR, SPOT, and to some extent high resolution NOAA images can be used to determine flood extent and areal coverage. Various precipitable water (PW) products have been developed and are available operationally for assessing the state of the atmosphere with respect to the magnitude of the moisture and its transport [53].

Hydrological models play a major role in assessing and forecasting flood risks. To be successful, however, these models often require several types of input data e.g., land use, soil type, soil moisture, stream/river base flow, rainfall amount/intensity, snow pack characterization, digital elevation model (DEM) data, and static data (e.g., drainage basin size) [53]. Furthermore, they need to be fine-tuned to effectively model a particular drainage basin correctly. Geospatial technologies can be employed to deliver most of these data sets and/or their derived products.

Model predictions or simulations of potential flood extent can help emergency managers develop contingency plans well in advance of an actual event to help facilitate a more efficient and effective response [53]. Specifically, flood forecast can be issued over the areas in which remote sensing is complementary to direct precipitation and stream flow measurements, and those areas that are not instrumentally monitored (or the instruments are not working or are in error). Evidently, high temporal resolution remote sensing data is fundamental to effective flood monitoring and forecasting.

16.4.3 Geospatial Applications to Flood Response and Mitigation

During floods, timely and detailed situation reports are required by the authorities to locate and identify the affected areas and to implement appropriate damage mitigation. It is essential that information be accurate and timely, in order to address emergency situations (for example, dealing with diversion of flood water, evacuation, rescue, resettlement, water pollution, health hazards, and handling the interruption of utilities etc.). This represents the most delicate phase of flood management as it involves rescue operations and the safety of people and property.

Jeyaseelan [53] presents the following lists of information used and analyzed in real time together with the sources for the same (outlined in the brackets): flood extent mapping and real time monitoring (satellite, airborne, and direct survey); damage to buildings (remote sensing and direct inspections); damage to infrastructure (remote sensing and direct inspection); meteorological (important real-time input from remote sensing data to show intensity/estimates, movement, and expected duration of rainfall for the next 0–3 h); and evaluation of secondary disasters, such as waste pollution, to be detected and assessed during the crisis (remote sensing and others). Adequate communication and logistical support is also important to ensure speedy delivery.

After the flooding, reconstruction of destroyed or damaged infrastructure, facilities and adjustments of the existing infrastructure will occur. At the same time, insurance companies require up-to-date information to settle claims. The time factor is not as critical as in the last stage. Nevertheless, both medium and high-resolution remote sensing images, together with an operational GIS, can help to plan many tasks. The medium resolution data can help establish the extent of the flood damages and can be used to establish new flood boundaries. They can also locate landslides and pollution due to discharge and sediments. On the other hand, high-resolution data employed in conjunction with GNSS are suitable for pinpointing locations and the degree of damages. They can also be used as reference maps to rebuild bridges, washed-out roads, homes and facilities [53].

16.4.4 Geospatial Support of Flood Management

Many examples on the operational use of space-based technology in the detailed monitoring and mapping of floods and post-flood damage assessment have been documented. Remote sensing information delivered from different sensors and platforms (satellite, airplane, and ground etc.) are used for monitoring floods worldwide. To estimate real time flood damages requires that a GIS be used. Besides mapping the flood and damage assessment, high resolution satellite data can be employed for operational management to map post flood river configuration, flood control works, drainage-congested areas, bank erosion and developing flood hazard zone maps.

GNSS could play a useful role of generating DEM (see e.g., [8]) and also find use in generating the drainage basin size. For example, GNSS was used to aid development of DEM of a flood prone area in Andhra Pradesh State of India that supported the assessment of spatial inundation at different water levels in the river. When the satellite derived land cover/use and ancillary ground-based socio-economic data was draped over the DEM, flood vulnerability was assessed to provide location specific flood warnings [52].

As discussed in Awange [8], GNSS meteorology promises to be a real boost to atmospheric studies with expected improvements on weather forecasting and climatic change monitoring, which directly impact on our everyday lives. One such area is in the monitoring of flash floods. Flash floods (see Sect. 16.4) are floods that come instantaneously and usually unexpectedly following heavy rains. Factors that lead to flash flood producing rains are summarized in Fig. 16.3. In India for example, deforestation, extension of agricultural activities in vulnerable areas, coupled with increased soil erosion and degradation of catchment areas have been cited as factors that have lead to frequent flash floods through reduction in natural storage capacity [52].

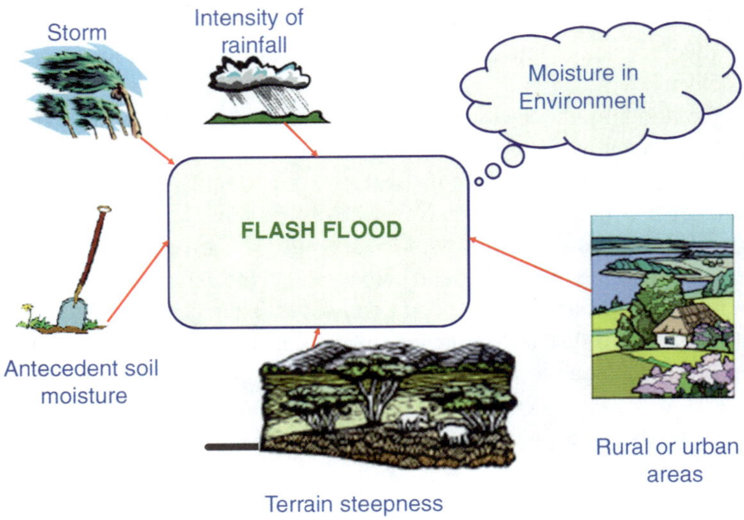

Fig. 16.3 Factors causing flash floods

In Awange and Fukuda [11], the possible use of IPWV for flash flood prediction is proposed, while Baker et al. [12] have outlined the potential of water vapor for meteorological forecasting. Let us consider Fig. 16.4[6] as an example where GNSS-meteorology could have assisted in flash flood forecasting and warning. In the figure, the amount of integrated water vapor in the atmosphere is related to the rainfall and used to monitor flash flood in Hawaii (see the left bottom corner of the figure). One observes that as it was raining, if the Integrated Precipitate Water vapor (IPWV) increased, more rain followed. By monitoring the intensity of water vapor, therefore, one can be able to predict the occurrence of flash flood. Station PGF1 and others were monitored and an increase noted after day 307. Shortly before day 308, the scientist reported the possibility of flash flood and indeed it occurred. Station PGF1 was washed away.[7]

GNSS could be used to map location of features, e.g., water tanks or perimeters of areas of potential danger such as contaminated dams, while in the second case, they could be used in conjunction with remote sensing methods to map inundated areas at large spatial coverage. In mapping flood events and damaged areas, GNSS could also be used to provide both horizontal and vertical controls as exemplified, e.g., in [103], where a constrained adjustment of a static GNSS survey (e.g., [8]) was performed to set basic horizontal and vertical controls for a flood control project near Guayama on the south coast of Puerto Rico. In this example, precise differential carrier-phase were applied and the end product was to obtain hydrographic and topographic data of project features to generate topographical maps for flood control.

[6] Source: Paroscientific Inc., http://www.paroscientific.com.

[7] More on information can be found by visiting http://www.paroscientific.com.

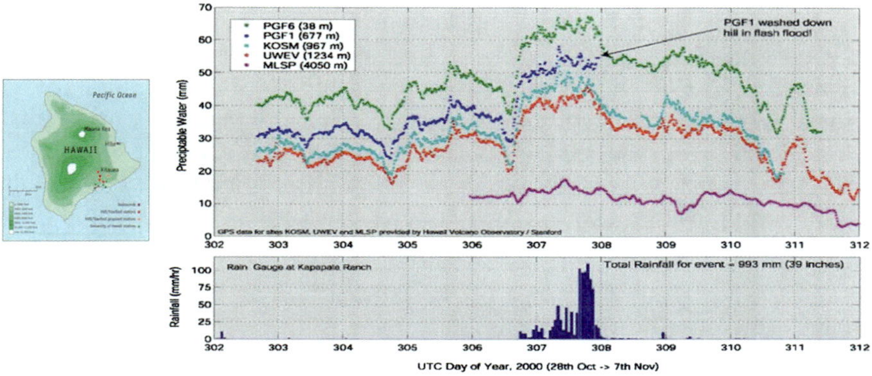

Fig. 16.4 An example of flash flood monitoring using GNSS-derived IPWV

In flood mitigation and response situations, location-based data and perimeter/area information could rapidly be provided by remote sensing and GNSS to assist in remedial measures. Furthermore, GNSS in conjunction with remote sensing satellite and GIS could be integrated to provide a unified system of monitoring and evaluating flood disasters. For example, Zhanga et al. [116] discusses an airplane-satellite-ground system that arose from a research program on flood monitoring promoted by the Chinese government, which integrates remote sensing, GNSS, data transmission, and image processing. Another example is in Brahmaputra river basin in India, where GNSS was used to provide information on flooded areas and damage to croplands, roads and rail-tracks [52].

Appropriate remote sensing data applied within a GIS environment could also be useful in identifying flood-prone river basins, assessment of expected risk levels, and selection of suitable land and water management measures to support management of flood plains. These can be employed to produce flood risk zone maps, which are essential in regulating the use of flood plains in a planned manner. For example, this approach was used in India to develop an integrated river basin development plan for major river basins, which was then applied to identify and monitor erosion-prone areas along river banks for the purpose of protecting land and habitat [52]. Maps prepared from remotely sensed data showing drainage congested areas could also be used effectively to avoid flooding [52]. In these cases, GNSS could provide location-based information for georeferencing of remote sensing images.

As an example, GNSS was used in conjunction with remote sensing satellites in flood management to map flood affected areas as demonstrated in the work of [30] who develop a method of wetland mapping and flood event monitoring based on a satellite multi-sensor data combination. Jayaraman et al. [52] state that because of the clear difference in the spectral signatures, it is quite possible to map areas under standing water, areas from where flood water had receded, submerged standing crop areas, sand casting of agricultural lands, breaches in the embankments, marooned villages and towns, etc., and that using multi-date satellite imageries enables the extent of damage due to crop loss, and the destruction of infrastructural facilities to be assessed.

Crétaux et al. [30] employed a series of airborne GNSS measurements to map horizontal areas covered by open water, aquatic vegetation, vegetation on dry land and then detect the limit zone between each type of terrain. This allowed the estimation of threshold values of the surface reflectance in different bands of frequency of the MODIS sensor that was used to characterize the land surface.

16.4.5 Monitoring of ENSO and IOD

The term "El Niño" originally applied to an annual weak warm ocean current that ran southward along the coast of Peru and Ecuador about Christmas time (hence Niño, Spanish for "the boy Christ-child") and only subsequently became associated with the unusually large warming that occur every few years and change the local and regional ecology [99]. The coastal warming, however, is often associated with a much more extensive anomalous ocean warming to the International Date Line (IDL), and it is this Pacific basin-wide phenomenon that forms the link with the anomalous global climate patterns [99]. The atmospheric component tied to El Niño is termed the "Southern Oscillation" forming El Niño-Southern Oscillation (ENSO), i.e., the phenomenon where the atmosphere and ocean collaborate together [98]. El Niño then corresponds to the warm phase of ENSO, while the opposite "La Niña" ("the girl" in Spanish) phase consists of a basin-wide cooling of the tropical Pacific and thus the cold phase of ENSO [99].

Satellite altimetry can play a crucial role in monitoring ENSO by observing the behaviour of the sea surface height (SSH). From the SSH obtained from TOPEX/Poseidon and Jason, Jet Propulsion Laboratory (JPL) use them to calculate the amount of heat stored below the ocean, from which annual and decadal changes in the ocean are used to monitor climate events e.g., ENSO.[8] Through the observation of SSH, the 1997–1998 ENSO event, which was probably the biggest in recorded history, and whose socio-economic impacts were felt in most places around the world could be monitored. Figure 16.5 shows the image of the Pacific Ocean produced using SSH measurements taken by TOPEX/Poseidon satellite, where the image shows SSH relative to normal ocean conditions on December 1, 1997.[9] In this image, the white and red areas indicate unusual patterns of heat storage; in the white areas, the sea surface is between 14 and 32 cm (6–13 inches) above normal while in the red areas, it

[8] see e.g., http://sealevel.jpl.nasa.gov/science/elninopdo/latestdata/.

[9] http://en.wikipedia.org/wiki/File:1997_El_Nino_TOPEX.jpg

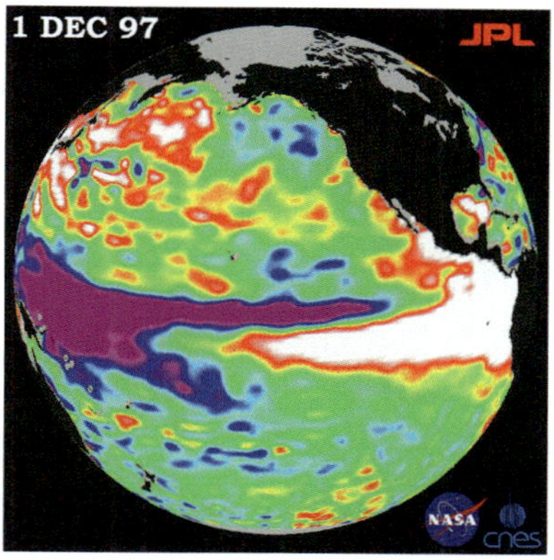

Fig. 16.5 1997–1998 ENSO monitored using TOPEX/Poseidon satellite. The *white-red* areas indicate unusual pattern of heat storage (*Source* http://en.wikipedia.org/wiki/File:1997_El_Nino_TOPEX.jpg)

is about 10 cm (4 inches) above normal. The green areas indicate normal conditions, while purple (the western Pacific) means at least 18 cm (7 inches) below normal sea level.[10]

The IOD (Indian Ocean Dipole) is an ocean-atmosphere interaction over the Indian Ocean, with alternative positive and negative sea surface height [20]. During positive events, anomalous cool (warm) waters appear in the eastern (western) Indian Ocean, associated with large-scale circulation changes that bring anomalous dry conditions to Indonesia and Australia, while East Africa experiences heavy rainfall [37]. By monitoring changes in stored water using GRACE (Sect. 13.4) and computing the correlation between these changes with ENSO-measured southern oscillation index (SOI) and the IOD-measured dipole mode index (DMI) as illustrated in Figs. 13.11–13.14 on pp. 213–216, it is possible to relate the effect of climate variability on water resources, see e.g., [20, 37, 101].

Garcia et al. [37] define SOI as being a measure of ENSO, which is based on the pressure difference between Tahiti and Darwin as provided by the Australian Bureau of Meteorology, while the DMI, a measure of IOD is estimated as the difference between the western and South-eastern tropical Indian Ocean sea surface temperature (SST) indices. SST data can be obtained from optical remote sensing sensors e.g., NOAA AVHRR and Terra MODIS as well as from microwave radiometer sensors (see e.g., [7]).

In conclusion, floods are evidently the most frequent (see Fig. 16.1) and most severe natural disaster occupying the highest rank among other natural disasters [50]. They have had a devastating impact in different parts of the world, resulting in the lose of many lives and destruction of numerous livelihoods in their wake. In this section, we have distinguished between different types of floods including; river-, coastal-, urban- and flash floods. Given their spontaneous nature, flash floods are evidently the most difficult type of floods to model. We have also discussed the different phases of flood management namely; flood prone/risk zone identification,

---

[10] http://en.wikipedia.org/wiki/File:1997_El_Nino_TOPEX.jpg

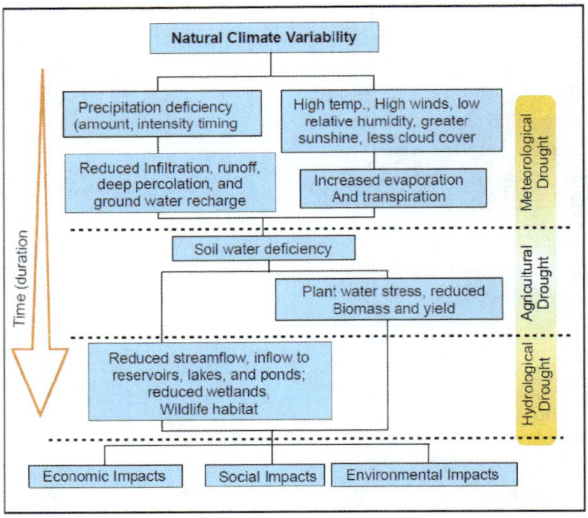

Fig. 16.6 Sequence of hydrological, agricultural, and meteorological drought impacts (*Source* [53])

flood monitoring and forecasting and lastly, flood response and mitigation. Using typical examples, different applications of geospatial in flood management have also been outlined. These range from the predominant use of remote sensing and GIS to the increasing application of GNSS. To wrap up things, the final subsection has focused on monitoring of the ENSO and IOD effects, both of which have a direct consequence on flooding and drought events at a global scale (e.g., Forootan [36]).

16.5 Droughts

The World Meteorological Organization (WMO)[11] classifies droughts on the basis of several parameters and considerations including: (i) rainfall, (ii) combinations of rainfall with temperature, humidity and/or evaporation, (iii) soil moisture and crop parameter, (iv) climatic indices and estimates of evapotranspiration, and finally (v) the general definitions and statements. Through such a classification regime, drought can be grouped into *hydrological drought*, *agricultural drought*, and *meteorological drought*, whose sequence of impacts is shown in Fig. 16.6. Its impacts (direct or indirect) include [53]: reduced crop, rangeland, and forest productivity; increased fire hazard; reduced water levels; increased livestock and wildlife mortality rates; and damage to wildlife and fish habitat.

Drought occurrences are subject to extreme weather and climate similar to floods. In contrast to floods, however, drought, which is a slow and progressive extreme climatic event, lasts longer and has been the least anticipated extreme in many regions. Drought is the single most important weather-related natural disaster often aggravated

[11] http://www.wmo.int/pages/index-en.html

by human action since it affects very large areas for months and years and thus has a serious impact on regional food production, life expectancy for entire populations and economic performance of large regions or several countries [53]. As a result, its impacts are more adverse than, for example, those of floods because of the inadequacy of existing coping mechanisms. Although drought is a natural, recurring phenomenon which cannot be controlled, prediction of its frequency, see e.g., [10, and the references therein], severity and probability is essential for users such as top level policy makers at the national and international organizations, researchers, middle level policy makers at the state, province and local levels consultants, relief agencies and local producers including farmers, suppliers, traders and water managers interested in reliable and accurate drought information for effective planning and management [53].

In the absence of such information, food security is threatened as exemplified in the 1999–2001 drought that affected most parts of Kenya, including some areas that normally receive high rainfall. At its peak, 4.5 million people in Kenya lost their livelihoods and ability to cope, and were subsequently entirely dependent on relief food provided jointly by the government and donors [35]. Evidently, arid and semi-arid regions are more prone to drought. For example, only recently (2011), drought yet again ravished the Horn of Africa affecting countries like Somalia, Ethiopia, and Northern Kenya. Adger et al. [1] discussed adaptation to climate change related phenomena in developing countries, while Barrett [19] described food assistance programs. The impacts of such extremes (in the context of climate change) on Lake Victoria Basin (LVB) has been assessed, e.g., in Phoon et al. [78]. On the average, statistics indicate that severe drought occurs once every five years in most of the tropical countries, though often they occur on successive years causing untold misery to human life and livestock [52]. For the Lake Victoria basin, Awange et al. [9, 10] obtained a cycle of 5–8 years.

To manage the effects of drought, therefore, measures should be taken, which involves both short-term that includes, e.g., early warning, monitoring and assessment of droughts; and long-term strategies aimed at drought mitigation measures through proper irrigation scheduling, soil and water conservation, and cropping pattern optimization [52]. Jeyaseelan [53] states that monitoring and assessment of drought using remote sensing and GIS are dependent on the factors that cause drought, and the factors of drought impact.

16.5.1 Early Warning of Drought

The first step in realizing a drought early warning system is to have a preparedness program, where drought-prone areas are identified well in advance, and drought-intensity and cycle predicted, e.g., [10]. Prediction is based on monitored parameters, e.g., rainfall anomalies, crop conditions, weather, and vegetation. Jayaraman et al. [52] list factors that can provide early indication of possible droughts as upper air winds, the development of hot low-pressure areas, ENSO phenomena, sea surface temperature (SST), snow cover, cloud patterns, wind velocity and direction, and atmospheric temperature and humidity profiles.

Geospatial tool of GNSS-meteorology presented in Awange [8] could be useful in providing temperature, pressure, and humidity profiles. Awange [8] discusses the possibility of using GNSS remote sensing techniques to monitor regional and global warming. Using the Australian example, Khandu et al. [57] showed how this technique, which has emerged over the past decade could prove to be an important tool for measuring global changes in tropopause's temperatures and heights, a valuable capacity given the tropopause's sensitivity to temperature variations. Similarly, remote sensing sensors like NOAA AVHRR and Terra MODIS could be employed to deliver SST image data and associated anomalies. Remote sensing could also be used to generate vegetation index, data that are useful in providing spatial information on drought-prone areas.

16.5.2 Geospatial Monitoring and Assessment of Drought

Monitoring and assessment of droughts are required for taking corrective measures at appropriate times in order to minimize the reduction in agricultural productivity in drought-prone areas, and also to provide objective information on the prevalence, severity level and persistence of drought conditions in a time-effective manner, which will be helpful to the resource managers in optimally allocating scarce resources to where and when they are most needed [52].

Normalized difference vegetation index (NDVI) and its associated derivatives, has traditionally been used for vegetation cover modelling [80], and is a good summary overview of the prevailing plant water stress as a function of the prevailing weather conditions [43]. It measures the amount of radiation absorbed by plants, where this radiation is directly related to evapotranspiration, hence relating NDVI to rainfall. This relationship is already being exploited by the Famine Early Warning Systems Network (FEWS NET)[12] to monitor crops and range lands in semi arid sub-Saharan Africa. In East Africa, there is a good correlation between NDVI and seasonal rainfall patterns [73], suggesting its possible use as a drought metric. In a recent study, Omute et al. [77] indicated that NDVI as a measure of vegetation vigour responded variably to *precipitation* and its *deficiency*. Its sensitivity to vegetation stress enables drought conditions to be continuously monitored on a real-time basis, often helping the decision makers initiate strategies for recovery by changing cropping patterns and practices [52].

Remote sensing techniques that provide NDVI could be integrated with GIS and GNSS to effectively forecast and monitor drought, where remote sensing would indicate areas of consistently healthy and vigorous vegetation, as well as stressed vegetation. On the other hand, GNSS would provide location-based information that would enable these vegetation to be located, and play the role of georeferencing

[12] http://www.fews.net/Pages/default.aspx

these remotely sensed satellite images. Finally, GIS would provide a modeling and decision support framework for analysis. In this regard, the generated NDVI data could be correlated with drought risk areas and integrated with both physical and meteorological data. Physical data could comprise data sets such as topography (slope), soil drainage, ground water resources etc., while meteorological data sets could include annual rainfall, annual frequency rainfall, annual evapotranspiration etc. Within a GIS environment all these different factors could be assigned appropriate weights, derived from expert opinion, and integrated through multi-criteria evaluation methods (see e.g., [7]) to give a realistic drought prognosis for a particular area.

The use of GRACE satellites discussed in Sect. 13.4 could provide spatial and temporal changes in terrestrial water storage, which could be used to monitor hydrological drought. Chen et al. [27] used GRACE satellites to measure a significant decrease in terrestrial water storage (TWS) in the central Amazon basin in the summer of 2005, relative to the average of the 5 other summer periods in the GRACE era. Their results demonstrated the unique potential of GRACE satellites to remotely sense large-scale severe droughts and flooding events, and in evaluating advanced climate and land surface models.

16.5.3 Combating Drought

Jayaraman et al. [52] indicate that while the construction of large reservoirs to ensure irrigation and drinking water contributed to a large extent towards mitigation of droughts in different countries, the poor and inefficient management of land and water resources in respective command areas have resulted in massive land degradation like salinity/alkalinity, water logging, etc., and thus causing serious concern to the conventional model of drought mitigation. In Chap. 15, ways in which geospatial techniques of remote sensing, GIS and GNSS could be helpful in land management are presented. Together with remote sensing, GNSS could also be useful in monitoring the large reservoirs.

16.6 Vector-Borne Diseases and Outbreak

Major epidemics of virulent disease have occurred with surprising frequency throughout human history, e.g., the numerous appearance of bubonic plaque in Europe in the late Middle Ages, the pandemic spread of influenza in the United States in 1918-19, and HIV-AIDS in our time [94]. Discovering and understanding the life cycle of a disease calls for painstaking research, literally years of trial and error, and for many biologists, geospatial could not have come at an opportune time [94]. In general, where temporal and spatial information is required, GNSS techniques could be integrated with remote sensing and GIS to enhance optimality and provide useful information applicable, e.g., in monitoring emerging infectious diseases and to studies of global change effects on vector-borne diseases, see e.g., [102]. For example, an integration of GNSS with GIS, satellite imagery, and spatial statistics tools have been used to analyze and integrate the spatial component in epidemiology of vector-borne disease into research, surveillance, and control programs based on a landscape ecology approach [102].

Defining landscape ecology as dealing with the mosaic structure of landscapes and ecosystems by considering the spatial heterogeneity of biotic and abiotic components as the underlying mechanism which determines the structure of ecosystems, Uriel [102] describes how the integration of GNSS with the tools above, and the landscape ecology-epidemiology approach could be applicable to vector-borne diseases. He supports his argument by presenting his work on malaria in Israel and tsetse flies in Kenya, and Lyme disease, LaCrosse encephalitis, and eastern equine encephalitis in the north-central United States as examples for application of the tools to research and disease surveillance [102].

On their part, Bonner et al. [23] consider the combination of GNSS and GIS geocoding in epidemiological research, while Jayaraman et al. [52] indicate the possible application of remote sensing in the assessment of crop pest/diseases by using the temporal and spatial distribution of desert vegetation and rainfall derived from NOAA data to identify potential locust breeding grounds. Once these locations have been identified, GNSS could help in providing the actual coordinates of these positions.

Environmental determinants such as presence of suitable habitats, temperature and climate, to a large extent, influence the outbreak and spread of vector-borne diseases. Consequently, it is unsurprising that techniques that are routinely used to map these environmental determinants, including remote sensing and GIS, have been employed extensively in public and environmental health to study and control these diseases, see e.g., [44, 46, 63] etc. Using these techniques, it is possible to identify the spatial patterns of outbreaks through epidemic curves as well as estimate the extent of the risk area through cluster analysis to support in mitigation measures.

The work of Dr. Snow in solving the problem of *cholera* outbreak in London in the 1850s (see Example 16.2) stands out as one of the most famous and earliest cases of maps being applied to understand the spread of a disease. Over the years, remote sensing and GIS have also been extensively employed to map different epidemiological cases including malaria, rift valley fever etc. Furthermore, with a better understanding of the environmental factors that drive the distribution, timing and abundance of pathogen carriers e.g., mosquitoes, tsetse flies etc., it is now possible to develop early warning systems. This can be used to flag regions at high risk of pathogen transmission, and more effectively inform and guide the distribution of surveillance and control measures. For instance, by studying indicators like NDVI it is possible to predict malaria seasons. Today, experts routinely use advanced mapping and computing technologies to understand the diffusion and spread of diseases such as AIDS and cancer. Hence, a map is no longer just an effective tool for finding locations, but it can also be a life saver!

Example 16.2 (Dr. Snow's Cholera analysis using GIS. Source: [93])

In 1854 a massive cholera outbreak occurred in Soho, London resulting in hundreds of deaths. In trying to diagnoze the source of the problem, Dr. John Snow plotted the locations of the deaths on a map and found they clustered around a pump in Broad Street. This provided strong evidence in support of his theory that cholera was a water-borne disease and additionally, that it was spread through dirty water. Specifically, Dr. Snow drew Thiessen polygons around the wells, defining straight-line least-distance service areas for each. A large majority of the cholera deaths fell within the Thiessen polygon surrounding the Broad Street pump, and a large portion of the remaining deaths were on the Broad Street side of the polygon surrounding the bad-tasting Carnaby Street well. Then Dr. Snow redrew the service area polygons to reflect shortest paths along streets to wells, and an even larger proportion of the cholera deaths fell within the Broad Street polygon or the Broad Street side of the Carnaby Street well's polygon. He suggested that the Broad Street pump be taken out of service—thus helping end the epidemic.

The above spatial analysis is outstanding as it is often considered to be the first:

(1) Epidemiological analysis which sought to understand the outbreak of disease by considering environmental factors; and
(2) Spatial analysis of disease data that involved plotting points on a map and looking for spatial relationships thereof.

End of Example 16.2

Example 16.3 (Leishmaniasis research [94])

Leishmaniasis is one of the infectious diseases causing significant health problems in Asia, Africa, India, North America, Latin America, and Europe, and is caused by a parasitic protozoan, which in turn can infect several hosts (i.e., an organism that harbors a parasite) and vectors (i.e., means of transmission, e.g., mosquitoes for malaria), making it difficult to control [94]. Researchers in Texas studied the relationship between the wood rats, which are the mammalian hosts for *Leishmania*, and the sand flies that act as vectors transmitting the disease from one host to another. To assist in their studies, the researchers employed GNSS to record the locations of the traps used to catch the rats, and also the locations of the Sand fly trapping stations used to catch and monitor the vector. GNSS also plays a subsidiary role of mapping roads and water features that are not found on a map, e.g., [94]. These information gathered from GNSS are integrated with other information in a GIS system through map overlaying to generate the information (maps) relating the trapping sites and the rats themselves, see e.g., [94, p. 46].

End of Example 16.3

16.7 Earthquakes

The outer layer of the Earth is made up of large pieces known as tectonic plates which are in motion. When these plates collide, as indicate in Fig. 16.14 in p. 324, an earthquake occurs. Other factors that can lead to earthquakes include fault raptures or forces such as stress and strain. But most causes result from plate collision as they try to move past one another. There is also a tendency of earthquakes to occur in "gaps" that are in places along an earthquake belt where strong earthquake had not previously been observed [52]. Earthquake risk assessment involves identification of seismic zones through collection of geological/structural, geophysical (primarily seismological) and geomorphological data and mapping of known seismic phenomena in the region (mainly epicenters with magnitudes) [52]. Kamik and Algermissen [55] have pointed out that knowledge of trends in time or in space helps in defining the source regions of future shocks, while Jayaraman et al. [52] point out that accurate mapping of geomorphological features adjoining lineaments reveal active movement or recent tectonic activities along faults.

GNSS satellites are emerging as a powerful geodetic tool for monitoring (geological) changes over time, which is the key for understanding the long-term geodynamical phenomena. They have been particularly useful in measuring the more complex deformation patterns across plate boundaries, where large and regional scale strain builds up, plate movements, and slips along faults, pre-seismic, co-seismic and post-seismic distortions [33, 48, 49, 52]. Indeed, GNSS through CORS (continuous operating reference stations, e.g., [8]) now enable detection of ground motion at the mm to cm-level accuracy before, during and after earthquakes [54], and have found use in seismological applications, see e.g., Larson [59].

Hammond et al. [41] point out that rapid detection and accurate characterization of earthquake related events can make a crucial difference during the minutes to hours that follow an earthquake event as was the case following the catastrophic 2004 Sumatra and 11th March 2011 Tohoku-oki earthquakes and tsunamis. In both of these events, the initial seismic notification of earthquake magnitude was available in minutes but was more than an order of magnitude smaller than the true event size, an uncertainty that can be addressed by GNSS in the rapid estimation of large earthquake magnitudes [41]. A review of the application of GNSS to geodynamics and earthquake studies is presented, e.g., in Segall and Davis [90].

Since natural disasters don't chose when to take place, prompt response is required, even when they occur at night. Cruz et al. [32] describe the application of nighttime remote sensors e.g., operational linescan system (OLS) which is a scan radiometer, for rapid detection of affected areas. When nightlights imagery acquired before and after disaster were compared, preliminary observations suggested the reduction of night-lights in the areas worst affected by the earthquake and the increase of ephemeral lights in the nearby areas originated by the concentration of survivors bonfires. This information can be useful in monitoring among other parameters; population movements, location of critical areas, area coordinates for high resolution imagery and dissemination of results to first responders.

Jia [54] documents three stages upon which measurements for the earthquake can be collected. These are pre-seismic, co-seismic and post-seismic defined as follows [54]:

Pre-seismic. These are measurements taken before the earthquake and normally serve as an early warning system of potential danger, especially when two plates indicate a possibility of collision (e.g., Fig. 16.14).

Co-seismic. These are measurements that provide direct information on the occurrence during an earthquake and are useful for providing data for estimating the likelihood of future earthquakes through, e.g., further investigations of fault slip models and of other seismic features of the earthquake.

Post-seismic monitoring entails using GNSS receivers to obtain time-series of location and plate movements long after the earthquake has occurred. Such post-seismic deformation information from all available GNSS sites in the earthquake region can help scientists analyze likely elastic, poroelastic and viscoelastic deformation, and plastic flow of the Earth's crust in the earthquake region, giving a better understanding of crustal relocation and redistribution after an earthquake.

By continuously sampling movements of permanent known GNSS locations, e.g., within a CORS network in a seismic area, accurate position differences between the stations can be established to an accuracy of a few millimeters in the horizontal and a centimetres in the vertical. These CORS networks are normally fitted with telemetry that relays the data to control stations for rapid processing (e.g., Fig. 16.14). The results are analyzed and used as monitoring data for possible future earthquakes. RTGPS presented in Awange [8] are suggested by Hammond et al. [41] to be useful in earthquake monitoring. Its integration with seismic time-series are expected to enhance seismic source monitoring through monitoring earthquake events occurring over very wide range of time scales. Its inclusion to extend measurements beyond typical seismic frequencies is essential to understanding the complete spectrum of fault slip behaviours associated with the earthquake cycle [41]. In what follows, two examples on the application of GNSS to monitor earthquakes is presented based on the works of Jia [54] and GEONET in Japan.

Example 16.4 (Sumatra-Andaman earthquake - Jia [54])

Jia [54] assesses the Sumatra-Andaman earthquake by analyzing data from more than 250 GNSS sites distributed through Australia, Malaysia, Thailand, Indonesia,

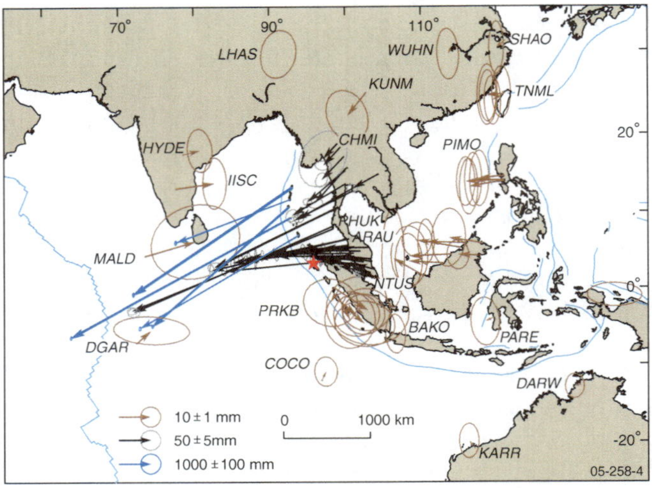

Fig. 16.7 Displacement field for the earthquake region determined by GPS. The *red star* represents the epicentre, and the *blue lines* show plate and fault boundaries. The displacements vary from 3 to 6 m in the Andaman and Nicobar Islands, indicated by the *blue arrows* while the *black arrows* show displacements at sites in Thailand, Malaysia and Sumatra. Almost 28 cm displacement was detected at the GNSS site PHUK (Phuket Island, southern Thailand near northern Malaysia), decreasing gradually towards the north and south. Displacements reduced to 2 cm at the NTUS site (Singapore) and 3 cm at the CHMI site (Northern Thailand). Deformation of around 10 mm was detected at large distances, indicated by *brown arrows*. Deformations from these sites, except for sites south-east from the epicentre, were also generally towards the epicentre or the great Sumatra fault, even though they were relatively small compared with their error ellipses (*Source* Jia [54])

the Philippines, China, India and the Maldives (Fig. 16.7). Co-seismic displacements from two combined seven-days' solutions, one before and one after the earthquake are computed. The displacements of the sites are calculated as the difference between the two solutions and the results presented in Fig. 16.8 for the co-seismic deformation at the GNSS site ARAU (Perlis, northern Malaysia) where a displacements of 15 cm in the east and 3 cm in the north is reported.

The results of Jia [54] indicate that the displacement detected by GNSS varies with location. Looking at the variation in GNSS coordinates computed by Jia [54] from stations near the earthquake every 30 s over 30 min periods before and after the earthquake (0:59, 26 December 2004) in Fig. 16.9, the progression of the rapture is noticed. Figure 16.9 indicates the deformation of 10 cm at a GNSS site of ARAU. In this particular case, Jia [54] points out that deformation was detected when the surface waves began to hit the site two minutes after the earthquake; four minutes later, positions at the site were relatively stable again.

Using a long-term GNSS time-series after the earthquake, Jia [54] then examined the post-seismic deformation process in Fig. 16.10 and showed that the deformation at the GNSS site LGKW (Langkawi Island, Malaysia) declined continuously over

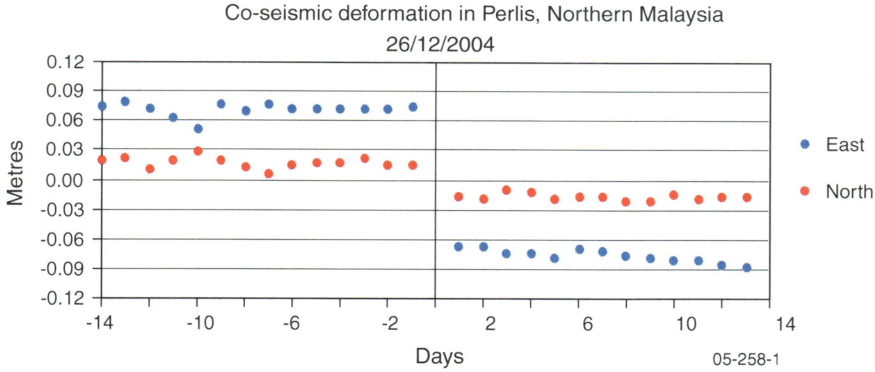

Fig. 16.8 Co-seismic deformation before and after the earthquake at the GNSS site ARAU (Perlis, northern Malaysia). A displacement 15 cm in the east and 3 cm in the north was detected (*Source* Jia [54])

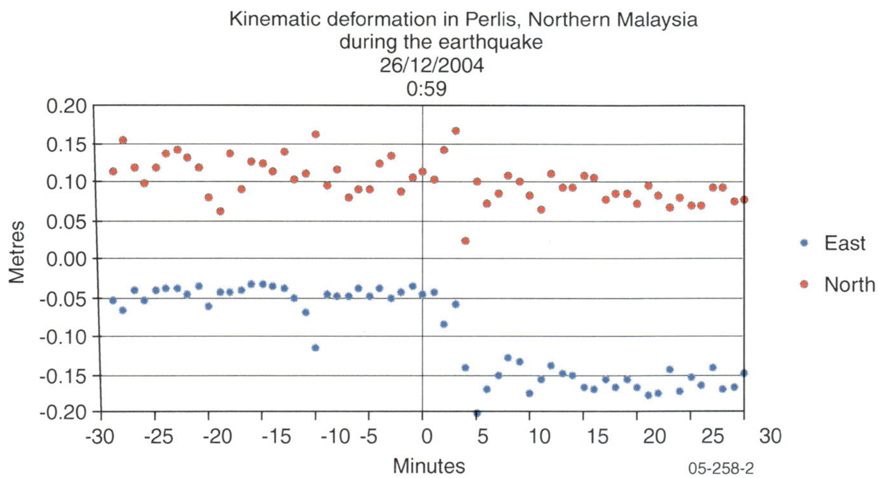

Fig. 16.9 The progression of rapture at a GNSS site ARAU (Perlis, northern Malaysia) during the earthquake, which occurred at about 0:59 on 26 December 2004. Measurements were taken every 30 s for a period of 30 min before and after the earthquake. Two min after the earthquake waves hit the site, a deformation of about 10 cm was detected (*Source* Jia [54])

time after the earthquake. An eastward deformation of more than 6 cm during the 80-day period after the earthquake is determined.

An example of such long-term GNSS time-series analysis in the work of Jia [54] shows that the Australian and Indian plates move towards Sumatra-Andaman at a rate of 5 cm and 4 cm per year, respectively (see Fig. 16.7) [54]. Jia [54] concludes that earthquake progressions of co-seismic type enable scientists to better understand the fault rupture process and that such studies could benefit tsunami early warning

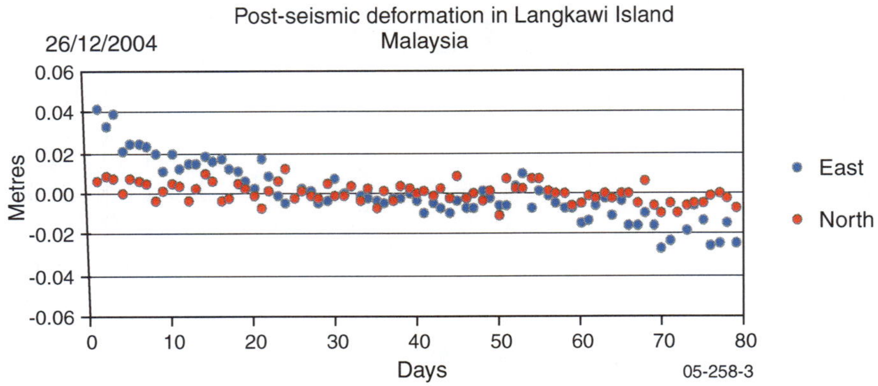

Fig. 16.10 Post-seismic deformation at a GNSS site LGKW (Langkawi Island, northern Malaysia) (*Source* Jia [54])

systems, such as the Australia Tsunami Warning System,[13] by allowing more reliable assessments of the likelihood of tsunami events to be made. For post seismic analysis, Jia [54] states that deformation information from all available GNSS sites in the earthquake region can help scientists analyze likely elastic, poroelastic and viscoelastic deformation, and plastic flow of the Earth's crust in the earthquake region, giving a better understanding of crustal relocation and redistribution after an earthquake.

End of Example 16.4

In post-seismic activities, GNSS are also useful in monitoring deformation in structures such as dams. Other structures that could be monitored for deformation include buildings, bridges, etc. This could be done by installing GNSS equipments at strategic locations on these features and comparing the locations reading before and after the seismic activity. In built environment, Hammond et al. [41] state that through high-sample rate GNSS data, e.g., from RTGPS discussed in Awange [8], the motion of large buildings and bridges in an inertial reference frame can be realized. Remote sensing and terrestrial photogrammetry can be used to document the damage caused immediately after an earthquake, with a view to supporting critical response and recovery operations. Specifically, high resolution satellite imagery (HRSI) (see, e.g.,[7]), thermal IR, video etc. are some of the sensors that can be employed at this stage.

Many studies have investigated the viability of HRSI in damage detection and assessment up to the individual building level, see e.g., [5, 58, 113, 114]. Other works have used aerial photographs and stereo aerial photographs, see e.g., [76, 100] etc, while very few have even employed unconventional datasets such as aerial

[13] http://www.bom.gov.au/tsunami/about/atws.shtml

television images, see e.g., [42]. The objective has been to classify building damage often using established macroseismic scales like the European Macroseismic Scale (EMS). Increasingly, automated image classification procedures have been employed (see e.g., [22, 68, 100] etc.) away from traditional visual interpretation methods, see e.g., [76, 114]. In the post event scenario, GIS can also use building damage data captured from remote sensing to evaluate the structural dynamics for buildings and assess their seismic vulnerability as reported in several works e.g., [71, 112] etc.

Example 16.5 (GNSS monitoring of the Tokai crustal movement)

In Japan, GEONET has been used to monitor crustal activities by analyzing daily solutions of GNSS CORS stations' coordinates [66]. For example, when an extraordinary data are detected, or a big earthquake or volcanic eruption occurs, emergency analysis is performed to get the solutions every 3 h [66]. This enables the determination of displacement rates and strain rates throughout Japan [86]. Matsuzaka [66] points to the fact that the more than 1200 GEONET CORS stations provide data that are used in various disaster related meetings and geophysical model estimation of crustal activities, and thus reflect the decision making process to cope with disasters, as well as, scientific researches. GEONET is also quite useful in earthquake studies, precisely detecting co-seismic, post-seismic, and inter-seismic deformation signals, with these observations used to infer physical processes at the earthquake sources [86]. GEONET operates by having its stations, which are well distributed throughout the country relay real-time data by a dedicated line to the control and analysis center at the Geospatial Information Authority of Japan (GSI) in Tsukuba (Ibaraki Prefecture), thereby enabling a real-time monitoring of crustal movements.

The Tokai crustal movement has been monitored through such analysis (see, e.g., Fig. 16.11). Figure 16.11 shows the data of horizontal displacement observed in the Tokai region due to crustal movements during the period of 27 March 2001 to 7 February 2004 with reference to the GPS station in Oogata, Niigata Prefecture.[14] From the Tokai region to Nagoya, an area of displacement in the south-east direction is identified thus further indicating the possibility of using GNSS as a possible key to understanding a mechanism of the Tokai crustal movement.[15]

Slow slip events on plate boundaries as demonstrated above have been found from GNSS data, thus providing an important constraint on the mechanism of faulting [86]. Sagiya [86], however, pointed to the fact that there had been no success in detecting pre-seismic deformation but highlights the fact that GEONET enabled a good linkage between monitoring and modelling studies, opening a possibility of practical data assimilation. He suggests that for further contribution to earthquake studies, it is necessary to continue GEONET with high traceability on the details in observation and analysis [86].

[14] http://www.gsi.go.jp/ENGLISH/page_e30068.html

[15] http://www.gsi.go.jp/ENGLISH/page_e30068.html

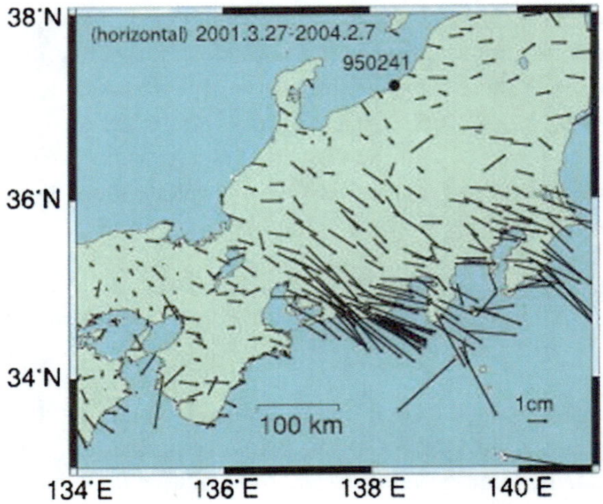

Fig. 16.11 Horizontal displacement observed in the Tokai region due to crustal movements during the period of 27 March 2001–7 February 2004 with reference to the GPS station in Oogata, Niigata Prefecture (950241) (*Source* http://www.gsi.go.jp/ENGLISH/page_e30068.html)

End of Example 16.5

Besides the use of GNSS for long term plate motion and plate boundary deformation monitoring as exemplified in the work of [86], other geodetic techniques that may be used are very long baseline interferometry (VLBI) that provides mm-level relative positioning accuracy and laser ranging methods (e.g., [7]).

16.8 Changing Sea Levels

Sea levels have been known to change for a variety of reasons, with the changes varying in temporal and spatial scales, see e.g., [28]. In time scales, some changes occur rapidly, e.g., sea level change due to the effects of tsunami, while others take long. Spatially, the changes can be local or global. The average sea level about which these changes occur is generally known as the *mean sea level* [81]. Global warming, fueled by the increase in greenhouse is thought to be having an impact on the sea level rise around the world. This is in part due to the melting of the polar ice on the one hand, and the increase in water temperatures causing expansion on the other hand. Today, one of the highest priorities is to understand and to reliably anticipate changes in the mean sea level and flood risks, particularly those that maybe due to global climate change [81]. In order to predict future changes of mean sea level, Pugh [81] suggests that it is necessary to have a full understanding of all the factors that influence sea levels at the coast and that the first step in realizing this is to measure the sea level over a long period of time to provide baseline data upon which scientific discussions could be made.

Measuring sea level entails the determination of the vertical height between the average surface of the sea and a fixed datum level (e.g., [8]), where the datum chosen depends on the application [81]. Some of the most commonly used datums such as tide gauge benchmark, chart datum, land survey datum (mean sea level), geocentric coordinates and geoid are discussed in [81]. As stated in Awange [8], the GOCE satellite mission is anticipated to significantly improve the accuracy of the static geoid, e.g., Hirt et al. [47], and in so doing contribute to improved monitoring of changes in sea level. Methods of measuring sea levels are listed by Pugh [81] as:

(1) Direct measurements by following the moving sea surface, e.g., using tide poles and float gauges. According to Pugh [81], for offshore, where there are no fixed structures, GNSS receivers have been placed on floating loosely moored buoys that measure sea levels to accuracies of few centimeters after averaging over a long period of time to remove the effect of waves.
(2) Use of fixed sensors (e.g., acoustic tide gauge and pressure measuring systems).
(3) Use of satellite altimetry discussed in Awange [8].

Precise measurements from GNSS combined with other methods are crucial in monitoring changing sea levels and provide early warning in case of evacuation. Since changes in sea level average about 1 to 3 mm per annum, any vertical motion of the crust must be monitored at this level of accuracy, and within short time spans of 3–5 years [38]. GNSS could also contribute indirectly to monitor changes in sea level through cryospheric measurements and also through reflected signals discussed in Awange [8].

16.8.1 Impacts of Rise in Sea Level

For sea level changes, climate models used to study the effects of atmospheric greenhouse gases predict an overall increase in the global temperature during the current century to be in the range of 1–3.5 °C [105]. Though sea level rise could also be attributed to salinity changes, see e.g., [6], increase in magnitude in temperature to the level predicted by [105] could easily lead to rapid melting of ice sheets and glaciers that could lead to a sea-level change that departs dramatically from the assumption of a uniform redistribution of meltwater [69]. For instance, Bamber et al. [14] reassessed the potential of sea level rise resulting from a collapse of the West Antarctic ice sheet and obtained a value for the global, eustatic (changes due to addition or removal of water mass) sea level rise contribution of about 3.3 m, with important regional variations. The maximum increase was found to be concentrated along the Pacific and Atlantic seaboard of the United States, where the value was about 25 % greater than the global mean, even for the case of a partial collapse [14]. Although the eustatic sea-level rise contribution for West Antarctic ice sheet collapse is uncertain (e.g., IPCC estimates 5 m), Mitrovica et al. [69] show that, whatever the value, sea level changes at some coastal sites will be significantly higher or lower than the predicted eustatic value.

Temperature rise due to global change in climate will have several effects, notably the global rise in sea level, an undesirable scenario given that a large fraction of the Earth's total population resides close to the sea. Catastrophic impacts on agriculture, tourism, industries, etc., of 1 m global rise in sea level have already been pointed out, e.g., by [34] who note that low lying regions that slope gently such as Florida and Indonesia with 15 % of the world's coastline or regions lying in the flood plains of large rivers such as Rhine, Nile, etc., are all in great risk, should the sea level rise by the predicted amount.

Titus et al. [97] estimated that in the United States, a global sea level rise by a similar magnitude would suffice to drown 20–85 % of the coastal wetlands resulting in an encroachment of up to 7,000 square miles (about 18,130 km^2), an area the size of Massachusetts. The foregoing discussion therefore clearly indicates the necessity of monitoring of the changes in sea/lake levels in order to provide early warnings and enable formulation of policies that will provide remedial measures.

Although Warrick et al. [105] report that tide gauge data show the sea level to be already rising at a rate between 1.0 and 2.5 mm/yr averaged over the past century, understanding and characterizing the sources of the rise of sea level remains a problem. The two prominent source of rise in sea level are;

(1) mass redistribution (from Antarctica, Greenland, and Glaciers and small ice caps), and
(2) thermal expansion and salinity variation.

In both sources, as already pointed out, global warming contributes to the rise in sea level by melting of the ice and glaciers that find their way into the ocean during mass redistribution. In the latter case, global warming causes the warming and expansion of the water during thermal expansion. Existing methods for determining secular rise in sea level is based on the tide gauge approach, which as already mentioned has its own limitations. For rise in sea level, therefore, it is vital for one to;

(a) characterize the sources of global rise in order to clearly distinguish between the thermal and mass redistribution contributions, and
(b) relate the rise in sea level to global warming.

These can be done using geospatial techniques of remote sensing satellites discussed in Awange and Kiema [7], which are specifically designed to provide solutions to problems related to global warming and the changes in sea/lake levels.

16.8.2 Tide Gauge Monitoring

The traditional means of monitoring sea and lake levels has been based on tide gauges, which are normally sensitive to water level changes. Tide gauge records collected over long periods indicate a rise in global sea level of 10–30 cm over the past century (cf. 10–25 cm from [105] above), with a large discrepancies in the rates of change of sea level indicated by different tide gauges, or groups of tide gauges spanning particular regions [38]. These differences are considered to be largely caused by vertical land motion resulting from phenomena such as glacial rebound, subsurface fluid withdrawal and sediment consolidation [38].

In order to decipher the signals observed by a network of tide gauges from vertical crustal movement so that absolute sea level changes could be monitored, it is necessary to establish the absolute positions (heights) of the tide gauges in an accurate, global, geocentric terrestrial reference frame. Currently, the International Terrestrial Reference Frame 2000 (ITRF2000) is used [38]. GNSS are used to georeference these tide gauges.

16.8.3 GNSS Monitoring

GNSS estimation of absolute and relative sea level changes: Due to their capability to provide three-dimensional coordinates $\{\phi, \lambda, h\}$, GNSS satellites play a crucial role in:

(1) the actual determination of the sea level changes by determining the phase center of the GNSS-equipped buoy receivers, and also through GNSS altimetry, see, e.g., Awange [8] [62, 82], and
(2) the realization of the reference frame upon which the tide gauges are referred.

For GNSS-equipped buoys capable of measuring dual frequency carrier-phases L1 and L2, their positions $\{\phi, \lambda, h\}$ relative to a fixed reference GNSS receiver can be computed as a function of time. The h component can be used to provide a time-series measurement of sea level. If the buoy is horizontally constrained and measurements are made at appropriate frequency, the mean sea level, the sea level change due to tides, and the wave amplitudes and frequencies can be determined to cm-level accuracy using GNSS [84].

Using kinematic positioning approach (e.g., Awange [8]), Kelecy et al. [56] demonstrated the applicability of GNSS by using two GPS-equipped buoys to collect data at two locations on two different days. Three-dimensional positions of the buoys were then computed using precise carrier-phase measurements relative to a GNSS reference station that was fixed 15 km away. From the derived positions, the height component h corrected for tilt and vertical displacement were then used to compute two mean sea level measurements at the buoy locations to 6 cm-level accuracy when compared to altimetry readings. Kelecy et al. [56] thus succeeded in demonstrating that accurate GNSS buoy measurements could successfully be used to detect changes in sea levels. They point out that such measurements could find use in complimenting altimetric data, e.g., in calibration of altimetric measurements, see e.g., [24, 61, 106], extending altimetric results to smaller scales by using an array of GNSS buoys for local studies, and enhancing the temporal resolution of altimetric data to resolve local uncertainties between satellite passes. Other application areas include measurement of absolute sea level, of temporal variations in sea level, and of sea level gradients (dominantly the geoid), while specific applications would include

Fig. 16.12 GNSS absolute sea level monitoring

ocean altimetric calibration, monitoring of sea level in remote regions, and regional experiments requiring spatial and temporal resolution higher than that available from altimetric data [56].

The variations of vertical crustal velocities at CORS sites near tide gauge stations may also be used to determine the "absolute" sea level change with respect to the International Terrestrial Reference Frame, what previously was impossible to conduct before the proliferation of CORS in coastal areas [92]. In the realization of the reference frame upon which the tide gauges should be referred, measurements of vertical crustal movement together with tide gauges in a global geocentric reference system has been shown, e.g., by Snay et al. [91] to have the potential of estimating absolute sea level. Denoting $S(p)$ as the rate of relative sea level change at a point p measured by a tide gauge and $U(q)$ the vertical velocity at a point q in the ITRF2000 system, Snay et al. [91] indicate that the absolute sea level is given by

$$S(p) + U(q) = \xi + \epsilon, \tag{16.1}$$

where ξ is the estimated absolute sea level rate for a sample of sites and ϵ the difference between the estimated values and the observations (i.e., the error vector). It is clear from the equation above that GNSS satellites will play a significant role in the determination of each ITRF2000 velocity $U(q)$ for each station using CORS observations. The quantity $S(p) + U(q)$ provides an estimate of the absolute sea level rate at p, denoted $A(p)$, when the distance between p and q (see Fig. 16.12) is small [91]. In addition to provision of the velocity observations, GNSS contribute to monitoring absolute sea level changes through a regular re-survey of local networks of benchmarks and reference marks [38].

GNSS remote sensing techniques discussed in Awange [8] further provide means by which GNSS could be useful in measuring changes in sea level. As an example, Lake Victoria's (Sect. 13.5.4) water levels for the period 1993 to 2006 are plotted for both TOPEX/Poseidon derived heights and tide gauge data in Fig. 16.13. The figure

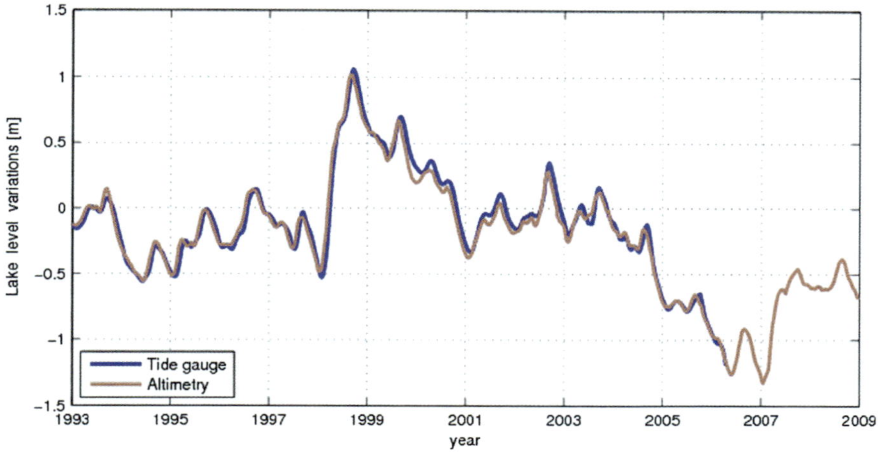

Fig. 16.13 A comparison of water gauge readings at Jinja station in Uganda (near Lake Victoria's outlet, see Fig. 13.15) and water Levels from Topex/Poseidon and Jason-1 altimetry satellites. The figure shows a close match between the tide gauge and satellite altimetry data (cf. Fig. 13.17 on p. 219 obtained from GRACE) (*Source* Awange [8])

indicates a close relationship between the two data sets. Crétaux et al. [29] compares water levels of Lake Victoria from the Jinja tide gauge and those from Jason-1 altimetry satellite and obtains a coefficient of correlation value of 0.99 with a standard deviation of 2.7 cm for the period 2004–2007. This supports the fact that satellite altimetry provides useful information on changes in sea level, and further shows the significant contribution of GNSS to satellite altimetry used for monitoring changes in water levels. Indeed, by averaging the few-hundred thousand measurements collected by the satellite in the time it takes to cover the global oceans (i.e., 10 days for TOPEX/Poseidon), global mean sea level can be determined with a precision of several millimeters [81]. Such information is vital for mitigation of disasters related to sea level changes. Detailed satellite altimetry study on East African lakes (e.g., Fig. 16.13) are given e.g., in [20, 29].

16.9 Tsunami Early Warning System

Boxing day, the 26th of 2004 will be remembered as the day tsunami caused havoc in Asia and Africa killing more than 280,000 people[16] and destroying property worth millions of dollars. In order to understand the cause of tsunami and how it could possibly be monitored using GNSS, use is made of Fig. 16.14. Tsunami is a series of waves created when a body of water such as ocean is rapidly displaced. Causes often range from land earthquakes, under ocean earthquakes, volcanic eruptions, under water

[16] see e.g., http://www.bom.gov.au/tsunami/about/atws.shtml.

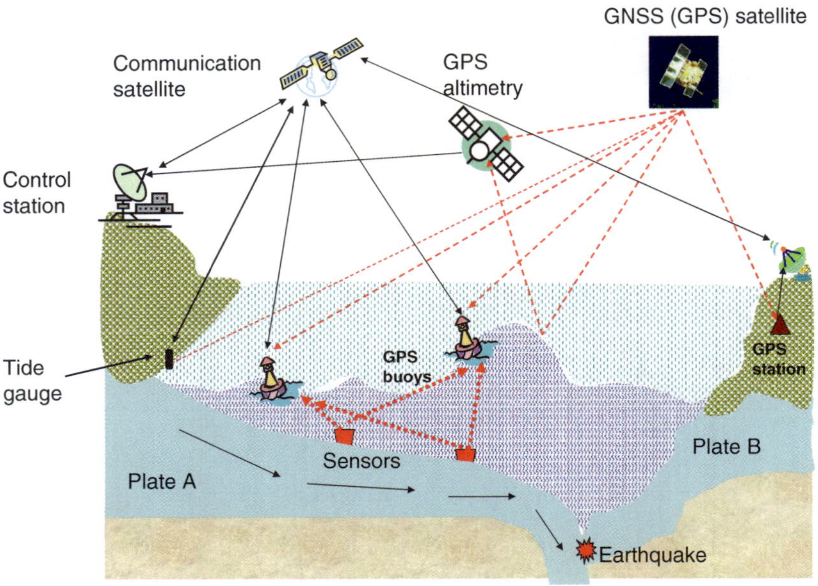

Fig. 16.14 A schematic illustration of the possible application of GNSS to tsunami monitoring (*Source* Awange [8])

explosions among others. In Fig. 16.14 for example, plate A is seen to move towards the right and as it does, it collides with plate B thereby causing under ocean earthquake with the epicenter at the point of collision as indicated in the figure. The effect of the earthquake immediately displaces the water mass causing rapid moving waves that rises to upto 30 m high [40]. As these approach land, the rapid speed enables the waves to penetrate deep into land causing massive destruction as witnessed by the 11th March 2011 tsunami at Tohoku-Oki in Japan. Monitoring the onset of Tsunami enables early warning system that could at least minimize the loss of lives.

Tsunami warning has particular requirements for calculating accurate earthquake magnitude, propagation direction, and vertical and horizontal motion of the seafloor, with the goal of rapidly recognizing when a tsunami event is occurring and improving predictions of where the wave will rise on near and distant coasts [41]. GNSS satellites are proving to be the key to such early warning systems in light of the fact that displacements at GNSS sites are useful in constraining a fault slip model, which predicts motion of the seafloor [41].

In the GITEWS (German Indonesian Tsunami Early Warning System),[17] which went into operation on 11th of November 2008 [40] with the ownership now in the hands of Indonesia, GNSS plays a leading role in key monitoring areas as demonstrated in Fig. 16.14. The indicators of this system are GNSS instruments, seismometers, tide gauges and GPS-buoys, as well as, ocean bottom pressure sen-

[17] see http://www.gtz.de/en/21020.htm.

sors. The augmentation of all these sensors ensures a rapid locating and analysis of the tsunami information to inform warning decisions.

Tide gauge sensors: Tide gauges are used to measure changes in sea level and therefore indicate exceptionally rapid rise in sea level, which signals unusual event. To enhance the tide gauges, the GITEWS project has equipped them with GNSS receivers that enable them to measure vertical and horizontal land displacements in addition to sea levels. Measurement of GNSS velocities provide an indication of the direction of horizontal displacement. The actual vertical and horizontal displacement are deduced from the GNSS positional time-series measurements. Any unusual displacement can serve to indicate an oncoming tsunami. During the catastrophic earthquake of 2004, a horizontal and vertical displacement of several decimeters to meters was evident even at a distance of some hundred kilometers from the earthquake. The direction of this resulting shift gives reference to the mechanism of the earthquake break and thus to the possible tsunami potential and the expected hazard [40]. This information is transmitted towards the communication satellite, which relays it to the control station in Indonesia.

GNSS Buoys: These help to relay information received from the underwater pressure sensors, which detect the effect of the tsunami. Besides relaying the information, the incorporated GNSS antenna can act as a measuring device for sea motion (horizontal motion) and sea level (heights). The buoys therefore provide an independent indicator of the onset of a tsunami. For the GITEWS, the buoys also detect tsunami waves, which with speeds of up to 800 km/h and wavelengths of 200 km in the open sea, and through proper filtering, achieve cm-level accuracy detection of rise in the sea level thus providing also an early detection of a tsunami wave [40].

Land based GNSS: These are combined with seismometers to provide location-based mapping of the ground motion. Horizontal and vertical displacements together with velocities are delivered by GNSS. All these, plus ground motion from the seismometers are relayed to the communication satellites and further transmitted to the control center in Indonesia where warning decisions are made.

GNSS altimetry: Low flying satellites such as GRACE are able to measure reflected signals from GNSS satellites (e.g., Awange [8]. They therefore would measure signals that hit the sea surface and are reflected to the low flying satellites. This helps in determining the sea levels as well as providing more information on tsunami indicators. Besides, the GNSS satellites also help in positioning these low flying satellites in space. Studies have been initiated to test the feasibility of this concept, see e.g., Helm et al. [45].

An example of a tsunami early warning system that was in test mode during the Tohoku-Oki earthquake and tsunami of 11th March 2011 was the GREAT (GPS Real-time Earthquake and Tsunami) Alert, see e.g., [41]. The GREAT Alert Project is a NASA-sponsored, multi-agency collaborative effort to develop an advanced Earthquake and Tsunami alert system that uses real-time GNSS to enable more accurate and timely assessment of the *magnitude* and *mechanism* of large earthquakes, as well as the *magnitude* and *direction* of resulting tsunamis.[18]

Another example is the Australian Tsunami Warning System (ATWS), which is a national effort involving the Australian Bureau of Meteorology (Bureau), Geoscience Australia (GA) and Emergency Management Australia (EMA) to provide a comprehensive tsunami warning system capable of delivering timely and effective tsunami warnings to the Australian population, and also support international efforts to establish an Indian Ocean tsunami warning system and contribute to the facilitation of tsunami warnings for the South West Pacific.[19]

16.10 Land Subsidence and Landslides

The various landslide types and underlying processes are discussed in different works, see e.g., [31, 64, 108] etc. Land-surface subsidence due to over-extraction of groundwater has long been recognized as a potential problem in many areas that have undergone extensive groundwater development [72]. This is particularly significant in expanding metropolitan areas in arid and semi-arid areas as witnessed in most parts of Australia and Iran. Surface subsidence has been observed worldwide in areas where withdrawal exceeds natural recharge thereby depleting the volume of the stored water. This has been evidenced in cities such as Mexico where subsidence of land has reached almost 9 m, Bangkok (Thailand), Shangai, Tanjin, Xi'an (China), Osaka, Tokyo (Japan), and Las Vegas (USA) [88]. Poland [79] outlines numerous case studies on land subsidence due to groundwater withdrawal. In 1991, the US National Research Council placed the annual costs from flooding and structural damage caused by land subsidence within the United States alone at over $125 million [72]. Another example of anthropogenic subsidence that can lead to property destruction and loss of life is that attributed to mining activities.

Remote sensing and photogrammetric techniques have long been applied in the mapping and monitoring of hazardous slope processes and landforms, see e.g., [67, 85, 87]. When integrated with height information from DEMs or InSAR[20] data discussed in Awange and Kiema [7], these methods have yielded relatively higher accuracy capable of monitoring land subsidence evolution (see e.g., [17, 18, 115]) etc.

Similarly, GNSS and other geospatial techniques, e.g., leveling, gravimetric, and InSAR have been widely used to detect the temporal and spatial pattern of surface deformation due to land subsidence. Of these methods, InSAR (see, e.g., [7]) provides a unique tool for detecting and monitoring deformation over regions of ongoing groundwater development with the advantage of having wide spatial coverage

[18] http://www.gdgps.net/products/great-alert.html. Accessed on 21/9/2011.

[19] http://www.bom.gov.au/tsunami/about/atws.shtml

[20] Interferometric Synthetic Aperture Radar.

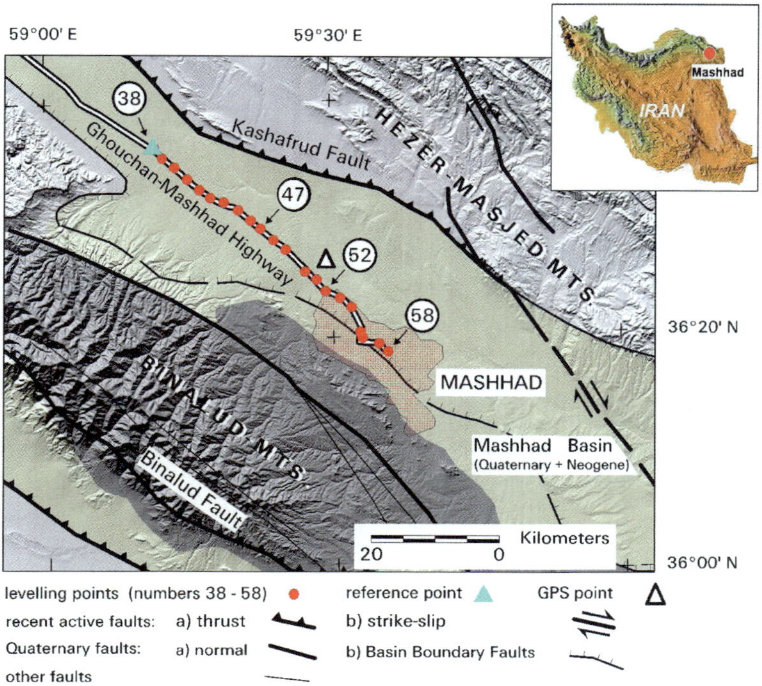

Fig. 16.15 Generalized geologic map of the Mashhad region based on the published 1:2,500,000-scale geological map from National Iranian Oil Company (NIOC), and the map of major active faults in Iran from International Institute of Earthquake Engineering and Seismology (IIEES). The map is draped with a shaded-relief topographic image generated from the 3-arcsecond Shuttle Radar Topography Mission (SRTM) data. The inset shows the location of the figure within Iran (*Source* Motagh et al. [72])

(10,000 km^2), fine spatial resolution (100 m^2), and high accuracy (~1 cm), and as such, offers new capabilities to measure surface deformation caused by aquifer discharge and recharge at an unprecedented level of detail never before possible with techniques like GNSS and levelling [72]. For this reason, Motagh et al. [72] demonstrate in their study how spatially dense InSAR results complement sparse geodetic measurements from GNSS and levelling, and thereby contributing to a better understanding of land subsidence associated with water extraction in the Mashhad area, northeast Iran (Fig. 16.15). Similarly, Schenk [88] was able to apply DInSAR[21] to interpret surface displacement in the Tehran/Iran region. He identifies inelastic deformation of the underlying acquifer systems to be the reason for the observed land subsidence affecting an area of 1350 km^2 and averaging subsidence rates of up 14 mm per month.

[21] Differential Synthetic Aperture Radar Interferometry.

Fig. 16.16 An example of surface fissures in Mashhad Valley caused probably by excessive groundwater withdrawal and associated aquifer-system compaction. The fissure is located northwest of Mashhad City and to the east of the Tous. A GNSS station is marked by a triangle in Fig. 16.15 near the city of Tous (*Source* Motagh et al. [72])

Example 16.6 (GNSS monitoring of subsidence in Mashhad [72])

The Valley of Mashhad is a northwest-southeast (NW-SE) trending valley in northeast Iran and is bounded to the south by the Binalud Mountains and to the north by the Hezar-Masjed Mountains, see Fig. 16.15 [72]. The basin encompasses the city of Mashhad, a provincial capital inhabited by over 2 million people and visited annually by millions of tourists [72]. Mashhad extends over an area of more than 200 km^2 across the basin, is the second largest city in Iran, and is one of the fastest growing metropolitan areas in Iran [72]. Water supply for the region mainly comes from groundwater, which is used for domestic, industrial and agricultural activities. However, substantial exploitation of this groundwater due to population growth, tourism and development, coupled with deficient natural recharge in recent decades, has resulted in severe depletion of the underground aquifer, leading to regional decline of water-table levels, lack of access to fresh water, and development of earth fissures and local sinking in many areas of the valley, see Fig. 16.16 [72]. From 2005 to 2006, a GNSS CORS station near the city of Tous provided continuous recordings that showed significant subsidence of approximately 22 cm at the station, e.g., Fig. 16.17 [72].

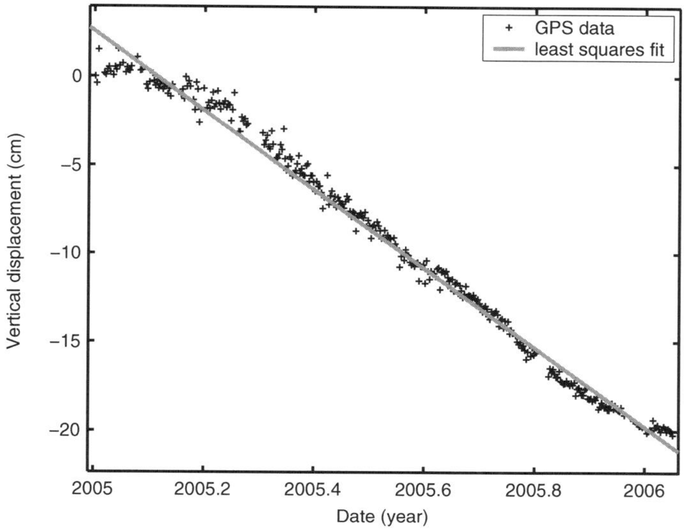

Fig. 16.17 A plot of GNSS vertical components (in cm) for the Tous GPS station (triangle in Fig. 16.15) with least-squares linear trend (slope ~22 cm yr.1) overlain (*Source* Motagh et al. [72])

End of Example 16.6

Landslides often occur on steep slopes following heavy rainfall. They often have a devastating effect such as the (2011) landslide disaster in Brazil, where more than 1,000 people died and thousands were displaced. Landslide monitoring can be undertaken in two ways [65] (i) analyzing and comparing various spatial maps and images, e.g., topographic maps, aerial photographs, remotely sensed imagery, cadastral maps, and digital elevation maps, which represent instantaneous views of an unstable site on different time epochs, and (ii) carrying out in-situ measurements of the surface displacements by combining a space and time resolution adapted to the evolution speeds of the phenomena.

In the context of the first approach, geospatial plays a pivotal role in map production that would aid in the analysis. With respect to the second approach, for example, GNSS has been applied to the case of monitoring the Vallcebre landslide in the Eastern Pyrenees, Spain [39]. This landslide had been periodically monitored since 1987 with terrestrial photogrammetry and total stations. The area of movement extended over 0.8 km^2 and showed displacements as large as 1.6 m during the period 1996–1997. Application of GNSS approach allowed greater coverage and productivity with similar accuracies (12–16 mm in the horizontal plane, and 18–24 mm in elevation) to classical surveying methods [39]. A general accuracy of GNSS monitoring of landslide is presented, e.g., in [65]. As elaborated above, GNSS has successfully been applied to monitor landslides. However, in cases where landslide prone areas lie in canyons like in many mountaneous regions GNSS signals are likely to be obscured and hence more innovative deformation monitoring is demanded for [111]. One possible solution is to integrate GNSS observations with height data obtained from remote sensing techniques e.g., InSAR [72].

16.11 Concluding Remarks

The introductory section of this Chapter has defined what constitutes natural disasters. It has gone ahead to point out the semantic controversy in the lexicon "natural", since disaster events are not hazards or disasters without some form of direct or indirect human involvement. This has led to the paradigm of human induced disasters being referred to as "man-made natural disasters". The Chapter has distinguished various natural disaster types and described how geospatial techniques can contribute to their monitoring and management.

However, it is important to state that what has been presented is by no means exhaustive on areas in which geospatial could contribute to monitoring and management of natural disasters. Moreover, it should be emphasized that no single method can provide a full proof monitoring of a disaster. What is discussed in this Chapter are simply ways through which geospatial could contribute, albeit marginally, to monitoring and management of disasters, which could lead to early warning and mitigation measures being taken. Geospatial could also contribute towards preparedness measures and also during post-disaster periods as discussed in various sections of the Chapter.

References

1. Adger WN, Huq S, Brown K, Conway D, Hulme M (2003) Adaptation to climate change in the developing world. Prog Dev Stud 3:179–195. doi:10.1191/1464993403ps060oa
2. Ailamaki A, Faloutsos C, Fischbeck P, Small M, VanBriesen J (2003) An environmental sensor network to determine drinking water quality and security. SIGMOD Record 32(4): 47–52. doi:10.1145/959060.959069
3. Alexander D (2002) Principles of emergency planning and management. Terra publishing, Harpended. ISBN ISBN 1-903544-10-6
4. Allenbach B, Andreoli R, Battiston S, Bestault C, Clandillon S, Fellah K, Henry JB, Meyer C, Scius H, Tholey N, Ysou H, de Fraipont P (2005) Rapid EO Disaster Mapping Service: Added Value, Feedback and Perspectives after 4 Years of Charter Actions. In: IGARSS05 Proceedings, pp 4373–4378
5. Al-Khudhairy DHA, Caravaggi I, Glada S (2005) Structural damage assessments from IKONOS data using change detection, object-oriented segmentation, and classification techniques. Photogram Eng Remote Sens 71:825837
6. Antonov JI, Levitus S, Boyer TP (2002) Steric sea level variations during 1957–1994: importance of salinity. J Geophys Res (Oceans) 107(C12): 8013. doi:10.1029/2001JC000964
7. Awange JL, Kiema JBK (2013) Environmental geoinformatics. Monitoring and management, vol. Springer, Berlin

8. Awange JL (2012) Environmental monitoring using GNSS. Global navigation satellite system. Springer, New York
9. Awange JL, Ogallo L, Kwang-Ho B, Were P, Omondi P, Omute P, Omulo M (2008) Falling Lake Victoria water levels: is climate a contribution factor? J Clim Change 89:287–297. doi:10.1007/s10584-008-9409-x
10. Awange JL, Aluoch J, Ogallo L, Omulo M, Omondi P (2007) Frequency and severity of drought in the Lake Victoria region (Kenya) and its effects on food security. Clim Res 33: 135–142. doi:10.3354/cr033135
11. Awange JL, Fukuda Y (2003) On possible use of GPS-LEO satellite for flood forecasting. In: Accepted to the international civil engineering conference on sustainable development in the 21st century "The Civil Engineer in Development" 12–16 Aug 2003 Nairobi, Kenya
12. Baker HC, Dodson AH, Penna NT, Higgins M, Offiler D (2001) Ground-based GPS water vapour estimation: Potential for meteorological forecasting. J Atmos Solar Terrestrial Phys 63(12):1305–1314
13. Ballesteros, LF (2008) What determines a disaster? 54 Pesos, Sep 2008:54 Pesos 11 Sep 2008. http://54pesos.org/2008/09/11/what-determines-a-disaster/. Accessed on 12/05/2011
14. Bamber JL, Riva REM, Vermeersen BLA, LeBrocq AM (2009) Reassessment of the potential sea-level rise from a collapse of the West Antarctic ice sheet. Science 324:901–903. doi:10.1126/science.1169335
15. Bancroft S (1985) An algebraic solution of the GPS equations. IEEE Trans Aerosp Electron Syst AES-21: 56–59
16. Bankoff G, Frerks G, Hilhorst D (eds.) (2003) Mapping vulnerability: disasters, development and people. ISBN 1-85383-964-7
17. Barlow J, Franklin SE (2007) Mapping Hazardous Slope Processes Using Digital Data. In Geomatics solutions for disaster management. Li J, Zlatanova S, Fabbri A (eds) Lecture Notes in Geoinformation and Cartography. Springer, NewYork, pp 74–90
18. Barlow J, Franklin S, Martin Y (2006) High spatial resolution satellite imagery, DEM derivatives, and image segmentation for the detection of mass wasting processes. Photogram Eng Remote Sens 72(6):687–692
19. Barrett CB (2002) Food security and food assistance programs. In: Gardner B, Rausser G (eds) Handbook of agricultural economics, vol 2. Elsevier Science, Amsterdam, pp 2103–2190
20. Becker M, Llowel W, Cazenave A, Güntner A, Crétaux J-F (2010) Recent hydrological behaviour of the East African Great Lakes region inferred from GRACE, satellite altimetry and rainfall observations. C Rendus Geosci 342(3):223–233. doi:10.1016/j.crte.2009.12.010
21. Bill R (2011) Precise positioning in ad hoc geosensor newtorks. http://www.ikg.uni-hannover.de/geosensor/Lecture/Wednesday/Session1/sess1_bill.pdf. Accessed on 22/01/2011
22. Bitelli G, Camassi R, Gusella L, Mognol A (2004) Image change detection on urban areas: the earthquake case. Proc ISPRS XXth Congr Istanbul 35(B7): 692–697
23. Bonner MR, Han D, Nie J, Rogerson P, Vena JE, Freudenheim Jo L (2003) Positional accuracy of geocoded addresses in epidemiologic research. Epidemiology 14:408–412. doi:10.1097/01.EDE.0000073121.63254.c5
24. Born GH, Parke ME, Axelrad P, Gold KL, Johnson J, Key KW, Kubitschek DG, Christensen EJ (1994) Calibration of the TOPEX altimeter using a GPS buoy. J Geophys Res 99(C12): 24517–24526
25. Brenner C (2011) Geo sensor networks—when and how? http://dgk.auf.uni-rostock.de/uploads/media/2_2-Brenner.pdf. Accessed on 22/01/2011
26. Buehler YA, Kellenber TW (2007) Development of processing chains for rapid mapping with satellite data. In: Li J, Zlatanova S, Fabbri A (eds) Geomatics solutions for disaster management. Lecture Notes in Geoinformation and Cartography. Springer, New York, pp 16–36
27. Chen JL, Wilson CR, Tapley BD, Yang ZL, Niu GY (2009) 2005 drought event in the Amazon River basin as measured by GRACE and estimated by climate models. J Geophys Res 114:B05404. doi:10.1029/2008JB006056

28. Church JA, Gregory JM, Huybrechts P, Kuhn M, Lambeck K, Nhuan MT, Qin D, Woodworth PL (2001) Changes in Sea Level. In: Houghton JT, Ding Y, Griggs DJ, Noguer M, Van der Linden PJ, Dai X, Maskell K, Johnson CA (eds) Climate change 2001: the scientific basis: contribution of working group I to the third assessment report of the inter-governmental panel on climate change. Cambridge University Press, Cambridge, pp 639–694
29. Crétaux J-F, Jelinski W, Calmant S, Kouraev A, Vuglinski V, Bergé-Nguyen M, Gennero M-C, Nino F, Cazenave A, Maisongrande P (2011) SOLS: a lake database to monitor in the Near real-time water level and storage variations from remote sensing data. Adv Sp Res 47:1497–1507. doi:10.1016/j.asr.2011.01.004
30. Crétaux J-F, Leblanc M, Tweed S, Calmant S and Ramillien G (2007) Combining of radar and laser altimetry, MODIS remote sensing and GPS for the monitoring of flood events: application to the flood plain of the Diamantina river. Geophys Res Abs 9: 07496. SRef-ID: 1607–7962/gra/EGU2007-A-07496
31. Cruden D, Varnes D (1996) Landslide types and processes. In: Turner K, Schuster R (eds) Landslides investigation and mitigation, transportation research board special report 247. National Academy Press, Washington, pp 36–75
32. Cruz, A, Laneve, G, Cerra, D, Mielewczyk, M, Garcia, MJ, Santilli, G, Cadau, E, Joyanes, G (2007) On the application of nighttime sensors for rapid detection of areas impacted by disasters. In: Li J, Zlatanova S, Fabbri A (eds) Geomatics solutions for disaster management. Lecture Notes in Geoinformation and Cartography. Springer, New York, 16–36.
33. Dalton R (2007) GPS could offer better fault line mapping. Nature News. doi:10.1038/news070521-9. http://www.nature.com/news/2007/070521/full/news070521-9.html. Accessed on 25/09/2011
34. Dickey JO, Bentley CR, Bilham R, Carton JA, Eanes RJ, Herring TA, Kaula WM, Lagerloef GSE, Rojstaczer S, Smith WHF, Van Den Dool HM, Wahr JM, Zuber MT (1996) Satellite gravity and the geosphere. National Research Council Report, 112 p, National Acadamy Press, Washington
35. DMCN (Drought Monitoring Centre Nairobi)(2002) Factoring of weather and climate information and products into disaster management policy. A contribution to strategies for disaster reduction in Kenya, UNDP, Government of Kenya and WMO, Nairobi
36. Forootan E, Awange J, Kusche J, Heck B (2012) Independent patterns of water mass anomalies over Australia from satellite data and models. Remote Sens Environ 124:427–443. doi:10.1016/j.rse.2012.05.023
37. Garcia-Garcia D, Ummenhofer CC, Zlotnicki V (2011) Australian water mass variations from GRACE data linked to Indo-Pacific climate variability. Remote Sens Environ 115:2175–2183. doi:10.1016/j.rse.2011.04.007
38. Geoscience Australia (2008) Need for the geodetic component for absolute sea level monitoring. http://www.ga.gov.au/geodesy/slm/spslcmp/. Accessed on 11/12/2008
39. Gili JA, Corominas J, Rius J (2000) Using global positioning techniques in landslide monitoring. Eng Geol 155(3):167–192
40. GITEWS (German Indonesian Tsunami Early Warning System) (2008) A new approach in Tsunami-early warning. Press-Information embargo: 11.11.2008, 10:00 CET. http://www.gitews.de/fileadmin/documents/content/press/GITEWS_operationell_eng_nov-2008.pdf. Accessed on 10/12/2008
41. Hammond WC, Brooks BA, Bürgmann R, Heaton T, Jackson M, Lowry AR, Anandakrishnan S (2011) Scientific value of real-time global positioning system data. Eos 92(15):125–126. doi:10.1029/2011EO150001
42. Hasegawa H, Yamazaki F, Matsuoka M, Seikimoto I (2000) Determination of building damage due to earthquakes using aerial television images. In: Proceedings of the 12th world conference on earthquake engineering, Auckland, CDROM, 8p
43. Hatfield JL, Prueger JH, Kustas WP (2004) Remote sensing of dryland crops. In: Ustin S (ed) Remote sensing for natural resources and environmental monitoring: manual of remote sensing, vol 4, 3rd edn. Wiley, New Jersey, pp 531–568

44. Hay SI, Lennon JJ (1999) Deriving meteorological variables across Africa for the study and control of vector-borne disease: a comparison of remote sensing and spatial interpolation of climate. Trop Med Int Health 4:58–71
45. Helm A, Montenbruck O, Ashjaee J, Yudanov S, Beyerle G, Stosius R, Rothacher M (2007) GORS— a GNSS occultation, reflectometry and scatterometry space receiver. In: Proceedings of the 20th international technical meeting of the satellite division of The Institute of Navigation ION GNSS 2007, Fort Worth, Texas. Sept 25–28:2011–2021
46. Herbreteau V, Salem G, Souris M (2007) Thirty years of use and improvement of remote sensing applied to epidemiology: from early promises to lasting frustration. Health Place 13:400–403
47. Hirt C, Gruber T, Featherstone WE (2011) Evaluation of the rst GOCE static gravity eld models using terrestrial gravity, vertical deections and EMG2008 quasigeoid heights. J Geodesy 85:723–740. doi:10.1007/s00190-011-0482-y
48. Hofman-Wellenhof B, Lichtenegger H, Collins J (2001) Global positioning system: theory and practice, 5th edn. Springer, Wien
49. Hofman-Wellenhof B, Lichtenegger H, Wasle E (2008) GNSS global navigation satellite system: GPS, GLONASS; Galileo and more. Springer, Wien
50. Istomina MN, Kocharyan AG, Lebedeva IP (2005) Floods: genesis, socioeconomic and environmental impacts. J Water Resour 32(4):349–358
51. James LF, Young JA, Sanders K (2003) A new approach to monitoring rangelands. Arid Land ResManage 17:319–328. doi:10.1080/15324980390225467
52. Jayaraman V, Chandrasekhar MG, Rao UR (1997) Managing the natural disasters from space technology inputs. Elsevier Science Ltd, Great Britain
53. Jeyaseelan AT (2003) Droughts and floods assessment and monitoring using remote sensing and GIS. In: Siva Kumar MVK, Roy PS, Harmsen K, Saha SK (eds) Satellite sensing and GIS applications in agricultural meteorology. Proceedings of the Trainging Workshop 7–11 july 2003, Dehra Dun, pp 291–313
54. Jia M (2005) Crustal deformation from the Sumatra-Andaman Earthquake. Geoscience Australia's analysis of the largest earthquake since the beginning of modern space geodesy. Ausgeo news, issue 80
55. Kamik V, Algermissen ST (1978) Seismic Zoning—chapter in the assessment and mitigation of earthquake risk. UNESCO, Paris, pp 1–47
56. Kelecy TM, Born GH, Parke ME, Rocken C, (1994) Precise mean sea level measuring using global positioning system. J Geophys Res 99(c4): 7951–7959
57. Khandu, Awange JL, Wickert J, Schmidt T, Shari MA, Heck B, Fleming K (2010) GNSS remote sensing of the Australian tropopause. Clim Change 105(3–4):597–618. doi:10.1007/s10584-010-9894-6
58. Kouchi K, Yamazaki F, Kohiyama M, Matsuaka M, Muraoka N (2004) Damage detection from Quickbird high-resolution Satellite images for the 2003 Boumerdes. Algeria Earthquake. In: Proceeding of the Asian conference on earthquake engineering, Manila, Philippines, CD-ROM, pp 215–226
59. Larson KM (2009) GPS seismology. J Geodesy 83:227–233. doi:10.1007/s00190-008-0233-x
60. Leavitt WM, Kiefer JJ (2006) Infrastructure interdependency and the creation of a normal disaster: the case of Hurricane Katrina and the City of New Orleans. J Public Works Manage Policy 10(4):306–314
61. Leuliette EW, Nerem RS, Mitchum GT (2004) Calibration of TOPEX/Poseidon and Jason altimeter data to construct a continuous record of mean sea level change. Mar Geodesy 27(1):79–94. doi:10.1080/01490410490465193
62. Lowe ST, LaBreceque JL, Zuada C, Romans LJ, Young L, Hajj GA (2002) First space borne observations of an earth reected GPS signal. Radio Sci 37(1):1007. doi:10.1029/2000RS002539
63. Lian M, Warner RD, Alexander JL, Dixon KR (2007) Using geographic information systems and spatial and space-time scan statistics for a population-based risk analysis of the 2002

equine West Nile epidemic in six contiguous regions of Texas. International Journal of Health Geographics 6, 42; available at www.ij-healthgeographics.com/content/6/1/42.
64. Malamud B, Turcotte D, Guzzetti F, Reichenbach P (2004) Landslide inventories and their statistical properties. Earth Surf Proces Land 29:687–711
65. Malet JP, Maquaire O, Calais E (2002) The use of global positioning system techniques for the continuous monitoring of landslides: application to the super-sauze earthflow (Alpes-de-Haute-Provence, France). Geomorphol 43(1–2):33–54. doi:10.1016/S0169-555X(01)00098-8
66. Matsuzaka S (2006) GPS network experience in Japan and its usefulness. In: Seventeenth United Nations Regional Cartographic Conference, Geographical Survey Institute, Bangkok Thailand
67. McKean J, Buechel S, Gaydos L (1991) Remote sensing and landslide hazard assessment. Photogram Eng Remote Sens 57(9):1185–1193
68. Mehdi R, Gruen A (2007) Automatic classification of collapsed buildings using object and image space features. In: Li J, Zlatanova S, Fabbri A (eds) Geomatics solutions for disaster management. Lecture Notes in Geoinformation and Cartography. Springer, New York, 135–148
69. Mitrovica JX, Tamisiea ME, Davis JL, Milne GA (2001) Recent mass balance of polar ice sheets inferred from patterns of global sea-level change. Nature 409:1026–1029. doi:10.1038/35059054
70. Mitrovica JX, Gomez N, Clark PU (2009) The sea-level fingerprint of West Antarctic collapse. Science 323(5915):753. doi:10.1126/science.1166510
71. Miura H, Midorikawa S (2006) Updating GIS building inventory data using high-resolution satellite images for earthquake damage assessment: application to metro Manila, Philippines. Earthq Spectra 22:151–168
72. Motagh M, Djamour Y, Walter TR, Wetze H, Zschau J, Arabi S (2007) Land subsidence in Mashhad Valley, northeast Iran: results from InSAR, levelling and GPS. Geophys J Int 168:518–526. doi:10.1111/j.1365-246X.2006.03246.x
73. Nicholson SE, Davenport ML, Malo AR (1990) A comparison of the vegetation response to rainfall in the Sahel and East Africa, using normalized difference vegetation index from NOAA AVHRR. Clim Change 17(2–3):209–241. doi:10.1007/BF00138369
74. Nittel S, Labrinidis A, Stefanidis A (eds) (2008) GeoSensor networks. Lecture Notes in Computer Science 4540. Springer, Berlin, pp. 1–6
75. Nittel S, Stefanidis A, Cruz I, Egenhofer M, Goldin D, Howard A, Labrinidis A, Madden S, Voisard A, Worboys M (2004) Report from the first workshop on Geo Sensor Networks. ACM SIGMOD Record 33(1)
76. Ogawa N, Yamazaki F (2000) Photo-interpretation of buildings damage due to earthquakes using aerial photographs. In: Proceedings of the 12th world conference on earthquake engineering, Auckland, CD-ROM, 8p
77. Omute P, Corner R, Awange JL (submitted) NDVI monitoring of Lake Victoria water level and drought. Water Resource Management
78. Phoon SY, Shamseldin AY, Vairavamoorthy K (2004) Assessing impacts of climate change on Lake Victoria Basin, Africa: people-centred approaches to water and environmental sanitation. In: 30th water engineering and development centre (WEDC) International Conference, Vientiane, Lao PDR, pp 392–397
79. Poland JF (1984) Guidebook to studies of land subsidence due to water withdrawal. Technical report. UNESCO
80. Privette JL, Fowler C, Wick GA, Baldwin D, Emery WJ (1995) Effects of orbirtal drift on advanced very high resolution radiometer products: normalized difference vegetation index and sea surface temperature. Remote Sens Environ 53(3):164–171. doi:10.1016/0034-4257(95)00083-D
81. Pugh D (2004) Changing sea levels. Effect of tides, weather and climate. Cambridge Univeristy Press, Cambridge

82. Rius A, Aparicio JM, Cardellach E, Martín-Neira M, Chapron B (2002) Sea surface state measured using GPS reflected signals. Geophys Res Lett 29(23):21–22. doi:10.1029/2002GL015524
83. Rizos C (2001) Alternatives to current GPS-RTK services and some implications for CORS infrastructure and operations. GPS Solution 11(3):151–158
84. Rocken C, Kelecy TM, Born GH, Young LE, Purcell GH, Wolf SK (1990) Measuring precise sea level from a buoy using the Global Positioning System. Geophys Res Lett 17(12): 2145–2148
85. Rood K (1984) An aerial photograph inventory of the frequency and yield of mass wasting on the Queen Charlotte Islands. British Columbia, BC Ministry of Forests, Land Management Report 34
86. Sagiya T (2005) A decade of GEONET: 1994–2003 The continuous GPS observation in Japan and its impact on earthquake studies. Earth Planets Sp 56: 29—xli
87. Sauchyn D, Trench N (1978) Landsat applied to landslide mapping. Photogram Eng Remote Sens 44:735–741
88. Schenk A (2006) Interpreting surface displacement in Tehran / Iran region observed by Differential Synthetic Aperture Radar Interferometry (DINSAR). Diplomarbeit, Technische Universität Berlin, Institut für Angewandte Geowissenschaften Fachgebiet Angewandte Geophysik
89. Scofield RA, Achutuni R (1996) The satellite forecasting funnel approach for predicting flash floods. Remote Sens Rev 14:251–282
90. Seidel DJ, Randel WJ (2006) Variability and trends in the global tropopause estimated from radiosonde data. J Geophys Res 111. doi:10.1029/2006JD007363
91. Snay R, Cline M, Dillinger W, Foote R, Hilla S, Kass W, Ray J, Rohde J, Sella G, Soler T (2007) Using global positioning system-derived crustal velocities to estimate rates of absolute sea level change from North American tide gauge records. J Geophys Res 112:B04409. doi:10.1029/2006JB004606
92. Snay R, Soler T (2008) Continuously operating reference station (CORS): history, applications, and future enhancements. J Surv Eng 134(4):95–104. doi:10.1061/(ASCE)0733-9453(2008)134:4(95)
93. Snow (2010) GIS Analyses of Dr. Snow's Map. http://www.udel.edu/johnmack/frec480/cholera/cholera2.html. Accessed on 02/04/2010
94. Steede-Terry K (2000) Integrating GIS and the global positioning system. ESRI Press, California
95. Stefanidis A (2006) The emergence of geoSensor networks. Directions magazine. http://www.directionsmag.com/articles/the-emergence-of-geosensor-networks/123208. Accessed on 22/01/2011
96. Terhorst A, Moodley D, ISimonis I, Frost P, McFerren G, Roos S, van den Bergh F (2008) Using the sensor web to detect and monitor the spread of vegetation fires in southern Africa. In: Nittel S, Labrinidis A, Stefanidis A (eds) GeoSensor Networks, Lecture Notes in Computer Science 4540. Springer, Berlin, pp 239–251
97. Titus JG, Park RA, Leatherman S, Weggel R, Greene MS, Treehan M, Brown S, Gaunt C, Yohe G (1991) Greenhouse effect and sea level rise: the cost of holding back the sea. Coast Manage 19:171–204
98. Trenberth K, Guillemot C (1996) Evaluation of the atmospheric moisture and hydrological cycle in the NCEP Reanalyses. NCAR Technical Note TN-430, December
99. Trenberth KE (1997) The Definition of El Niño. Bull Am Meteorol Soc 78:2771–2777
100. Turker M, Cetinkaya B (2005) Automatic detection of earthquake-damaged buildings using DEMs created from pre- and post-earthquake stereo aerial photographs. Int J Remote Sens 26(4):823–832
101. Ummenhofer C, England M, McIntosh P, Meyers G, Pook M, Risbey J, Gupta A, Taschetto A (2009) What causes southeast Australiaś worst droughts? Geophys Res Lett 36:L04706. doi:10.1029/2008GL036801
102. Uriel K (1998) Landscape ecology and epidemiology of vector-borne diseases: tools for spatial analysis. J Med Entomol 35(4):435–445

103. US Army Corps of Engineers (2007) NAVSTAR Global Positioning System surveying. Engineering and Design Manual, EM 1110–1-1003
104. Voigt S, Riedlinger T, Reinartz P, Knzer C, Kiefl R, Kemper T, Mehl H (2005) Experience and perspective of providing satellite based crisis information, emergency mapping and disaster monitoring information to decision makers and relief workers. In: Zlatanova S, van Oosterom P P, Fendel E (eds) Geoinformation for Disaster Management. Springer, Berlin, pp 519–531
105. Warrick RA, Le Provost C, Meier MF, Oerlemans J, Woodworth PL (1996) Changes in sea level. In: Houghton JT, Meira Filho LG, Callander BA, Harris N, Klattenberg A, Maskell K (eds) Climate change 1995, the science of climate change. Cambridge University Press, Cambridge, pp 359–405
106. Watson C, Coleman R, White N, Church J, Govind R (2003) Absolute calibration of TOPEX/Poseidon and Jason-1 Using GPS buoys in Bass Strait. Australia. Mar Geodesy 26(3–4):285–304. doi:10.1080/01490410390256745
107. Webster TL, Forbes DL, Dickie S, Shreenan R (2004) Using topographic LiDAR to map flood risk from storm-surge events for Charlottetown, Prince Edward Island, Canada. Can J Remote Sens 30:64–76
108. Wills C, McCrink T (2002) Comparing landslide inventories: the Map depends on the method. Environ Eng Geosci VIII 4:279–293
109. Wisner B, Blaikie P, Cannon T, Davis I (2004) At risk—natural hazards, people's vulnerability and disasters. Routledge, Wiltshire
110. Worboys M, Duckham M (2006) Monitoring qualitative spatiotemporal change for geosensor networks. Int J Geograph Inf Sci 20(10):1087–1108. doi:10.1080/13658810600852180
111. Xue Z, Li G, Li Z, Wu X, Wei J (2007) Monitoring Xian Land Subsidence Evolution by Differential SAR Interferometry. In: Li J, Zlatanova S, Fabbri A (eds) Geomatics solutions for disaster management. Lecture Notes in Geoinformation and Cartography. Springer, New York, pp 427–437
112. Yamaguchi N, Yamazaki F (2001) Estimation of strong motion distribution in the 1995 Kobe earthquake based on building damage data. Earthq Eng Struct Dyn 30(6):787–801
113. Yamazaki F, Kouchi K, Kohiyama M, Muraoka N, Matsuoka M (2004) Earthquake damage detection using high-resolution satellite images. In: Proceedings of IEEE 200 international geoscience and remote sensing symposium, IEEE, CD-ROM, 4p
114. Yamazaki F, Yano Y, Matsuoka M (2005) Visual damage interpretation of buildings in Bam City using QuickBird images. Earthquake Spectra 21(1):329–336
115. Zhang Q, Zhao, C, Ding, X, Peng, J (2007) Monitoring Xian Land Subsidence Evolution by Differential SAR Interferometry. In: Li J, Zlatanova S, Fabbri A (eds) Geomatics solutions for disaster management. Lecture Notes in Geoinformation and Cartography. Springer, New York, pp 91–102
116. Zhanga J, Zhoub C, Xua K, Watanabe M (2002) Flood disaster monitoring and evaluation in China. Environ Hazards 4:33–43. doi:10.1016/S1464-2867(03)00002-0

Chapter 17
Environmental Impact Assessment

> *"Distances and locations are the important determinants of many choices that economists study. Economists often rely on information about these variables that are self-reported by respondents in surveys, although information can sometimes be obtained from secondary sources. Self-reports are typically used for information on distances from households or community centers to roads, markets, schools, clinics, and other public services. There is growing evidence that self-reported distances are measured with errors and that these errors are correlated with the outcomes of interest. In contrast to self-reports, global positioning systems (GPS) can determine locations to within 15 m in most cases. The falling cost of Global Navigation Satellite Systems (GNSS) receivers makes it increasingly feasible for field surveys to use GNSS to more accurately measure locations and distances."*
>
> J. Gibson and D. MacKenzie [23]

17.1 Role of Geospatial in EIA, SEA, and SA

17.1.1 Impact Assessments and the Need for Monitoring

Environmental Impact Assessment (EIA) is defined by Munn [49] as the need to *identify* and *predict* the impact on the environment and on man's health and well-being of legislative proposals, policies, programs, projects, and operational procedures, and to interpret and communicate information about the impact. EIA is thus a process, a systematic process that examines the environmental consequence of development actions in advance [25, p. 4]. Glasson et al. [25] have defined the purpose of EIA as an *aid to decision making*, *an aid to the formulation of the development actions*, and *an instrument to sustainable development*. In order to achieve these goals, EIA requires monitoring data that can be used to identify and predict impacts, and also to evaluate the impacts of a given project once approved. Whereas EIA has been traditionally restricted to projects that are deemed to have significant impacts on the environment, it has recently expanded to include *strategic environmental assessment* (SEA) discussed in Sect. 17.4 and *sustainability assessment* (SA) presented in Sect. 17.5.

N. O. Agola and J. L. Awange, *Globalized Poverty and Environment*,

DOI: 10.1007/978-3-642-39733-2_17,

Monitoring involves the measuring and recording of physical, social and economic variables associated with development impacts. The activities seek to provide information on the characteristics and functioning of variables in *time*, *space*, and *scale*, and in particular in the occurrence and magnitude of impacts [25, p. 185]. It offers the possibility of determining or assessing the extent of human impacts on the environment and also compares human impacts with natural variation in the environment. The advantages of monitoring following project implementations are that it can improve project management, it can be used as an early warning system to identify harmful trends in a locality before it is too late to take remedial action, it can help to identify and correct for unanticipated impacts, and it can also be used to provide acceptable data and information, which can be used in mediation between interested parties [25, p. 185]. Glasson et al. [25, p. 186] defines environmental impact auditing as the comparison between the impacts predicted in *environmental impact statements (EIS)* and those occurring after implementation in order to assess whether the impact prediction performs satisfactorily. EIS is the document that contains the information and estimates of impacts derived from the various steps of the EIA process.

17.1.2 Geospatial Applications

GNSS satellites could be used to support the processes of project-based EIA, SEA and SA in provision of location-based data that support *monitoring* and *auditing*. As an example, in March of 2009, Kelly Core Salmon (KCS) Ltd filed an application with the Noca Scotia Department of Fisheries and Aquaculture (NSDFA) to relocate and expand the boundaries of the existing three aquaculture sites (Sand Point, Boston Rock, and Hartz Point) located in Shelburne Harbour, Nova Scotia [63]. The desire to relocate and expand was motivated by the need to improve the environmental performance of the three sites by allowing a greater flow and depth on the sites, easier access to the sites, and increased production, ensuring greater economic stability for KCS production in Nova Scotia [63].

For the relocations and expansion to take place, EIA was undertaken in order to satisfy the criteria of the New Brunswick Department of Agriculture and Aquaculture (NBDAA), Nova Scotia Department of Agriculture and Aquaculture (NSDAA), and Fisheries & Oceans Canada (DFO) [63]. In support of provision of location-based data, Global Navigation Satellite Systems (GNSS)'s method of Differential Global Positioning System (DGPS) (see [7]) was employed to provide the relocated boundary co-ordinates.

GNSS could also be useful in supporting impact assessments in the following ways:

(a) Provide location-based data useful in identification of features of interest, which could be impacted during the undertaking of the project-based EIA. For example, GNSS could be used to provide the locations of boreholes in a given region where a project that has the potential of contaminating groundwater has been proposed.

(b) Providing distance information that is useful in measuring access to infrastructure and social services such as health care. Gibson and McKenzie [23] discuss how GNSS-based information on spatial distribution of population and services can lead to improved understanding of access to services. Understanding access to services is essential in spatial multi-criteria selection (e.g., Sect. 17.3.2), where a decision to choose an option from various alternatives is to be made. For instance, Perry and Gessler [50] applied GNSS to measure access from communities to health-care facilities in Andean Bolivia, and used the results to propose an alternative model of health distribution in the study area.
(c) Its distance and travel time data can be useful in identifying barriers to the use of services [23]. Often, such hidden barriers can lead to poor decision leading to the selection of a given alternative at the expense of the other methods, which might be optimum. Knowledge of these hidden barriers could thus enable policy and decision makers to make informed decisions.
(d) A combination of GNSS-based location data and Geographical Information System (GIS) would be very effective in illustrating access to services in a form that would be easily understood by the community during participatory stage of discussing environmental impact statement (EIS), and also for policy and decision makers during the selection of an option from given alternatives.
(e) In support of collection of socio-economic data, e.g., household surveys. Here, GNSS could be useful in improving the quality and cost-effectiveness of the survey data. GNSS locations could for instance be used to provide sampling boundaries as opposed to cases where such boundaries are arbitrarily selected or regular grids used where they are not useful (e.g., in monitoring variable features irregularly distributed over space such as air pollution). For example, Kumar [38] show how a combination of GNSS and remote sensing was useful in drawing samples in a survey of 1,600 households spread across different air pollution zones in Dehli (India).
(f) For SEA and SA, GNSS can be of use in providing data for econometric modeling of casual impacts of policies [23]. In this regard, it could provide data that could enable practitioners to better control the geographical and regressional characteristic of their models, e.g., by comparing individuals who are subjected to a given policy and those who are not.
(g) Further, for SEA and SA, its integration with GIS can prove particularly useful in supporting the evaluation of cumulative impacts (see e.g., Sect. 17.4.1). This is achieved through the ability of GIS-GNSS to consider spatial component and allow the analysis of the temporal evolution, see, e.g., [64].

17.2 Impact Monitoring to Detect Change

In defining monitoring (e.g., in [7]), *impact monitoring* is noted to focus on identifying possible impacts of human activities on environment and to distinguish them from the non-human environmental processes, while *compliance monitoring* has the objective of supporting stipulated legislations that aim at protecting and conserving the environment. According to Downes et al. [13], both compliance and impact assessments have a key objective of detecting change in selected variables, with impact assessment relying on comparisons within the collected data to assess whether an impact has occurred and the magnitude of such impact. Because impact assessment monitoring tend to be defined relative to natural conditions rather than being pegged to external criteria, Downes et al. [13] propose a monitoring design model, which if properly implemented, could support *change detection.*

The model is location-based taking into consideration the fact that in most cases, variables are measured at a specific impact location or locations, i.e., the *impact location(s)*. They then argue that a change being monitored in a variable should be seen to have occurred by comparing the variable's status prior to the activity (baseline data) which they call "*Before*" and after or during the activity (operational data) which they call "*After*". This Before-After model takes place at the impact location. In order to distinguish between natural and impact induced changes, a location outside the activity (impact area) is suggested, i.e., the "*control*" upon which data is to be simultaneously sampled together with the impact location "before" and "after". This before and after, control and impact locations form the BACI (Before-After-Control-Impact) model. The model proceeds as follows [13]:

- Data are collected at some ***impact locations*** over some period ***before*** the activity starts.
- Data are collected at some ***impact locations*** over some period ***after*** the activity starts.
- Data are collected at some ***control locations*** over the same period ***before*** the activity starts.
- Data are collected at some ***control locations*** over the same period ***after*** the activity starts.

In the BACI model above, the control location provides proxy data that are used to remotely sense the impact locations in the absence of a triggering activity. The assumption of the model is that if similar changes occur at both the control and impact locations, then the trigger for this changes would be natural causes since the control location does not have the activity. On the contrary, if the changes are only noticeable at the impact's location and not at the control location, then the activity at the impact location would be the most likely suspect. Because of the varying dynamics of the impacts and control locations, Downes et al. [13] suggest that several control locations and possibly impact locations be used, thus extending the BACI model to MBACI model, where multiple locations are considered.

Within these BACI and MBACI models, GNSS could be useful in providing the positions of control and impact locations upon which environmental impact assessment monitoring could be collected simultaneously before and after the activity. Remote sensing could be employed to collect variable spatio-temporal data, with GIS being used to manage the various datasets, in addition to providing the platform for detecting the change being monitored.

Example 17.1 (Illustration of tourism impact on groundwater)

Consider that a particular hotel utilizes groundwater and due to increased number of tourists, plenty of water is used, and that the impact of groundwater abstraction on the hotel is to be monitored to avert the potential danger of the building collapsing. Using relative GNSS positioning technique (e.g., [7]), coordinates of the hotel being monitored could be measured *before* it started operating to provide base data. During the operational phase, GNSS could be used to provide continuous coordinates of the building *after* the groundwater abstraction started. These observations are simultaneously observed to an established GNSS control points on stable locations some distance far away from the hotel both before, and after the groundwater abstraction started. The *relative positions* obtained will indicate the spatial variation of the hotel's position relative to the GNSS control (reference) before-and-after-the-impact. If no variation is noticed at the control location, but visible at the hotel (impact) location, then the variation could be attributed to groundwater abstraction. In such case, GNSS would have played a double role of providing locations of both impact (hotel) area and the control area, and also provision of time-variable data useful in generating relative motion (both horizontal and vertical) of the hotel useful in assessing the impact of groundwater abstraction.

♣ End of Example 17.1

17.3 Project EIA

17.3.1 Geospatial in Support of EIA Process

EIA generally goes through various stages, see e.g., [25, pp. 88–184] and [48, p. 8]. Some of these stages, and possible areas in which geospatial could be useful are discussed. The first of these stages is *screening*, where a project is assessed as to whether it requires EIA or not. GIS is the basic tool that could be employed to support screening in EIA. For example, in the work of Geneletti [22], GIS was combined with a decision aiding tool known as Multi-criteria analysis (MCA) to produce thematic nature conservation layer maps used to support decisions on whether to undertake EIA for a proposed project and also to choose the most suitable locations for new projects in the alpine area located in Trentino (northern Italy). Antunes et al. [3] propose a GIS approach for computing scores for criteria for use in MCA. Since GIS brings visual capability, its combination with MCA analytical tools will play a significant role in screening EIA projects as discussed in the next section. As a matter of fact, virtually all commercial GIS software in operation today have built and integrated MCA functionality into their systems.

The next stage of EIA after screening is the *scoping* stage, where the impacts and issues to be considered are identified. The process of scoping is that of deciding, from all of a projects possible impacts, and from all the alternatives that could be addressed, which ones are the most significant [25, p. 91]. Identification of significant alternatives requires comparison to be made at the scoping phase. Usually, at the initial phase of scoping, a small number of alternatives will be selected for further analysis from many potential alternatives, and in the final evaluation, these alternatives are subjected to more detailed evaluation. An example is presented in the rare earth case where 15 sites were selected in the initial case from which 6 sites were chosen for further analysis [4].

In evaluating possible alternatives, GIS is an attractive proposition given its ability to consider different factors within an integrated framework. In this respect, remote sensing could be employed to develop various factor maps. GNSS satellites could also play a vital role of not only providing the coordinates (i.e., positions) of alternative locations, but could be used to provide rapid field measurements of factors such as distances to environmental sensitive locations (e.g., groundwater or conservation parks), and the actual spatial coverage of areas of each alternative.

Example 17.2 (GNSS in support of choosing from alternative locations)

Consider Fig. 17.1 where three alternative perimeter locations are to be considered for the purpose of setting out a project such as sugar processing factory. *First*, the areas of these locations are to be established so that the smallest parcel of land is chosen to accommodate the factory and at the same time minimize on the land purchasing cost. *Second*, the distances of the sites to the nearest water source is required so as to asses the potential of the sugar factory contaminating the groundwater source. GNSS could be used to establish the corner positions of the various sites A, B, and C from which the perimeter and area of each parcel of the land could be rapidly calculated. Further, distances from each site to the nearest water source can rapidly be obtained in the field by measuring baselines of two receivers, one stationed at a given site and the other stationed at the water source as illustrated in Fig. 17.1. Another possibility would be to use a hand-held GNSS (e.g., Fig. 14.1) to obtain the direct distance measurements from each site to the water source using the navigation functions of these receivers by simply walking from the water source to the proposed sites. It is worth mentioning that these analyzes could be best undertaken within a GIS environment.

♣ End of Example 17.2

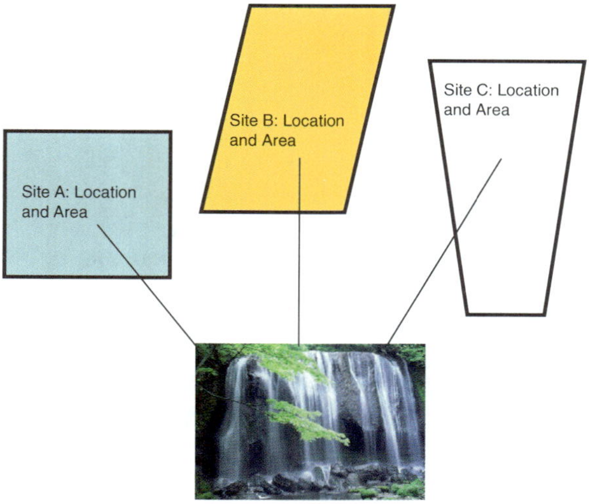

Fig. 17.1 GNSS support of site selection from three different alternatives. GNSS could be useful in providing positions, distances, perimeters and areas of each site, information that could inform decision markers choice of the correct site

Analysis stage of EIA consists of *identification*, *prediction* and *evaluation* [1]. Impact identification brings together project characteristics and baseline environmental characteristics with the aim of ensuring that all the potential significant environmental impacts (adverse or favorable) are identified and taken into account during EIA process [25, p. 107]. Remote sensing, photogrammetry and GNSS could help in provision of environmental baseline data before the project is established. This could then be used in impact prediction which requires that it be based on available environmental baseline data and proper use of technology to identify environmental modification, forecast the quantity and/or *spatial dimension of change* in the environment, and estimation of the probability that the impact will occur [1].

Techniques for impact identification and prediction are discussed, e.g., in Glasson et al [25, pp. 88–184]. Several methods of impact identification exist and are generally divided into the following categories [25, p. 108]; *checklist*, *matrices*, *quantitative methods*, *networks* and *overlay maps*. These methods have been discussed in detail, e.g., by Shopley and Fuggle [62] and Westman [73]. Incidentally, the last impact identification method i.e., overlay of maps provides spatial visualization and is best carried out within a GIS environment as discussed in Chap. 2. Evaluation in EIA looks mostly at the cost and benefits of a proposed projects to the users, assesses the impact on environment and compares various alternatives that will yield benefit of the project with minimum environmental and social impacts. Such alternatives could be evaluated through methods such as multi-criteria analysis (see Sect. 17.3.2). Remote sensing and GNSS satellites can play a vital role in this aspect of EIA with regard to documenting and identifying impacts associated with spatial changes. As illustrated in Fig. 17.1, GNSS could also assist in the determination of positions, distances, perimeters, and areas needed by decision makers to make informed choices.

17.3.2 Geospatial and Multi-Criteria Analysis

17.3.2.1 Spatial Multi-Criteria Analysis

The vast majority of environmental management decisions are guided by multiple stakeholders' interests. These decisions are often characterized by *multiple objectives, multiple alternatives* and *considerable uncertainties* [26]. Alternatives are means for accomplishing particular goals [68, pp. 109–110] and their evaluation is a requirement in EIA of some countries. For example, the *National Environmental Policy Act* (NEPA) 1969 (US) requires that alternatives be considered while undertaking EIA. When multiple stakeholders with varied interests are involved, and multiple objectives and alternatives have to be considered, the situation often turns out to be very complex. In such cases, *multi-criteria analysis* (MCA), a framework for evaluating decision alternatives against multiple objectives comes in handy. MCA is currently emerging as a popular approach for supporting multi-stakeholder environmental decisions as reported, e.g., in Regan et al. [56].

MCA are methods that seek to allow for a pluralist view of society, composed of diverse stakeholders with diverse goals and with differing values concerning environmental changes [25]. According to Munier [48, p. 132], MCA are tools that are used for the analysis of projects, plans, programmes and options either with single objective or with several objectives with many different attributes or criteria. Besides being a tool for aiding the selection of the best preferred alternative, Marttunen and Hamalainen [44] suggests that it could also be used to increase the understanding of the problem by value structuring (i.e., identification of the objective and the analysis of values). The components of MCA are listed by Annandale and Lantzke [2] as;

- a given set of alternatives,
- a set of criteria for comparing the alternatives, and
- a method for ranking the alternatives based on how well they satisfy the criteria.

Spatial multi-criteria decision problems typically involve a set of geographically-defined alternatives (events) from which a choice of one or more alternatives is made with respect to a given set of evaluation criteria [34, 43]. For spatial multi-criteria decision analysis, two considerations that are of utmost importance are [11, 34]:

(1) A GIS component such as data acquisition, storage, retrieval, manipulation, and analysis capability
(2) Spatial analysis component such as aggregation of spatial data and decision maker's preferences into discrete decision alternatives.

MCA can help decision makers to choose between several alternatives by comparing the advantages and disadvantages of each alternative, one against the other, see e.g., [35]. The significant advantage of most MCA methods as stated by Annandale and Lantzke [2] is the capability to allow the evaluation criteria to be measured in either quantitative and/or qualitative terms, thus providing flexibility compared with other techniques such as cost-benefit analysis that require quantification of all values. Cost benefit analysis techniques are used, e.g., in economics to evaluate different alternatives, see e.g., [48, pp.106–114].

There are several MCA techniques in operation in various countries, see e.g., [39, 48]. Examples of these techniques include Analytical Hierarchy Process (AHP), Mathematical Programming (MP), Additive Weighting and Concordance Analysis presented, e.g., in [2, 46, 48]. Malczewski [43] conducts a survey of GIS-based multi-criteria decision analysis. A principled problem in choosing a decision aid method for a real-life problem is that, for the same data, different methods may produce different results [39]. This problem is further compounded by the difficulty of objectively identifying the best alternative or method in view of these differing results. In realization of this shortcoming, Lahdelma et al. [39] lists the requirements of MCA methods for use in environmental problems as:

1. Being well defined and easy to understand, particularly regarding the essential tasks such as setting of criteria and definition of weights.
2. Being able to support the necessary number of decision makers.
3. Being able to manage the necessary number of alternatives and criteria.
4. Being able to handle the inaccurate or uncertain criteria information.
5. Due to time and money constraints, the need of preference information from the decision makers should be as small as possible.

Clearly, it is difficult to have a method that satisfies all these requirements. All MCA methods have their strengths and weaknesses. The Additive weighting and Concordance analysis presented in the Example of Sect. 17.3.2.3 fulfil requirements 1, 2 and 3.

MCA does not actually provide an absolute answer by specifying a particular alternative, instead, it provides a process that ranks various alternatives and leaves the final decision to the policy makers. On the one hand, several studies indicate the success of MCA in ranking alternatives and therefore aiding in decision making, see e.g., [56]. On the other hand, researchers are still learning how it impacts on what could otherwise be an intuitive or ad-hoc group decision-making process [31]. As an example, Bojorquez-tapia et al. [10] report that some researchers have found out that MCA can alienate decision makers or experts in multi-stakeholder problems due to its complexity and 'black box' nature.

To address the shortcoming of alienating stakeholders, who most often comprise of the community (e.g., conservation groups and people likely to be directly affected by the project), CWP (community weighting process) in MCA is currently gaining momentum as a possible solution that attempts to carter for the community's interests. The increasing role played by CWP in environmental decision making with MCA as a processing tool is captured, e.g., by Hajikowicz [31] who states that the common reasons for applying MCA in multi-stakeholder decisions are to provide a transparent, structured, rigorous and objective evaluation of options.

Some examples of applications of MCA in EIA:

As already discussed, EIA processes involve several stages, see, e.g., in [25, pp. 88–184], and [48, p. 8] many of which may utilize MCA. At the screening stage, for example, where a project is assessed whether or not it requires EIA, MCA could be used, e.g., where one alternative location is to be chosen from several, see, e.g., [37]. The scoping stage of EIA is that of deciding, from all of a project's possible impacts, and from all the alternatives that could be addressed, which are the significant ones [25, p. 91]. Identification of significant alternatives requires comparison to be made at the scoping phase. Usually, at the initial phase of scoping, a number of alternatives are selected for further analysis, and in the final stage, a small number of alternatives are chosen and subjected to more thorough evaluation. An example is provided by the EIA performed for Ashton Mining Ltd, which required a selection of the best location for iron ore processing from six possible locations [4]. MCA could be used in such scenario during scoping stage. This example is discussed further in Sect. 17.3.2.3.

Evaluation in EIA looks mostly at the cost and benefits of a proposed project to users, assesses the impacts on environment, and compares various alternatives that will yield benefits to the project, while at the same time minimizes environmental and social impacts. MCA plays a vital role in evaluation in EIA as exemplified in the work of Janssen [35].

17.3.2.2 Decision Making and Alternatives

Steinemann [67] considers alternatives as means to accomplish ends, and that from the perspective of EIA; these ends include not just a particular agency's goals, but also broader societal goals such as the *protection* and *promotion* of environmental quality. Steinemann [67] further opines that developing the set of alternatives that become the choice set and the center of analyzes is the most important part of the EIA process. Decision makers can then chose from these choice sets rather than simply having to rubber stamp a proposal. However, two problems that confront the development of alternatives are cited by Steinemann [67]. *First*, the public involvement often occurs too late to influence the development of the alternatives, and *second*, the alternatives are frequently eliminated from further consideration based on weak evaluations, which are not well-documented in the environmental impact statements (EISs). The first problem is associated with the very nature of project based EIA where the outcomes are almost always predetermined. In contrast to the project based EIA, SEA (Sect. 17.4) and SA (Sect. 17.5) enable earlier participation of the public. In evaluating alternatives, decision making is often based on some selected *criteria* and the desired objectives. Criteria are aggregate values computed from a much larger amount of so-called primary factors, which form the lowest level of information, also known as the assessment level [39].

The problems with environmental decision making, however, are that they are intrinsically complex because they almost always involve many alternatives and multiple attributes (e.g., biological, economical, and social), the relative importance of which has to be determined by subjective evaluations [44]. In an effective EIA process, alternatives will be sought that attempt to balance the data set with multiple attributes. The balancing act becomes even more crucial in SEA or SA where the desire is to balance the diverse ecological, social, and economic values over space, time and scale. These values are usually represented in the form of multiple criteria and *indicators* that sometimes express conflicting management objectives [69].

In SEA or SA, complex projects are often involved which present many alternatives to choose from, necessitating the need for MCA for comparison. The situation is worsened when many stakeholders are involved and they conflict over the relative importance of the different comparison criteria. Annandale and Lantzke [2] state that "when decisions become this complex, there is a need for special tools or techniques to help in making sense of what can be a large amount of information". In addition, complex environmental planning problems will almost always include value judgements, public opinion, and controversies. So, the techniques need to deal with more than just technical information [2].

In such complex situations, MCA provides the means for comparing the advantages and disadvantages of each alternative, one against the other. By doing this, decision makers are provided with the means of choosing between several alternatives [35]. One of its advantages is that it permits public involvement in the process by allowing their voices to be heard through weighting of the criteria according to their preferences. Community weighting process (CWP) in MCA therefore leads to the community participating in decision making as already stated, and enhances public confidence in the final decision as opposed to where decisions are made using weak evaluation tools as already pointed out by Steinemann [67]. Its vital role is captured by Sheppard and Meitner [61] who state that "public involvement needs more effective, defensible techniques usable by managers at the sharp end of decision making, rather than just in the scoping of public concerns and in setting broad strategies".

Specification of alternatives: Alternatives are different ways of achieving an objective. For example, if the objective is to find a suitable waste dumping site, the alternatives would be the various possible locations that can serve as dumping sites at a minimal cost and minimize environmental and social impacts. In real life, there will be, almost always, people with vested interest(s) in these locations, thereby complicating the task of objectively identifying a suitable site. Specification of alternatives is helpful in such situations as they account for as many of the stakeholder opinions as possible. Annandale and Lantzke [2] suggest that the best approach in determining alternatives for a decision aiding exercise is the involvement of stakeholders and allowing them to offer as many alternatives as possible.

Specification of comparison criteria: In comparing alternatives, decision makers look for those alternatives that would be less costly in implementing but at the same time satisfy the environmental and social benefits. Criterion offers a possibility of comparing alternatives. Munier [48, p. 48] defines criteria as parameters used to evaluate the contribution of a project to meet the required objectives. Desirable properties for criteria are presented, e.g., in Annandale and Lantzke [2].

Scoring the alternatives: Annandale and Lantzke [2] discuss the three types of measurement scales; *ordinal*, *interval* and *ratio*. According to Annandale and

Lantzke [2] ordinal scales provide information on order only and are unsuitable for mathematical manipulations (addition, subtraction, multiplication and division). It can only indicate that one alternative scores higher than another alternative, but does not indicate by how much (i.e., magnitude). Ordinal scales favour qualitative attributes and are often used interchangeably with quantitative reserved for ratio or interval scales [2]. The interval scale indicates the difference between two alternatives without giving the actual magnitude. Its advantage over the ordinal scale is that it permits addition and subtraction only. The ratio scale has a natural origin (zero value) and provides a measure of both difference and magnitude [2]. It permits the mathematical operations and as such, favours scores obtained when the attributes are directly measured. Glasson et al. [25] suggests that scoring may use qualitative or quantitative scales according to the availability of information. Both qualitative and quantitative scales could be used simultaneously as demonstrated in Annandale and Lantzke [2].

Weighting the criteria: Commonly, in MCA methods, a number is assigned to each criterion describing its importance relative to other criteria. These numbers are called weights, and they model the decision maker's subjective preferences [39]. The interpretation of weights depends completely on the decision model used. Therefore, it is essential that the decision model be chosen prior to collecting weights, see e.g., [70]. The primary purpose of weighting the criteria is to develop a set of values which indicate the relative importance of each criterion as valued by the community. These values are then used in ranking algorithms to determine the relative value of each alternative [32].

There are several ways of assigning weights. For example, weights could be assigned directly by the individuals undertaking the analysis to represent hypothetical point of view, or they could be based on the data collected from opinion polls, focus groups, public meetings or workshops, or other direct forms of sampling public or expert opinion [2, 40]. Weights can also be assigned using some mathematical functions as indicated, e.g., in [48, p. 53]. This is therefore the part of MCA which takes into consideration divergent views of stakeholders on a project.

This is captured by Glasson et al. [25, p. 145] who states that MCA seeks to recognize plurality of views and their weights. Weights thus allow different views and their impacts on the final outcome to be expressed explicitly [2]. Several techniques for weighting are presented in literatures, e.g., direct assessment and pair-wise comparison methods such as AHP, see e.g., [57, 58]. In general, there exist no right weights that would allow comparisons between different alternatives. The weights obtained depend on the technique used [39].

17.3.2.3 Application of Geospatial in Support of MCA

The following example illustrates how geospatial could be used together with MCA to assist in the selection of alternatives for sitting of the secondary processing plant of a high-grade rare earth's deposit at Mt. Weld reported in Ashton Mining

Ltd [4, 5]. This example uses both ratio and ordinal scales to score the alternatives relative to the criteria.

Background of the Mt. Weld project: In 1991, a 2-year study program was undertaken by Ashton Mining Ltd [4] to determine the feasibility of commercial development of a high-grade rare earth's deposit at Mt. Weld, near Laverton in the Eastern Goldfields in Western Australia. The project was to involve the mining and beneficiation of ores at Mt. Weld and the secondary processing of rare earth concentrates to produce rare earth chemicals at a site that was to be determined [4]. The evaluation of the sites was undertaken in two stages. In the first stage, 15 sites assessed to have the potential for the sitting of the secondary processing plant were evaluated. These were [4]: Collie, East Rockingham, Esperance, Kalgoorlie, Karratha, Kemerton, Koolyanobbing, Kwinana, Moore River, Mt. Weld, Muchea, Geraldton, Picton, Pinjarra and Northam in Western Australia.

Ashton Mining Ltd [4] adopted *qualitative* and *semi-quantitative* approaches to compare each of the sites. The *semi-quantitative* method focused on the economic considerations, i.e., capital and operating costs, while the qualitative assessments included environmental considerations namely; public health, town planning, flora and fauna, and groundwater. It also included social considerations such as community infrastructure, availability of skilled labor, road and road-rail transport, and social acceptance.

Five appraisal categories adopted for each of the factors were; little or no constraint; manageable constraint, significant constraint, requiring detailed evaluation, and overriding constraint with the potential to preclude development. Out of the 15 sites, 6 (East Rockingharn, Collie, Kalgoorlie, Kemerton, Geraldton and Northam) were selected and subjected to further evaluation [4].

The results of the second evaluation stage indicated *Northam* as the preferred site. Five alternative sites in the Northam region were then evaluated, and the proposed Meenaar Industrial Park was assessed as being the site with the greatest potential [5]. Between road only and road-rail options considered for transporting the ore concentration, residues and chemicals, the road option was preferred. The proposal was then submitted for environmental impact assessment (EIA) and was subjected to a public environmental review (PER) in 1992, see [4].

Now, let us apply two multi-criteria analysis (MCA) methods (*Additive weighting* and *Concordance analysis*) together with geospatial to assist in the selection of alternatives and show that the same results, i.e., Northam could have been reached. Six alternative sites for the Mt. Weld EIA case study are evaluated using these MCA methods. For each of the 6 alternatives, 11 criteria were compared and scored using ratio and ordinal scales and processed.

Application of MCA:

Site Evaluation Criteria: In the site evaluation by Ashton Mining Ltd, a number of general and specific site requirements were identified and used to develop appropriate criteria which were applied to each site, see e.g., [4, pp. 24–25]. The site evaluation criteria considered were those most suitable for the establishment of the secondary processing plant. Ashton Mining Ltd [4] adopted *economic*, *environmental* and *social* criteria to evaluate the sites. In these criteria, which we discuss below, GNSS could play the role of providing site locations and the distances of various environmental features, e.g., groundwater source or community infrastructure from a given site.

The main economic criteria considered were to minimize the capital and operating cost to establish and operate the plant. Capital cost was needed for the construction of the secondary processing plant and to establish infrastructure (i.e., supplying power, water, natural gas and housing). Operating cost was to cover the cost of power, water, natural gas, land rates, transport of concentrates, residues, chemicals and products. All the assumptions made in calculating capital and operating costs are presented in Ashton Mining Ltd [4, p. 52].

The environmental and social criteria adopted were those which minimized a site's potential for [4, p. 24]: Off-site effects on the public and to public health; conflict with surrounding (and future) land use, impact on the existing flora and fauna, impact on high-quality groundwater resources or other significant components of the physical environment, and inefficient utilization of land.

Social criteria were those which would ensure a site [4, p. 25]; is close to established and well developed community infrastructure, is near a suitably sized labor force with appropriate skills, minimizes the disruption and risks to the public from the transportation of materials, and is likely to be acceptable to the public. The assumptions made in deriving the environmental and social criteria are presented in [4, p. 25]. Both the environmental and social factors can be modeled in a GIS environment with appropriate cost layers developed for the different factors.

The results of the example when MCA was applied indicated both additive and concordance methods ranked Northam as the top site followed by East Rockingham, and demonstrated the suitability of Concordance analysis for evaluating alternatives when the criteria are scored using mixed ratio and ordinal scales, thus underscoring the usefulness of MCA in assisting decision makers to chose between alternatives during the evaluation process of environmental impact assessment (EIA). Care should however be taken to know the limitations of each method (e.g., Additive weighting), use proper weights, and agreeable threshold.

17.3.3 Example of Gnangara Mound Groundwater Resources

During 1992–1995, a review was undertaken in Western Australia on the proposed changes to environmental conditions of Gnangara Mound groundwater resources under Sect. 46 of the *Environmental Protection Act (EPA)* 1986 (WA). Using it as an example, a theoretical examination of the possible areas of EIA process that could have benefited from using geospatial is presented.

17.3.3.1 Background

The Gnangara Mound is Western Australia's largest source of groundwater, supplying up to 60 % of Perth's drinking water [6, 14]. Its area is estimated to be 2,356 km^2 and comprises Gnangara, Yanchep, Wanneroo, Mirrabooka, Gwelup, Perth and Swan Groundwater Management Units (GMUs). Gnangara Mound supports local wetlands and lake ecosystems and supplies irrigation for horticulture and agriculture [6]. It is also a major water source supporting a number of groundwater abstraction schemes operated by the Water Authority [71]. It is bounded to the north by Gingin Brook and Moore river, to the East by Ellen Brook, to the south by Swan River, and Indian Ocean to the West.[1]

Physical environment: Gnangara Mound is characterized by hot dry summers and mild wet winters with an average annual rainfall of about 800 mm [71]. Department of Water [15] gives an average annual value of 814 mm. The hottest month of the year is reported as February with an average maximum temperature of 34°, while August is the coldest month with an average maximum of 18° [71]. Water Authority [71] state that the area does not have natural surface runoff due to the porous nature of soil in the area. Most of the water that falls as rainfall recharges the groundwater and that any surface water is due to discharge from groundwater. Recharge of groundwater depends largely on rainfall pattern, vegetation cover, and the water table.

Groundwater flows westerly from the top of the Mound following the terrain slope. Wetlands are generally found in the low areas where the water table reaches above the ground surface much of the year. Due to the presence and absence of water above the ground in these wetlands, soil and vegetation have adopted to the pattern of groundwater. In general, groundwater quality is reported to be excellent [71]. It is however widely recognized that sustainability of the Mound as a water resource is under threat due to *climate change* and excessive drawing of water.

Biological environment: Water Authority [71] reported the dominant terrestrial vegetation as the candle Banksia (*Banksia attenuate*) and firewood Banksia (*B. Mennziesii*). Vegetation of high significant conservation value was also reported in the area [71]. Vegetation, soils and land forms of Gnangara have been mapped, e.g., in [42]. Fauna survey of 1977 and 1978 recorded 12 native mammals, 70 reptiles and amphibians and 223 bird species. Five caves out of the 273 documented caves in the Yanchep National Park were reported to be the most species rich subterranean ecosystem ever recorded, supporting 30 and 40 species compared to caves elsewhere in the world, which rarely have five animal species [71]. GNSS could be useful in providing the locations of these five caves.

Social environment: Water Authority [71] reported a general increase in urbanization in the Gnangara Mound area that led to incremental approach to planning, subsequently having significant implications for the future of the area. Increase in urbanization comes along with changes in land use, which in turn impacts on the groundwater level. In the rural areas, common land uses reported at the time included market gardening and poultry farming. Specialized activities included

---

[1] See, e.g., http://www.water.wa.gov.au/sites/gss/ggs.html.

flower, mushroom, and strawberry growing, and gourmet pheasant production, all of which required groundwater. Large areas of Gnangara Mound are State Forest under the management of Conservation and Land Management (CALM). Approximately 20,000 ha of this land was Pine plantation with the remainder of the State Forest being natural bushland [71].

Water Authority (ibid) further reported 14 archaeological sites registered with the Western Australian Museum. Specifically, McNess, Lake Mariginiup, Lake Joondalup, Lake Goollelal, and Lake Gnangara among others were said to be sites of Aboriginal mythology and/or historical Aboriginal use. According to Water Authority [71], it was also likely that most of the wetlands in the western linear wetland chain are potential areas of Aboriginal significance.

17.3.3.2 Review of Allocation and Management of Groundwater Resource

Under the guidance of the Environmental Protection Authority (EPA), Water Authority manages groundwater resources of the Mound. Private groundwater abstraction is managed through area allocation and licensing of users [71]. Water Resource Authority, therefore, has the task of ensuring that the environmental impacts from users and its own activities are minimized. This is achieved, e.g., through assessing the impacts of proposed land use changes on groundwater levels and in providing advice to land management and planning organizations. In 1986, Water Authority submitted the Gnangara Mound Water Resources Environmental Review and Management Program (ERMP) to Environmental Protection Authority (EPA) for;

(1) approval to develop the Pinjar Groundwater Scheme, and
(2) approval for changes to private groundwater allocations.

In 1988, the Minister of Environment approved development of Pinjar Stage 1 Groundwater Scheme and the changed private groundwater allocation quotas, subject to a number of environmental conditions [71]. The approval allowed for increased abstraction of groundwater by the Water Authority and other users. The conditions to be met included measures to protect the environment through; *Maintenance of water level in the wetlands*, *limits on private groundwater allocations*, *establishment of a management and monitoring program*, and *setting in place a range of administrative mechanism regarding inter agency interaction on groundwater management* [71]. In 1992, Water Authority identified the need to review the management of the southern portion of the Mound. Factors which necessitated the requirement for the review were [71]:

- Identification of other ecosystems, which had been, or had the potential of being affected by groundwater abstraction. These included shallow cave streams and phreatophytic vegetation.
- Rapid increase in knowledge of Environmental Water Requirements (EWR), which suggested that water levels set by EPA in 1988 should be reviewed.
- Increase in demand of groundwater by private users called for an assessment of the potential impacts that would result from further groundwater allocation.

- There was a need by Water Authority to further develop groundwater schemes (e.g., Pinjar Stages 2 and 3) on the Gnangara Mound and as such, a review of allocation and management was essential before development of the schemes could commence.
- The recognition that land use on the Mound could significantly affect groundwater availability required that the impacts of likely future land use scenarios be considered in allocating and managing groundwater.
- Since the outcome of the review was likely to involve changes in some of the environmental conditions which applied to the management of Gnangara Mound, notably wetland water levels, allocation quotas and land use issues, consideration of any changes required the review to take the form of Environmental Impact Assessment (EIA).

A formal referral was submitted by Water Authority to the EPA[2] in late 1992, and it was decided that the conditions should be reviewed under Section 46 of the *Environmental Protection Act (EPA)* 1986 (WA). This then led to the review of allocation and management of groundwater resource by Water Authority. EPA guidelines for this review are presented in Water Authority [71, Appendix 2].

Focus of the environmental review: Environmental conditions reviewed focused on three main areas; *wetland water level*, *allocation quotas* and *land use issues*. EPA acknowledged that little information was available to determine EWR[3] for the wetlands and that there maybe changes to the set levels in future and required the initiation of research to provide an improved understanding of wetland ecology, which could then be used as a basis to review the wetland water level criteria.

With the continuing urban development in the Wanneroo region, evolving patterns of land use led to considerable changes in the pattern for demand for private water. Water demand in some areas, e.g., Flynn Drive could not be met. There was need to review groundwater availability with the view of allocating further resources to high demand areas. This could be achieved through further development of groundwater schemes within Gnangara Mound comprising Pinjar Stage 2 Part 1 groundwater scheme, which was scheduled for December 1996 and had been approved by EPA subject to the outcome of the allocation and management review.

Since also allocation of Pinjar Stage 2 Part 2 and Stage 3 groundwater schemes were being sought at the time, Water Authority believed in reviewing the allocation and management of water resource before further development of groundwater schemes so as to ensure equitable distribution between the public water supply and private use while minimizing environmental impacts.

17.3.3.3 Possible Areas of Geospatial Support to the Gnangara EIA

Impacts identification: Water Authority adopted the checklist method in identifying the impacts. This is the common procedure used in Western Australia where the

---

[2] Environmental Protection Authority.

[3] Environmental water requirement.

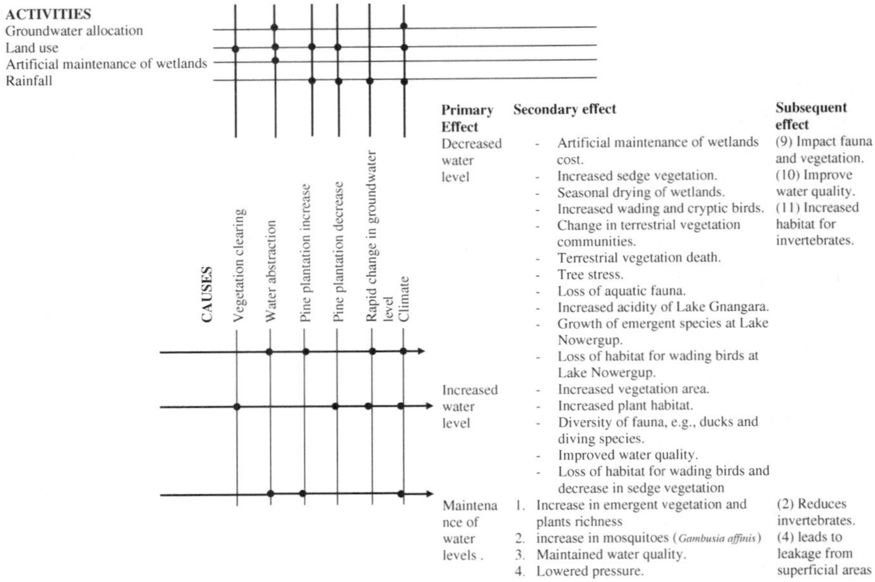

Fig. 17.2 Sorensen network for Gnangara Mound impact identification

proponent is required to complete a referral form [15]. In what follows, a network approach based on Sorensen [65] and a GIS method are compared in order to demonstrate how geospatial could have been useful in enhancing impact identification.

The Sorensen Network Approach: Recreating the impacts in Water Authority [71], first the activities to be undertaken in the Gnangara Mound review are specified. In this case, three major activities are identified from Water Authority [71] as; *groundwater allocation, land use*, and *artificial maintenance of wetland's water levels*. Let us add rainfall to this list as a climate variable that has the potential of impacting on the water level. The causes of environmental changes associated with the activities above are then identified and a matrix format applied to trace its impact. In Fig. 17.2, use is made of the Sorensen [65] principles to identify the impacts.

For instance, land use activity potentially results in clearing of vegetation, water abstraction, increase and decrease in density of pine plants, and climate change. These environmental changes, in turn, results in increased water level (e.g., when vegetation is cleared) as the primary effect. Increase in water level in turn leads to increased wetland vegetation areas and plant habitats, diversity of fauna, e.g., ducks and diving species, and improved water quality all of which are positive secondary impacts. A negative secondary impact is the loss of habitat for wading bird species and sedge vegetation that rely on seasonal drying of the wetlands.

Water abstraction will occur when land use activities involve irrigation of farms, maintenance of golf course, and other uses which require water. This in turn leads to low wetland water levels as primary effects. Secondary effects as a result of the primary impact are presented in Fig. 17.2. All the potential impacts of land use can be traced in a similar way as demonstrated in Fig. 17.2. As demonstrated for the case of land use activity, the Sorensen network is used to identify the primary, secondary and subsequent impacts associated with the allocation of groundwater, artificial maintenance of wetlands and rainfall. Figure 17.2 summarizes the identified impacts using this method. It is evident that the identified impacts compare well with those reported in [71].

GIS Approach: For identification of environmental impacts having spatial distribution in nature, GIS with the assistance of GNSS satellites is an ideal tool. The potential of GIS in environmental impact assessments has been demonstrated, e.g., by Antunes et al. [3] who applied it to evaluate the impacts of a proposed highway in Central Portugal. Antunes et al. [3] suggested identification of environmental components (e.g., ecosystem) and receptors (e.g., a particular species likely to be affected by the component) using GIS. Another example of application of GIS to EIA is presented by Haklay et al. [33] who advances a GIS-based scoping method and discusses the conditions necessary for its utilization.

For the Gnangara Mound example, the environmental components that were likely to be affected by groundwater allocation and management were *pine trees*, *vegetation* and *wetlands*. Using Gnangara Mound Map of 1987 as a base for example, annual map layers of pine trees, vegetation, wetlands and urbanization can be overlaid on the base map in a GIS environment to produce a composite map, which can be used to identify hot spots (areas where land use are clearly identified to impact on wetland water levels and wetland vegetation). In this example, 1987 is selected as a base since environmental conditions issued by the Minister became operational in 1988. Annual groundwater level for specific wetlands are entered as attributes or produced in maps as contours. From the hot spots, areas and contours indicating water level changes and potential impacts can be identified. Where there is intense land use and sharp reduction in wetlands vegetation area, that specific land use could be said to impact on wetland water level, and subsequently vegetation. Linear trends can also be obtained on, e.g., the rate of pine growth/decline, vegetation clearing and urbanization by comparing annual values from 1997 to 1995. These could then be correlated with the groundwater levels to further identify the impacts. Negative linear trends will indicate adverse impact, while positive trend will indicate positive impact. Besides the trend analysis, visual examination of the layers could also indicate the spatial distribution. In this method, GNSS satellites provide location-based data to which the attributes, such as impacts on wetlands, are related.

Compared to the Sorensen method, the GIS approach has the advantage of being able to identify pertinent environmental effects on the basis of readily available information under stringent time and budget constraints [33]. Since it is best suited for *spatially distributed impacts*, it can analyze cumulative impacts better than the checklist or Sorensen network approach. It also provides friendly visual presentations, which are easily understandable by non-experts. Its drawbacks, however, are that it does not consider the likelihood of an impact, secondary impacts of the difference between reversible and irreversible effects [25], and that it may require initial capital to establish.

Impacts prediction: Impact prediction requires that it be based on available environmental baseline data. In this example, the baseline data were readily available since regular water level monitoring had been taking place as part of the initial Ministerial conditions set out in 1988. Geospatial could have supported impact prediction of this EIA in several ways namely:

1. Provision of baseline data, and
2. Mapping boundaries of changing spatial features, e.g., wetland boundary changes as illustrated in Fig. 17.3 (see, e.g., [8]).
3. Visualization of various "what if scenarios" that are likely to result from some prior conditions being fulfilled, e.g., understanding impact of drought. This can be implemented using GIS modeling techniques.

By having permanent reference marks set around the wells, GNSS satellite could be used to provide continuous measurements of positions and elevations of these reference marks. The measured depths of the wells could then be referred to these reference marks and thus help in monitoring the state of the groundwater levels (Fig. 17.3). This can be related to the state of vegetation and fauna. To predict the impacts of groundwater abstraction on wetlands and other vegetation, similar techniques to those discussed in [7] could be used to document changes in the wetlands' boundaries (i.e., perimeters and areas), as illustrated in Fig. 17.3 (for years 1 and 2). By analyzing annual trend of these boundary changes, it is possible to predict the impacts of groundwater abstraction on e.g., wetlands, assuming that the changes are unrelated, e.g., to evaporation. This will require some control location (see e.g., discussions on BACI model in Sect. 17.2).

Impacts of groundwater variation on terrestrial vegetation were reported by Mattiske [45] as ranging from small change in community structure in favour of more drought tolerant species, through to deaths of Banksia woodland vegetation. Indeed, that the death of Banksia vegetation were triggered by groundwater variation was supported by the findings of Water Authority [72], which suggested that Banksia trees that occurred where depth to groundwater was less than 6 m were most vulnerable to groundwater reduction. The study further suggested a general stress on vegetation due to reduced groundwater level.

Groom et al. [28] deduced that a lowering of groundwater level by 2.2 m at a station P50 between the summers of 1990 and 1991, resulting from the cumulative effects of abstraction and below average annual rainfall (low groundwater recharge), coincided with a loss of between 20 and 80 % of mature Banksia species within 200 m of the bore. Over a similar time period, no significant decreases in the abundance of species were recorded in the monitored site to have been influenced by groundwater abstraction. They concluded that negative impact of groundwater draw-down on Banksia populations made it an important indicator of decreasing groundwater levels on the Gnangara groundwater Mound. GNSS monitoring of groundwater abstraction (Fig. 17.3) therefore could be useful in predicting impact on Banksia trees if the proposed changes in environmental condition would impact on groundwater by similar level, i.e., lowering of groundwater level by more than 2.2 m.

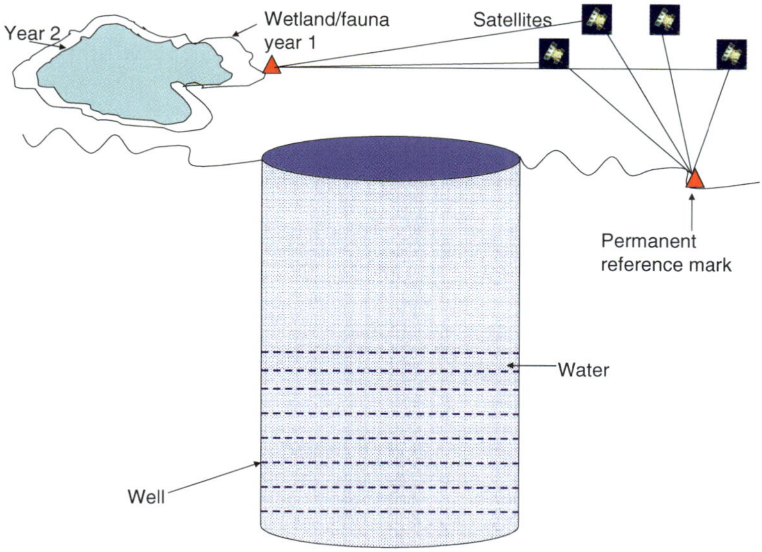

Fig. 17.3 GNSS monitoring of the impacts of well water abstraction using, e.g., techniques discussed in [7] spatial changes maps of wetlands from years 1 to 2. Also, GNSS provides reference points upon which measured groundwater levels can be referenced

Geneletti [22] demonstrates the capability of using GIS method to compute spatial indicators to predict and quantify critical impacts, such as ecosystem loss and fragmentation, soil erosion, geomorphologic hazards, interference with flora and fauna, and visibility. Since GIS has successfully been used by Geneletti [22], it could be applied together with GNSS to predict variation in spatial distribution of environmental components caused by groundwater level variation.

Once groundwater changes have been obtained using, e.g., piezometric readings, and boundaries of impacted features (e.g., wetland in Fig. 17.3) mapped using GNSS, overlaying the land use and vegetation cover maps could then be performed using GIS for the same time period. Correlation between the land use, terrestrial vegetation and groundwater level could then be developed and predictions made on the impact of land use and terrestrial vegetation on groundwater level, and impacts of groundwater level on wetland vegetation. Linear and cyclic trends analysis could then be developed to give predictions at various temporal resolutions.

To support the prediction of impact of groundwater abstraction on fauna in caves, hand-held GNSS receivers can be used to provide locations of these caves, which can be related to groundwater level. Jasinska and Knott [36] listed about 100 caves in Yanchep National Park and reported that little information was known about the biology of the aquatic fauna within these caves. They found aquatic species of high conservation value and concluded that one of the greatest threats to these species would be the permanent or temporary drying of the caves streams in which they occur. Since these aquatic fauna are of high conservation value, they could be seriously affected by drying of the streams within the cave to a point of extinction. GNSS could be useful in the prediction of the effect of regional warming on groundwater through the analysis of the GNSS derived tropopause heights as discussed in [7].

Comparison of the prediction methods: The model based approach adopted by the Water Authority [71] is the most commonly used method in most EIA of groundwater impacts. Models rely on the input data and the assumptions that are taken into consideration. The more they fit in the model, the more reliable are the output. The disadvantage of using models, however, is that they require some expert knowledge during their development and operation stages. Any wrong assumptions, input data, and usage can lead to false information and interpretation.

The Field experiment using GNSS and GIS has the advantage of using real data in their predictions as opposed to simulated values as is the case of models. They also provide easy visual interpretation of the results. The disadvantage is that it comes at a cost. The initial cost of installing a GIS may be high. Besides, there is the cost of validating the data using GNSS. Another disadvantage is the incapability to predict higher order impacts. In the Gnangara Mound example, it was difficult to use GNSS and GIS to predict the impacts of variation of wetland vegetation to fauna which may require other methods for enhancement.

In summarizing this example, in identifying and predicting the impacts for Gnangara Mound using alternative GIS-based methods to those adopted in the Water Authority [71], it has been pointed out that identification and prediction models are labor intensive and require knowledge of the system. More often, they are based on assumptions which may not fit the model leading to delivery of meaningless results. Field experiments, though straight forward, requires some validation which may increase the cost of EIA. Finally, baseline environmental parameters should be established upon which judgment of an impact can be made. Identification method should be a combination of methods that are simple to use, but which are capable of identifying higher order impacts and their inter-relations. Where models are adopted, they should be well understood by the analyst and assumptions must be clear and meaningful.

In particular, when used in conjunction with GIS and field data from the GNSS, a suitable approach for identifying and predicting impacts, which are spatially distributed could be obtained. This example indicates the possibilities of geospatial techniques to support identification and prediction of environmental impacts associated with the proposed change in environmental conditions of Gnangara Mound and highlight the limitations of the methods.

17.4 Strategic Environmental Assessment

Strategic environmental assessment (SEA) is the process that aims at integrating environmental and sustainability considerations in strategic decision-making [68]. In so doing, the goal is to protect the environment and promote sustainability. Sadler and Verheem [59] define SEA as a systematic process for evaluating the environmental consequences of a proposed policy, plan or program initiative in order to ensure that they are fully included and appropriately addressed at the earliest appropriate stage of decision making at par with economic and social considerations. Wood and Djeddour [74] define a policy as inspirational and guidance for action, a plan as a set of coordinated and timed objectives for the implementation of the policy, and a programme as a set of projects in a given area, see also [68, p.12].

The basic principles of SEA have been presented, e.g., by Therivel [68] as being a tool for improving the strategic actions, promoting participation of stakeholders in decision making process, focusing on key environmental/sustainability constraints, identifying the best option; minimizing negative impacts, optimizing positive ones, and compensating for the loss of valuable features and benefits; and ensuring that strategic actions do not exceed limits beyond which irreversible damage from impacts may occur. Its advantages include [68]:

(1) Being able to shape the projects at an earlier stage through the appraisal of strategic action. This offers the chance to influence the kinds of projects that are going to happen, not just the details after the projects are already being considered.
(2) SEA deals with impacts that are difficult to consider at project level. It deals with *cumulative* and *synergistic* impacts of multiple projects, e.g., cumulative impacts of various mining sites on the development of an entire area.
(3) SEA can deal with large-scale environmental impact such as those of biodiversity or global warming more effectively than individual EIA.
(4) Unlike project based EIA, which formulate goals around an already selected approach, SEA promotes better consideration of alternatives, thereby ensuring a strategic approach to action.
(5) It incorporates environmental and sustainability consideration in decision making thus adding an additional dimension to decision making.
(6) It enables public participation in decision making thus making the whole process inclusive and transparent.
(7) It has the potential to promote streamlined decision making.

SEA has also benefited from MCA as illustrated by Noble [51] who presents five scenarios that were evaluated within SEA to determine the most suitable option for power generation to be developed to cover the Canadian need up to the year 2050. The role and contribution of geospatial in supporting global warming monitoring is treated in [7]. In what follows, the cumulative impact aspect of SEA and the possible role and contribution of geospatial is discussed.

17.4.1 Geospatial and Cumulative Impacts Assessments

Cumulative effects refer to the phenomenon of *temporal* and *spatial* accumulation of change in environmental systems in an additive or interactive manner and may originate from either an individual activity that recurs with time and is spatially dispersed, or *multiple activities* (independent or related) with sufficient spatial and temporal linkage for accumulation to result [66]. The attributes of cumulative effects are classified by Spaling and Smit [66] into three categories; temporal accumulation, which occurs if the interval between perturbation is less than the time required for an environmental system to recover from each perturbation, *spatial accumulation, which results where spatial proximity between perturbation is smaller than the distance required to remove or disperse each perturbation*, and the nature of human induced activities or perturbations, which also affect accumulation of environmental change provided the perturbations are sufficiently linked in time, space and scale.

Cumulative impact assessment is thus defined by the Commonwealth Environmental Protection Agency [12] as predicting and assessing all other likely existing, past and reasonable foreseeable future effects on the environment arising from perturbations. In some legislations, e.g., in Canada, EIA regime has made it specific and mandatory, where consideration of cumulative effects assessment has been made explicit and mandatory both federally and in several provinces [25]. In USA, *National Environmental Policy Act* (1967) requires the assessment of cumulative impacts, while in Australia, assessments have largely been carried out by regulatory authorities, rather than project proponents. In Western Australia for example, the EIA process does not come out forcefully on cumulative impacts assessment.

Spaling and Smit [66] provide an in-depth look at the contributions and shortcomings of EIA to assessing cumulative impacts. Three key factors in favor of EIA are theoretical understanding of environmental change through empirical analysis and modeling of responses of environmental systems through human induced perturbations, the development of various analysis methods for projecting and assessing the various environmental changes associated with the proposed human activities, and regulatory and administrative mechanism contributed by EIA in the integration of environmental consideration in decision making. In EIA, cumulative impacts can be identified at the *scoping stage* where issues to be examined are pruned. It is at this stage where the spatial and temporal effects of cumulative impacts can be considered [60].

Geospatial could be useful in *providing the locations of multiple activities*, *mapping the changes in spatial coverage*, and *monitoring variation in groundwater* as a result of cumulative impacts.

17.4.2 Example of Marillana Creek (Yandi) Mine

Background: Marillana Creek (Yandi) Mine operated by BHP Billiton Ore Pty Ltd (BHPBIO) is located approximately 90 km north-west of Newman in the Pilbara region of Western Australia [9]. The mine is situated within lease ML 270SA, and is operated under the Iron Ore (Marilliana Creek) Agreement Act 1991 [9]. BHPBIO also has a smaller lease (M 47/292) located to the immediate north of ML 270SA. The Yandi ore body occurs within an ancient channel iron deposit (CID). This deposit is subdivided into a series of mine areas, i.e., central pits (C1 to C5), eastern pits (E1 to E8), and the western mesa pits (W1 to W6) [9]. The CID is about 80 m thick and the majority of mining is within the upper 60 m. BHPBIO (referred to as proponent in this example) operates dewatering bores that lower the water table in the vicinity of each pit by approximately 30 m [9].

In May 1988, an approval was granted by the Minister for Environment to mine E2 and C5 at a rate of 5 million tonnes per annum [20] and in 1991, mining commenced. In 1992, 1994 and 1995, EPA assessed modifications to the original proposal, which involved increased rates of production and mining of additional pits [17–19]. At the time of application for approval by the proponents, mining was taking place in the E2, C1/C2 and C5 areas. In 2004, the proponent sought approval under Part IV of the *Environmental Protection Act* (WA) 1986 to concurrently mine from pits across the leases (ML 270SA and M 47/292), and in addition update, assess, and agree on closure concepts for the whole of the deposit. During the mining of individual pits, the proponents proposed to partially fill the voids with overburden (waste) material from other pits, and to use the same open cut mining techniques and ore processing methods over the remaining life of the mine [9]. The agreed concepts were to be documented through closure specific conditions that were issued by the Minister of Environment. The project was called the Marillana Creek (Yandi) Life of Mine and was expected to deliver 40 million tonnes per annum with a lifespan of 30 years [9].

The expansion of delivery capacity from an initial 5 million tonnes per annum to 40 million tonnes per annum resulting from concurrent pit mining had the potential to *significantly impact* on the environment. Besides, Hamersley Iron also holds a mining lease (274SA) over the CID, east of BHPBIO's lease, and was mining up to 34 million tonnes per annum at the same time. The potential impact on environment, therefore, was not only likely to come from the proposed project but also *cumulative* taking Hamersley into consideration. Thus, there existed a need for EIA under Part IV of the *Environmental Protection Act (EPA) 1986* (WA).

BHPBIO [9] produced an EIS[4] that documented the environmental objectives, potential impacts, proposed environmental management measures and predicted outcomes. Environmental Management Plan and Decommissioning and Final Rehabilitation Plan were also presented as key supportive documents to the Environmental Protection Statement (EPS). EPA was advised of the proposal in January 2004 and based on the information provided, considered that the proposal had the potential to impact on the environment, but could be managed to meet the EPA's environmental objectives [16]. Consequently, EPA determined, under section 40(1) of the *EPA* 1986 (WA), that the level of assessment for the proposal was EPS.[5] EPA's advice and recommendations were then forwarded to the Minister of Environment in accordance with section 44(1) of *EPA* 1986 (WA) [16]. The Minister of Environment granted approval with a set of conditions on 6th of July 2005.

Cumulative impacts: Since the EPS proposed to concurrently mine the pits within the leases as opposed to the previous pit by pit mining, there existed a potential to lead to *cumulative impacts* as discussed in Spaling and Smit [66]. Cumulative

---

[4] environmental impact statements (EIA).

[5] Environmental protection statement.

impacts were likely to be felt on the surface and groundwater. As part of the effort to manage cumulative effects accrued from surface water, the proponent proposed to integrate the surface water monitoring program to a wider monitoring initiative in the Marillana Creek catchment. This was to be achieved by adding flow gauge stations on the Marillana Creek and its tributaries within, upstream, and downstream of ML 270SA and 47/292 [9].

On groundwater resource, BHPBIO [9] identified the potential of cumulative impacts given the presence of Hamersly Iron operation in the neighborhood. The proponents took into account the impacts in their regional groundwater model and de-watering license. Both surface and groundwater monitoring to be undertaken by the proponents during mining was expected to provide a mechanism for monitoring cumulative impacts [9]. GNSS could play a role in providing location-based information on test sites (flow gauge station) and also provide perimeter/area information that could help in monitoring of the cumulative impacts in the entire mining region (e.g., Fig. 17.3). This information could be integrated with a GIS to support management decisions.

17.5 Sustainability Assessment

Sustainability has been defined as meeting the needs of current and future generations through integration of environmental protection, social advancement and economic prosperity [27]. Sustainability assessment (SA) can be performed when a proponent requests a regulator to do so (external) for the purpose of approval or internally as a mechanism for improving internal decision-making and the overall sustainability of the final proposal, see e.g., [52, 53].

Pope et al. [54, 55] classify SA into objective-led (strategic) and EIA based (narrow) approaches. Morrison-Saunders and Therivel [47] rank the various SA approaches with the EIA-led approach on bottom and the integrated (objective-led) approach at the top, see also [29]. Between them are various approaches, e.g., the win-win-win. Citing the dangers inherent in using the separate findings of the three sustainability pillars (environment, social, and economics) at the decision stage, Gibson [24] proposes adoption of an integrated approach. Caution should, however, be observed when using the term "integration" as it is used variedly by different authors, see e.g., [41, 47].

SA involves (i) *Sustainable decision making protocol*. A sustainable decision making protocol is a process of setting objectives, criteria and targets that underpin SA. Hacking and Guthrie [30] present several sources of sustainability development objectives and propose the use of threshold as one of the means, (ii) *alternative approaches*, which are options, choices, or courses of action. They are means to accomplish particular goals [67]. Alternatives have been shown to be affected by the formulation of the decision question. SA, similar to SEA, has also benefited from MCA. In the province of Reggio Emilia (Northern Italy), Ferrarini et al. [21] used MCA to rank 45 municipalities based on 25 state of the environment indicators.

Their results provided information on the state of sustainability in the province as a whole. GIS could support SA in choosing the best alternative as already discussed in Sect. 17.3.2.2.

17.6 Concluding Remarks

The use of GIS to support impacts' assessment is still developing and certainly that of GNSS is a new concept. This chapter attempted to motivate EIA, SEA and SA experts dealing with impact assessments to exploit the full potential of GNSS, especially with regard to its superb provision of location-based information and measurement of spatial variations. GNSS satellites could for example be used to provide information related to distances, e.g., distances of the alternative sites from established and well developed community infrastructures. The distances can readily be computed and incorporated into the MCA criteria and used to compute the desirable alternative that will inform decision making. GNSS could also be useful in providing information on spatial coverage of the proposed sites and also in environmental audit to evaluate compliance where location and spatial variation data are required. GNSS data together with remote sensing data can then be integrated in GIS to support the EIA process as discussed in the chapter.

References

1. Al-Rashdan D, Al-Kloub B, Dean A, Al-Shemmeri T (1999) Environmental impact assessment and ranking the environmental projects in Jordan. Eur J Oper Res 118:30–45. doi:10.1016/S0377-2217(97)00079-9
2. Annandale D, Lantzke R (2000) Making good decisions: a guide to using decision-aiding techniques in waste facility siting. Institute for Environmental Science, Murdoch University, Perth
3. Antunes P, Santos R, Jorda OL (2001) The application of geographical information systems to determine environmental impact significance. Environ Impact Assess Rev 21:511–535
4. Ashton Mining Ltd. (1991) Mt Weld rare earths project: Sites evaluation study for a rare earths Secondary Processing Plant, report prepared by Kinhill Engineers
5. Ashton Mining Ltd. (1992) Mt Weld rare earths project: Public environmental review, report prepared by Kinhill Engineers
6. Australian Water Resource (2005) Combined water management area: Gnangara Mound. http://www.water.gov.au/RegionalWaterResourcesAssessments/SpecificGeographicRegion/TabbedReports.aspx?PID=WA_GW_51x. (Accessed on 8/03/2008)
7. Awange JL (2012) Environmental monitoring using GNSS. Springer, Berlin, 354 pp
8. Awange JL, Kiema JBK (2012) Environmental geoinformatics. Monitoring and management. Springer, Berlin, 354 pp
9. BHP Billiton Iron Ore Pty Ltd (2005) Marillana Creek (Yandi) Mine - Environmental Management Plan, Revision 3
10. Bojorquez-tapia LA, Sanchez-colon S, Martinez AF (2005) Building consensus in environmental impact assessment through multicriteria modeling and sensitivity analysis. Environ Manage 36:469–481

11. Carver SJ (1991) Integrating multi-criteria evaluation with geographical information systems. Int J Geogr Inform Syst 5(3):321–339. doi:10.1080/02693799108927858
12. Commonwealth Environmental Protection Agency (1994) Review of Commonwealth environmental impact assessment. Canberra: CEPA 10
13. Downes BJ, Barmuta LA, Fairweather PG, Faith DP, Keough MJ, Lake PS, Mapstone BD, Quinn GP (2002) Monitoring ecological impacts: concepts and practise in flowing waters. Cambridge University Press, Cambridge
14. Department of Water (2008) Gnangara Mound—a unique water resource. http://portal.water.wa.gov.au/portal/page/portal/WaterManagement/Groundwater/Gnangara/ (As of 14/03/08)
15. Department of Water (2007) Environmental Management of Groundwater Abstraction from the Gnangara roundwater Mound 2004–05, annual compliance report to the Environmental Protection Authority, July 2004 to June 2005
16. EPA (Environmental Protection Authority) (2005) Marillana Creek (Yandi) Mine Life of Mine proposal, mining leases 270SA and 47/292, 90 km north-west of Newman. Report and recommendation of the EPA, assessment no 1555, Bulletin 1166. Perth, WA
17. EPA (1992) Yandicoogina (Marillana) iron ore project—change of ministerial condition due to increase in rate of production. Report and Recommendations of the Environmental Protection Authority, Bulletin 622. Environmental Protection Authority, Perth, Western Australia
18. EPA (1994) Yandicoogina (Marillana) iron ore project—change of environmental conditions due to increase in rate of production. Report and Recommendations of the Environmental Protection Authority, Bulletin 738. Environmental Protection Authority, Perth, Western Australia
19. EPA (1995) Duplication of iron ore mining operation, Yandi mine ML 270SA, Hamersley Range, 90 km north west of Newman. Report and Recommendations of the Environmental Protection Authority, Bulletin 802. Environmental Protection Authority, Perth, Western Australia
20. EPA (1988) Yandicoogina (Marillana) iron ore project. Report and Recommendations of the Environmental Protection Authority, Bulletin 323. Environmental Protection Authority, Perth, Western Australia
21. Ferrarini A, Bodini A, Becchi M (2001) Environmental quality and sustainability in the province of Reggio Emilia (Italy): using multi-criteria analysis to assess and compare municipal performance. J Environ Manage 63:117–131. doi:10.1006/jema.2001.0465
22. Geneletti D (2007) An approach based on spatial multicriteria analysis to map the nature conservation value of agricultural land. J Environ Manage 83:228–235
23. Gibson J, MacKenzie D (2007) Using global positioning systems in household surveys for better economics and better policy. World Bank Res Obs 22(2):217–241. doi:10.1093/wbro/lkm009
24. Gibson R (2006) Beyond the pillars: sustainability assessment as a framework for effective integration of social, economic and ecological considerations in significant decision-making. J Environ Assess Policy Manage 8(3):259–280. doi:10.1142/S1464333206002517
25. Glasson J, Therivel R, Chadwick A (2005) Introduction to environmental impact assessment, 3rd edn. Routledge, New York
26. Gough JD, Ward JC (1996) Environmental decision making and land management. J Environ Manage 48(1):1–15
27. Government of Western Australia (2003) Hope for the Future: The Western Australian State Sustainability Strategy. Department of the Premier and Cabinet Perth, Western Australia.http://www.sustainability.dpc.wa.gov.au/docs/Final%20Strategy/SSSFinal.pdf (Accessed on 2/03/2008)
28. Groom BPK, Froend RH, Mattiske EM (2000) Impact of groundwater abstraction on woodland, Swan Coastal Plain, WA. Ecol Manage Restor 1:117–124
29. Hacking T, Guthrie P (2008) A framework for clarifying the meaning of triple bottom-line, integrated, and sustainability assessment. Environ Impact Assess Rev 28(2–3):73–89. doi:10.1016/j.eiar.2007.03.002
30. Hacking T, Guthrie P (2006) Sustainable development objectives in impact assessment: why are they needed and where do they come from? J Environ Assess Policy Manage 8(3):341–371. doi:10.1142/S1464333206002554

31. Hajikowicz SA (2007) Supporting multi-stakeholder environmental decisions. J Environ Manage. doi:10.1016/j.jenvman.2007.03.020
32. Hajikowicz SA, McDonald GT, Smith PN (2000) An evaluation of multiple objective decision support weighting technique in natural resource management. J Environ Planning Manage 43(4):505–518
33. Haklay M, Feitelson E, Doytsher Y (1998) The potential of a GIS-based scoping system: An isreali proposal and case study. Environ Impact Assess Rev 18:439–459
34. Jankowski P (1995) Integrating geographical information systems and multiple criteriadecision making methods. Int J Geogr Inform Syst 9(3):251–273
35. Janssen R (2001) On the use of multi-criteria analysis in environmental impact assessment in The Netherlands. J Multi-Criteria Decis Anal 10:101–109
36. Jasinska EJ, Knott B (1991) Stability of root mat ecosystem in groundwater stream: Cabaret cave, Yanchep National Park, WA, University of Western Australia
37. Kiker GA, Bridges TS, Varghese A, Seager TP, Linkov I (2005) Application of multicriteria decision analysis in environmental decision making. Integr Environ Assess Manage 1(2): 95–108. doi:10.1897/IEAM_2004a-015.1
38. Kumar N (2007) Spatial Sampling for Collecting Demographic Data. Paper presented at the Annual Meeting of the Population Association of America, March 29–31, New York
39. Lahdelma R, Salminen P, Hokkanen J (2000) Using multicriteria methods in environmental planning and management. Environ Manage 26:595–605
40. Lantzke R (2006) MCA: the weighting process and overall comments on a weighting process. Environmental Science, Murdoch University. Unpublished. Available at password restricted URL: 16 http://lms.murdoch.edu.au/webct/urw/lc152224987001.tp152225013001/RelativeResourceManager/Template/content/10_13evaluation/three-mca-wt-mthds.pdf (Accessed 15/04/2008)
41. Lee M (2006) Bridging the gap between theory and practise in integrated assessment. Environ Impact Assess Rev 28:57–78. doi:10.1016/j.eiar.2005.01.001
42. Macarthur R (2004) Geographical information systems and their use for environmental monitoring. In: Artiola J, Pepper IL, Brusseau ML (eds) Environmental monitoring and characterization. Elsevier Academic Press, New York, pp 85-100
43. Malczewski J (2006) GIS-based multicriteria decision analysis: a survey of the literature. Int J Geogr Inform Sci 20(7):249–268
44. Marttunen M, Haimailaiinen RP (1995) Decision analysis interviews in environmental impact assessment. Eur J Oper Res 87:551–563
45. Mattiske EM (1994) Monitoring the effects of groundwater extraction on native vegetation on the Nortrhern Swan Coastal Plain. EM Mattiske and associates
46. Ministry of Environment and Energy, Government of Ontarion (1990) Evaluation methods in environmental, assessment, pp. 3–12, 33–51, 112–137
47. Morrison-Saunders A, Therivel R (2006) Sustainability integration and assessment. J Environ Assess Policy Manage 8(3):281–298. doi:10.1142/S1464333206002529
48. Munier N (2004) Multicriteria environmental assessment. A practical guide. Kluwer Academic Publishers, Dordrecht
49. Munn RE (1979) Environmental impact assessment: principles and procedures, 2nd edn. Wiley, New York
50. Perry B, Gessler W (2000) Physical access to primary health care in Andean Bolivia. Social Sci Med 50(9):1177–1188. doi:10.1016/S0277-9536(99)00364-0
51. Noble B (2002) Strategic environmental assessment of Canadian energy policy. Impact Assess Project Appr 20(3):177–188. doi:10.3152/147154602781766681
52. Pope J (2006) What's so special about sustainability assessment? J Environ Assess Policy Manage 8: v–x. doi:10.1142/S1464333206002505
53. Pope J, Grace W (2006) Sustainability assessment in context: issues of process, policy, and governance. J Environ Assess Policy Manage 8:373–398. doi:10.1142/S1464333206002566
54. Pope J, Annandale D, Morrison-Saunders A (2005) Applying sustainability assessment models. Impact Assess Proj Appr 23(4):293–302. doi:10.3152/147154605781765436

55. Pope J, Annandale D, Morrison-Saunders A (2004) Conceptualizing sustainability assessment. Environ Impact Assess Rev 24:595–616. doi:10.1016/j.eiar.2004.03.001
56. Regan HM, Colyvan M, Markovchick-Nicholls L (2006) A formal model for consensus and negotiation in environmental management. J Environ Manage 80(2):167–176
57. Saaty TL (1980) The analytic hierarchy process. McGraw-Hill, New York
58. Saaty TL (1987) Analytical hierarchy process what it is and how it is used. Math Model 9(3):161–176
59. Sadler B, Verheem R (1996) SEA: status, challenges and future directions, Report 53, Ministry of Housing, Spatial Planning and the Environment, The Hague, Netherlands
60. Scace R (2000) Vital relationship: complementary responsibilities of government and proponent in scoping. In: Kennedy A (ed) Cumulative environmental effects management: tools and approaches. Alberta Association of Professional Biologists, Edmonton, pp 31–41
61. Sheppard SRJ, Meitner M (2005) Using multi-criteria analysis and visualisation for sustainable forest management planning with stakeholder groups. Forest Ecol Manage 207:171–187
62. Shopley J, Fuggle R (1984) A comprehensive review of current EIA methods and techniques. J Environ Manage 18:25–47
63. Sweeney International Management Corp (2009) Environmental Impact Assessment for Kelly Cove Salmon Ltd. proposed aquaculture site relocation for Sand Point, Boston Rock, and Hartz Point. Submitted to NS Department of Fisheries & Aquaculture. SIMCorp. File #SW2008-016, 017 & 018
64. Smit B, Spalding H (1995) Methods for cumulative effects assessment. Environ Impact Assess Rev 15(1):81–106. doi:10.1016/0195-9255(94)00027-X
65. Sorensen JC (1971) A framework for the identification and control of resource degradation and conflict in multiple use of the coastal zone. Department of Landscape Architecture, University of Carlifornia, Berkeley
66. Spaling H, Smit B (1993) Cumulative environmental change: conceptual frameworks, evaluation approaches, and institutional perspectives. Environ Manage 17(5):587–600. doi:10.1007/BF02393721
67. Steinemann A (2001) Improving alternatives for environmental impact assessment. Environ Impact Assess 21:3–21
68. Therivel R (2004) Strategic environmental assessment in action. Earthscan, Sterling
69. Varma VK, Ferguson I, Wild I (2000) Decision support system for the sustainable forest management. Forest Ecol Manage 128:49–55
70. Vincke P (1992) Multicriteria decision-aid. Wiley, Chichester
71. Water Authority of WA (1995) Review of proposed changes to environmental conditions-Gnangara Mound groundwater resources (Section 46), Water Authority of WA
72. Water Authority of WA (1992) Gnangara Mound vegetation stress study results of investigation, Report No WG 127, Water Authority of WA
73. Westman WE (1985) Ecology, impact assessment, and environmental planning. John Wiley, New York
74. Wood C, Djeddour M (1991) Strategic environment assessment of policiees, plans and programmes. Impact Assess Bull 10(1):3–22

Index

DOI: 10.1007/978-3-642-39733-2, © Springer-Verlag Berlin Heidelberg 2014

Printed by Printforce, the Netherlands